T0143588

5G Mobile Communications

Concepts and Technologies

5G Mobile Communications
Concepts and Technologies

Saad Z. Asif

CRC Press
Taylor & Francis Group
Boca Raton London New York

CRC Press is an imprint of the
Taylor & Francis Group, an **informa** business

CRC Press
Taylor & Francis Group
6000 Broken Sound Parkway NW, Suite 300
Boca Raton, FL 33487-2742

© 2019 by Taylor & Francis Group, LLC
CRC Press is an imprint of Taylor & Francis Group, an Informa business

No claim to original U.S. Government works

Printed on acid-free paper

International Standard Book Number-13: 978-1-4987-5155-1 (Hardback)

Library of Congress Cataloging-in-Publication Data

Names: Asif, Saad Z., author.
Title: 5G mobile communications : concepts and technologies / Saad Asif.
Description: First edition. | Boca Raton, FL : CRC Press/Taylor & Francis
Group, 2018. | "A CRC title, part of the Taylor & Francis imprint, a
member of the Taylor & Francis Group, the academic division of T&F Informa
plc." | Includes bibliographical references and index.
Identifiers: LCCN 2018008690| ISBN 9781498751551 (hardback : acid-free paper)
| ISBN 9780429466342 (e-book)
Subjects: LCSH: Mobile communication systems--Technological innovations. |
Wireless communication systems--Standards.
Classification: LCC TK5103.2 .A47 2018 | DDC 621.3845/6--dc23
LC record available at https://lccn.loc.gov/2018008690

Visit the Taylor & Francis Web site at
http://www.taylorandfrancis.com

and the CRC Press Web site at
http://www.crcpress.com

To my late father Muhammad and late younger brother Babar

Contents

SECTION III

SECTION IV

SECTION VI

Preface

Telecommunication is a fast, ever-changing field of STEM (Science, Technology, Engineering, and Mathematics). I have been associated with this field for the past 20 years and have had the opportunity to work for the technology organizations of three leading service providers—Sprint (U.S.), Jazz (Veon), and Telenor Pakistan (Telenor Group). I am currently working at the ICT (Information and Communication Technology) policy-making institute—Ministry of Information Technology and Telecommunication, Pakistan. Based on these two distinct experiences, I made an attempt in this book to cover not only the technical attributes but also the policy elements of telecommunication.

The purpose of this book is to enable the readers to understand the technological ecosystem along with certain key policy/regulatory elements of mobile telecommunication. The key phases of this ecosystem include research and technology development, standardization, product development, network development, device and application development, and burning challenges and best practices. A separate section has been allocated to each of these phases.

- Section I looks into the concepts of basic and applied research along with their applicability to telecommunication. Next, the allocation/identification of the radio frequency spectrum at the global level along with some solutions to address the spectrum needs of 5G (fifth generation mobile communications) are discussed.
- Section II describes the standardization processes of key SDOs (Standard Development Organizations). The key concepts of 5G that are either under standardization or have been standardized are also described. Additionally, a high level governmental/regulatory policy for 5G is also provided.
- Section III describes the domain of semiconductors and product development which is the following step after standardization. Internet-of-Things is described as a potential next growth engine for the semiconductor industry.
- Section IV presents the end-to-end architecture of a mobile network. All the key areas of a network including radio access network, transport network, core network, and operation support systems are discussed. The radio access aspects of 4G (Long Term Evolution [LTE]-Advanced) and 5G NR (New Radio), transport network technologies, 5G NGC (Next Generation Core), end-to-end 5G network architecture, network development, and many other characteristics of mobile networks are described.
- Section V—Smart phones and smart applications have brought seismic change to the way we communicate and conduct business. To support ITU's (International Telecommunication Union) requirements for 5G, certain components of devices such as batteries, processors, and antennas may require major upgrades. These potential enhancements are discussed from the perspectives of theoretical standards and state-of-the-art innovations. Application delivery mechanisms along with exciting applications such as mobile money and Internet of Things (IoT) are explained.
- Section VI explains some of the industry's key challenges as well as the best practices that can be later addressed/used in the subsequent generation of mobile telecommunication.

The book also presents a number of case studies that may help the readers to learn about ICT from the perspectives of both developing and developed economies. To assist academia, every chapter is provided with a mix of simple and thought-provoking problems. A solution manual is also available from the publisher for the assistance of instructors.

I concluded the book with certain suggestions, particularly on the importance of basic research, departure from the traditional 10-year evolution cycle, and having a 20 to 30-year plan which may help in further strengthening the longevity of the telecommunication sector.

From the bottom of my heart I would like to thank my mother for her prayers, my wife for her support, and my daughters for sparing precious time from their studies.

In conclusion, I hope you will enjoy reading this book and realize enormous benefits from it.

Saad Z. Asif

Disclaimer: The responsibility for the content of this book rests upon the author, not with his employers (current and any former).

Author

Saad Z. Asif has been associated with the field of telecommunication for more than 20 years. He gained experience through working with three of the top-tier telecommunication operators—Sprint (US), Jazz (Veon), and Telenor Pakistan, and also through the Ministry of Information Technology and Telecommunication, Pakistan, which is a public policy-making institute.

Asif is a strategic thinker, researcher, and telecommunications policy expert. He has been at the forefront of technology research and standardization and in providing strategic guidance throughout his career. He has led one of the first teams across the globe to evaluate state-of-the-art technologies such as 3G, 4G, Smart Antennas, e-band microwave radios, dense wavelength division multiplexing (DWDM), and many more. He played a key role in designing Sprint's wireless high-speed data strategy, Jazz's Broadband strategy, and Telenor's Pakistan's 3G and Transmission Network strategies.

He is one of the main authors of GSMA's (Global System for Mobile Association) award winning Pakistan's National Telecommunications Policy 2015. He was also one of the key players in the successful execution of two frequency spectrum auctions that generated more than US$700 million for the national exchequer. He was also instrumental in defending the country's scarce resource (frequency spectrum) at the World Radiocommunication Conference 2015. He has produced a Policy Directive on 5G and developed Pakistan's first 3-year rolling Frequency Spectrum strategy and Cyber Governance policy.

Asif has written two books and numerous peer-reviewed technical papers on telecommunications. He has been granted five patents as a co-patentee by the United States Patent & Trade Office. He has also been listed as a scientist in the *Productive Scientists of Pakistan* directory since 2009. He has also been a senior member of the IEEE (Institute of Electrical and Electronics Engineers) since 2004. He also served on the board of directors of Pakistan's National Radio & Telecommunication Corporation.

Asif earned a BS and an MS in electrical engineering from Oklahoma State University in 1996 and 1997, respectively. He also earned an MS in engineering management from the University of Kansas in 2001.

1 The Beginning

Mobile Cellular Telephony is one of the greatest innovations of the twentieth century and today in the twenty-first century it can be safely said that it has brought nothing less than a revolution in the way communications take place across the globe.

The Mobile Cellular Telephony is enabled through a combination of cellular networks and mobile devices which communicate to each other by means of radio frequency spectrum (i.e. wirelessly). A cellular network consists of thousands of nodes that assist mobile device users in performing plethora of tasks. Mobile device has become the Third Screen after Television and Computer and is becoming more economical and powerful with continual technological advancements. There are more than 5 billion mobile subscribers in the world and for the very vast majority the mobile device has become a necessity and without it they can't go about with their daily routine lives.

1.1 MOBILE CELLULAR TELEPHONY EVOLUTION

The cellular concept was conceived by Bell Laboratories in 1947 and enabled companies to provide wireless communications to a large population [1]. Like any other field of science and technology, mobile communications is continuously evolving and the sector* has made astonishing progress in the last 70 years.

The first generation (1G) cellular networks was deployed in the 1980s, the second generation (2G) in the 1990s, while the third generation (3G) in the 2000s. Today, 4G (fourth generation) cellular networks are being deployed and the world is getting ready to embrace the fifth generation (5G) of mobile cellular telephony.

The 1G analog systems are no longer operational, which only provided voice services and had no support for data. The 2G digital systems are currently operational and support voice and limited data services. The 3G systems support voice, low speed data, and enable a number of data services. The 4G systems enable mobile broadband in the true sense, targeting 100 Mbps or higher on the move.

5G systems are expected to provide an enhanced mobile broadband targeting peak data rate of 20 Gbps, extend 4G's Internet of Things capability, and enable mission-critical applications that require ultra-high reliability and low latency. 5G networks are expected to be designed by taking a user-centric approach.

1.2 HEXAGON BASED MOBILE CELLULAR TELEPHONY

A cellular network or mobile network is a wireless network spread over the land through a web of cell sites. Each of these sites or cell towers is comprised of a transceiver (transmitter/receiver) for communications with mobile devices. From a technological perspective, mobile devices† rely on die hard cellular towers for communications and these cell sites or cell towers are designed to keep a hexagonal shape in mind. The use of hexagonal cells was invented by Bell Laboratories in the 1970s [2]. This shape was selected over other geometrical shapes since by using it the cells can be laid next to each other with no overlap, thus providing coverage theoretically to the entire service area without any gaps [3]. The hexagon design has been at least so far remained as necessary for mobile communications as cement for the construction of buildings or coal tar for carpeting the roads.

---

* The sector includes both private and public entities associated with mobile communications, that is, governments, ministries, regulators, and private industry.
† Mobile devices can include mobile phones, tablets, data modems, iPads, and similar devices.

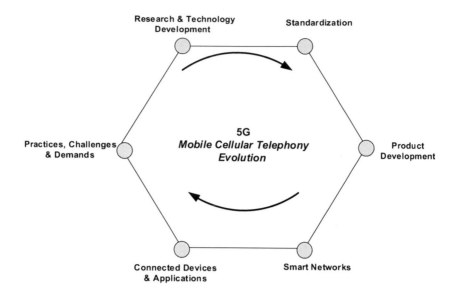

FIGURE 1.1   Key steps in mobile cellular telephony.

If we look at it from an end-to-end perspective, the key phases of the mobile cellular telephony can nicely fit on the six corners of the hexagon shape (Figure 1.1). These generic six phases are research and technology development, standardization, product development, network development, device and application development, and the sector's practices and challenges.

In a nutshell, though not necessary, research and technological development is the first step that leads into standardization followed by product development. Once telecom products are ready, they get deployed in the cellular networks. Device development usually lags behind network equipment production. Once networks are up and running and users are connected to networks through their devices, applications start to pour in; from there, the sectors begin to see the good or not-so-good practices, bottle necks and challenges, and new business demands which then leads back to the first step to start all over again.

1.3   MANUSCRIPT OVERVIEW

This manuscript has been divided into the following six sections corresponding to the hexagonal cell.

Section I is on research, technology development, and the frequency spectrum. Like any field of science and technology, mobile communications also rely on research and technology development for progress and evolution. Chapter 2 discusses how these are currently capitalized in mobile networks and how these can be strengthened in the future. The topic of the radio frequency spectrum, which is controlled by governments, is discussed in Chapter 3. The ITU-R (International Telecommunications Union Radiocommunication Sector) process of frequency allocation and identification and some potential technical and financial solutions to address the spectrum needs of 5G are discussed in this chapter.

Section II looks into the element of standardization. Mobile communications is governed by global standards which are developed to achieve economies of scale and to attain many other benefits. Chapter 4 looks into the standardization of 5G along with the standardization processes of certain key standard development organizations (SDOs), ITU-T (ITU Standardization Sector) guidelines for establishing SDOs in developing nations, and a case study on the lack of research and standardization in OIC (Organization of Islamic Conference) member states. Chapter 5 summarizes certain key elements of 5G such as multiple access techniques, cognitive radio, massive cloud radio access network, vehicular communications, and network slicing. 5G is currently in the standardization phase and ITU is expected to approve the 5G standard in 2020.

Section III describes the key aspects of semiconductor development and product development. After standardization, the next step is to develop integrated circuits (ICs) conforming to standards, which is described in Chapter 6. Every piece of telecom equipment is equipped with ICs which is a set of electronic circuits on a small plate (chip) of semiconductor material, normally silicon. Various technologies for IC development along with a perspective on semiconductor business in Pakistan are presented in this chapter. The ICs along with components (shelves, chassis, nut, bolts, and so on) go into full scale product development. Chapter 7 presents a few such products, namely multimode base stations, 5G enabled base stations/small cells, and so on.

Section IV is all about networks covering the areas of radio access network, transport network, core network, and operation support systems. The section starts with defining the end-to-end cellular network architectures of 2G, 3G, 4G, and 5G in Chapter 8. The radio access network of 4G along with several advanced features of LTE-Advanced and LTE-Advanced Pro are discussed in Chapter 9. The requirements, architecture, and air interface of 5G NR (new radio) are also described in the same chapter. Chapter 10 starts off by defining traffic capacity requirements for 5G. It also discusses key technologies in three subareas of transport networks, namely mobile backhaul, metro transport network, and core transport network. A case study is also presented on the applicability of 80 GHz e-band microwave radios for the cellular and broadband providers of Pakistan. Chapter 11 provides an overview of the 4G Evolved Packet Core and evolving 5G NGC (Next Generation Core). CDN (Content Development Network), evolution of IMS (IP Multimedia Subsystem), and OSS (Operations Support Systems) are also discussed in the chapter.

Section V deals with connected devices and certain key mobile applications that are described in Chapters 12 and 13, respectively. Chapter 12 discusses the three key components of devices. namely batteries, processors, and antennas that perhaps need major as upgrades to meet the requisites of 5G. Chapter 13 describes application delivery mechanisms such as SDP (Service Delivery Platform) and IMS, and the role of cloud computing in the arena of applications. In the latter part of the same chapter, advanced value added services such as mobile financial services, mobile health, and IoT (Internet of Things) are described in detail.

Section VI, which is the last section, describes the industry's key burning challenges and best practices which can later be formulized in the following generation of mobile communications. Chapter 14 looks into the challenges due to signaling storms (massive signaling traffic due to smart devices), an abundance of HetNets (heterogeneous networks), device-to-device communications, and big data. Chapter 15 discusses some key practices of the industry, namely spectrum management, energy management, and patent portfolio management where each practice is further elaborated with a case study.

The book concludes with a futuristic and thought provoking discussion on how to go beyond the traditional way of doing business. Chapter 16 urges the telecom sector to look at the bigger picture and into a much more distant future in collaboration with the overall ICT (Information and Communication Technology) sector.

REFERENCES

1. Engineering and Technology History Wiki 2016. Milestones: List of IEEE Milestones. http://ethw.org/Milestones:List_of_IEEE_Milestones
2. Engineering and Technology History Wiki 2015. The Foundations of Mobile and Cellular Telephony. http://ethw.org/The_Foundations_of_Mobile_and_Cellular_Telephony
3. Rappaport, S.T. 1996. *Wireless Communications Principles and Practice.* Prentice Hall PTR, Upper Saddle River, NJ.

2 Research and Technology Development

Like any other field of science and technology, mobile telecommunication requires thorough and perpetual research to succeed. Research is the backbone of the telecommunications world and a critical enabler for future achievements.

Today's research in telecommunications primarily falls into the category of applied research. Applied research is business driven, focusing on solving practical problems for gaining profits in short to midterm. The basic research, on the other hand, is conducted without boundaries and without business justifications and focuses on long term. The research is forward looking, drives innovation, enables socio-economic prosperity and helps to move from one generation of technology to the next.

There are as such no hard-coded steps but normally research takes into technological development. In more tangible terms the R&D from academia goes into industrial research labs and from there to standard development organizations. The standardization process is executed, within certain boundaries, considering economical and deployment constraints and interoperability challenges. Standardization is followed by product development and then to network development/deployment/operations.

The focus of this chapter is on research and technology development (R&D/R&TD), which is the fundamental block for making continuous and substantial progress in telecommunications.

2.1 BASIC VERSUS APPLIED RESEARCH

Basic research is curiosity driven. It is performed without a thought for practical ends as defined in the report "Science, the Endless Frontier" [1] by Vannevar Bush to the President of the United States in 1945 and quoted by the National Science Foundation in one of its articles published in 1953 [2].

Basic research leads to new knowledge and provides scientific capital. The knowledge provides the means of addressing a good number of practical problems, though at the same time, it may not give a complete specific answer to any of them [2]. The goal of basic research needs to be defined by the researchers themselves rather than by certain individuals/organizations that are mainly concerned with short term improvements in existing products and services [3]. At the end of the day, it is important that the unrestrained exploration of new knowledge results in discoveries and inventions. A perfect example of this fact is the discovery of gravitational waves in 2016 after a century of expectation.

Applied research, on the contrary, is designed to answer specific questions aimed at solving specific practical problems. The knowledge acquired through applied research has specific commercial objectives in the form of products, processes or services. A good example could be the use of single base transceiver stations (BTS) to support various radio access technologies in the late 2000s as compared to having a one-to-one mapping between BTS and radio technology.

It may be noted that both types of research complement each other. The basic research provides scientific know-how. Applied researchers can use this knowledge to develop new technologies/products and also to make improvements on existing products, technologies, and processes. This is followed by a step in which basic researchers take advantage of improved products and services to answer new fundamental questions. Overall, it is an important cycle for the progress and advancement of all the fields of science and technology [4].

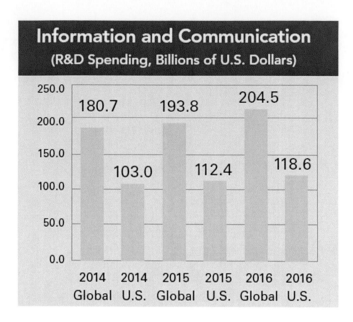

FIGURE 2.1 ICT industry R&D spending. (From Battelle and R&D Magazine 2016. 2016 Global R&D Funding Forecast [7].)

The telecom industry focuses on applied research, as the emphasis is on short term practical gains rather than on researcher-driven investigations. There is immense competitive pressure to focus on the short term rather than on longer terms in the corporate world. The horizons are getting shorter and are driven by fast moving and incremental gain loving industry and society in general. However, there is a need for basic research for the longer term future of telecommunications that needs to be produced at both the national and global levels as illustrated by the following case.

A well known example of the relationship between basic and applied research is the invention of the laser. This activity was substantially funded by U.S. federal research dollars. The work of inventors was criticized at the time by some of their colleagues as "a solution looking for a problem" [5]. However, history proved otherwise, making the laser a solution for many problems. The laser has ultimately revolutionized not only the telecommunications industry through fiber optics, but also provided an unlimited set of applications from laser printers to laser guided weapons. In a nutshell, the U.S. government provided the much-needed support upfront for basic research and later various industries funded the applied research to produce countless benefits [6].

2.2 R&D IN ICT

At the global ICT (Information and Communication Technology)* level, the U.S. continues to be the dominant R&D force. The U.S. accounted for about 58% of this sector's global R&D share in 2016 as tabulated in Battelle and R&D Magazine (Figure 2.1). The vast majority of the R&D investors are suppliers as shown in [7,8], where the investment activity is led by Intel and Microsoft, with each investing more than US $13 billion in R&D during 2016.

ICT heavily relies on the semiconductor industry that has laid the foundation for modern electronics and the modern lifestyle. Every piece of ICT equipment contains integrated circuits which are a set of electronic circuits on a small plate (chip) of semiconductor material. The

---

* The ICT industry provides hardware, software, and services that make up the modern information age, spanning semiconductors, telecommunications, computers, phones, and tablets [7].

TABLE 2.1
Top 10 Semiconductor R&D Spenders

2016 Rank	2015 Rank	2014 Rank	Company/ Headquarters	2014 R&D Exp ($M)	2014 R&D/Sales	2015 R&D Exp ($M)	2015 R&D/Sales	2016 R&D Exp($M)	2016 R&D/Sales
1	1	1	Intel, USA	11,537	22.4%	12,128	24%	12,740	22.4%
2	2	2	Qualcomm, USA	3,695	19.2%	3,702	23.1%	5,109	33.1%
3	4	4	Broadcom, USA	2,373	28.2%	2,105	25.0%	3,188	20.5%
4	3	3	Samsung, South Korea	2,965	7.8%	3,125	7.5%	2,881	6.5%
5	7	6	Toshiba, Japan	1,853	16.8%	1,655	17.0%	2,777	27.6%
6	5	5	TSMC[a], Taiwan	1,874	7.5%	2,068	7.8%	2,215	7.5%
7	8	9	MediaTek, Taiwan	1,430	20.3%	1,460	21.8%	1,730	20.2%
8	6	7	Micron, USA	1,598	9.6%	1,695	11.4%	1,681	11.1%
9	–	–	NXP[b], Netherlands	–	–	–	–	1,560	16.4%
10	9	12	SK Hynix, South Korea	1,340	8.2%	1,421	8.4%	1,514	10.2%
Top 10 Total				**28,655**	**–**	**29,359**	**–**	**35,395**	**–**

Source: Electronic Specifier. The Top 10 Semiconductor R&D Spenders in 2015. http://www.electronicspecifier.com/around-the-industry/the-top-10-semiconductor-r-d-spenders-in-2015; IC Insights 2017. Research Bulletin—Intel Continues to Drive Semiconductor Industry R&D Spending; IC Insights 2016. Research Bulletin—Semiconductor R&D Growth Slows in 2015 [9–11].

[a] TSMC: Taiwan Semiconductor Manufacturing Company.

[b] NXP: Next Experience.

semiconductor is a substance, usually a solid chemical element or compound, that supports electrical conductivity between a conductor, such as copper, and an insulator, such as glass. Elements like silicon, germanium, and compounds of gallium are the most widely used in electronic circuits. From the perspective of semiconductor research and development spending, Intel leads the chart followed by Qualcomm and Broadcom as shown in Table 2.1 [9–11]. The advancements in semiconductors are a prerequisite in providing higher data speeds and energy efficiency in telecom networks and devices.

2.2.1 R&D in Telecom

A similar trend is also followed in telecommunications (a subset of ICT) where the vendor community is primarily driving R&D investments. The vast majority of the service provider community, on the other hand, does not conduct any sizeable research and relies on vendors' R&D results. Vendors and operators sometimes jointly set up innovation centers to bring research more closely align with practicality. The results from such joint ventures speed up the stages of product development and deployment. The vendor community, in general, consists of network infrastructure providers, semiconductor developers, device manufacturers, and perhaps application developers that annually spend billions of dollars on research to gain a competitive advantage and greater profits in the longer run. Figure 2.2 shows the R&D spending of some key players of the telecom sector. It can clearly be seen that vendors are in the drivers' seat and spend billions on the applied R&D. DOCOMO (a Japanese operator), which has one of the largest if not the largest R&D budget, still lags behind the vendor community [12–19].

Some examples of R&D portfolios for vendors and operators are as follows [20–22].

Apple is the most valuable company in the world, the world's largest information technology company by revenue [23], the world's largest technology company by total assets [24], and the

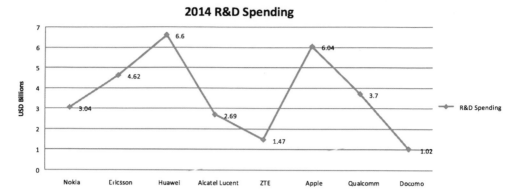

FIGURE 2.2   Telecom R&D spending.

world's second largest mobile phone manufacturer [25]. In relation to telecommunications, it designs, develops, and sells consumer electronics like the famous iPhone branded products. It spends about 3% of its revenue on R&D which skyrocketed from US $782 million in 2007 to US $8.1 billion in 2015 [26,27].

Ericsson's research and development function is a part of Group Function Technology and focuses on a variety of research areas, including but not limited to wireless access networks, networking, radio, and cloud technologies. Group Function Technology funds research activities with several major universities and research institutes all over the world. It also leads and supports research programs within the European Union and substantially contributes to standards development activities. Ericsson has been granted 39,000 patents, and is the largest holder of standard-essential patents on mobile communications. It spends about 15% of its sales on R&D expenditures [28].

Huawei has a large pool of product and solution R&D employees, comprising more than 45% of its total workforce worldwide. It runs 16 R&D centers in several countries that include Germany, Sweden, the U.S., France, Italy, Russia, India, and China. It has also set up 28 joint innovation centers with leading telecom operators to bring solutions to the market earlier and with fewer implementation challenges. Huawei participates in more than 150 industry standards organizations. It has been granted 30,240 patents and spends about 13.7% of the company's annual revenue on R&D [29].

Bell Laboratories is the R&D unit of Nokia founded in 1925. Its primary activities are directed from its Murray Hill, New Jersey, U.S. based headquarters, and it also operates R&D facilities in a few additional countries. Bell Labs participates in approximately 100 standards organizations, making contributions to more than 200 different working groups. It collaborates with more than 250 universities across the globe. It has more than 33,000 patents and has received seven Nobel Prizes which are shared among its 12 researchers. It spends about 16% of the company's annual sales on R&D [30,31].

Qualcomm is a fabless semiconductor company and its business model is built around technology licensing. The U.S. Telecommunications Industry Association adopted Qualcomm's CDMA (Code Division Multiple Access) technology, which is used in all the 3G standards, as a cellular standard in 1993. It spends about 20% of the company's annual sales on R&D. It holds one of the largest standard-essential patents' portfolios in mobile communications [32]. It is actively engaged with academia and collaborates heavily with standards development organizations.

Orange Labs is the R&D arm of France Telecom that consists of 18 laboratories in (the UK, France, Poland, China, Korea, Japan, Jordan, Egypt, and the US) and a few technocenters. A technocenter is an incubator for projects with high innovation potential. Moreover, the goal of technocenters is to reduce the time required to bring a concept to market. It actively contributes to regional and global groups of standardization bodies. Orange devotes about 2% of its revenue to research and development [33].

NTT DOCOMO R&D (Nippon Telegraph and Telephone do communications over the mobile network) has been in the forefront of mobile technology, not only in Japan, but also at a global scale. During 2014, it created one of the first labs dedicated to conducting R&D on 5G and to speeding up work on its standardization. It contributes about 2.5% of its revenue to research and development [34].

It is evident from the above examples and illustrations that the vendor community spends over 10% of its revenue on R&D, whereas operators' spending hovers around 2% (if any). R&D spending by some of the largest operators is on the order of a few hundred million dollars, whereas it runs into billions for large-scale manufacturers. Orange Labs and DOCOMO are the two key R&D spenders in the operator community, whereas the vast majority of the service providers do not set aside a budget for R&D.

2.2.2   R&D Process (Telecom)

Established vendors and startups are always seeking to bring research and innovations to their customers, that is, the operators. In the majority of cases, operators have the final say as to whether or not to take the supplier's innovation to the next level. The established suppliers, big or small, have deeper pockets and more leverage over operators as compared to startups. The startups spend considerable time seeking funding from venture capital firms and endorsements from operators for their technology incubation.

As the research starts to mature, suppliers start working to sell their ideas, which leads to technology development (or technology incubation for startups). Technology development enables development and enhancement of practical solutions. Technologies are first normally standardized and then utilized during product development.

2.2.2.1   Example

Consider the case of antenna diversity in mobile phones which has become a fundamental component of 3G, LTE, 4G, and eventually 5G networks and devices. Antenna diversity (a form of smart antennas[*]) is about having more than one antenna either in the transmit or receive direction or in both to increase the coverage and capacity of telecom networks.

Twenty or so years ago, in the 1990s, that was a novel and commercially unproven idea. The academia, suppliers, and startups were highly involved in putting their arms around smart antennas, including MIMO (multiple-input multiple-output). During that time, some startups such as Metawave Communications, even disappeared due to a lack of interest from the operators.

Finally, the inception of 3G and data in the early 2000s brought smart antenna techniques to a more practical level. Tests and evaluations were happening at a faster pace with the involvement of operators. One such investigation [36] in the form of having two receiving antennas (mobile receive diversity) in the 3G handset revealed voice and data capacity gains in excess of 3 dB[†] in the forward link (base station to mobile). Such investigations later resulted in full scale standardization of smart antenna solutions or antenna diversity solutions, that is, MIMO (Multiple Input Multiple Output). MIMO involves adding antenna elements at both the base station and at the terminal. MIMO is a key feature of LTE and LTE-Advanced and is expected for 5G systems as well.

2.3   STRENGTHENING THE ICT R&D ECOSYSTEM

Over the last several decades, the development of the ICT industry has had a positive impact on nearly every facet of the global economy and global family. The continuation of this positive

---

[*] A smart antenna is a digital wireless communications antenna system that takes advantage of the diversity effect at the source (transmitter), the destination (receiver) or both [35].

[†] 3 dB or 3 decibels represents a ratio of two to one or a doubling of power.

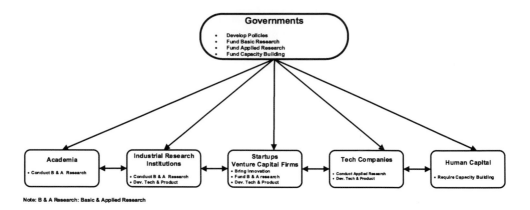

FIGURE 2.3   ICT R&D ecosystem.

trend highly depends on the health of the ICT research ecosystem. A robust research ecosystem (Figure 2.3) comprises strong government policies, energetic university and industrial research institutions, emerging startups, mature technology companies, funding for both basic and applied technologies from government and private institutions (including venture capital firms), and a large pool of talented researchers. Only a few countries have all of the ingredients required to establish and maintain a dynamic ICT research ecosystem. Thus, strong collaboration is required at the global level so that one country can address the missing elements of another nation [6,37–40].

A critical element for continuation of this ecosystem is investment in basic research, which is almost nonexistent, particularly in the ICT sector, worldwide. Most of the investment since the last decade has been pouring into applied research for immediate returns. This trend is expected to continue as ICT companies operate in a highly competitive and commoditized environment, which is forcing them to quickly bring products to market with razor thin margins.

2.3.1   EXAMPLE

A perfect example of such a trend can be witnessed in Bell Laboratories (the R&D organization of Nokia). Bell Labs was heavily involved in basic, long term ICT research prior to restructuring in the 1980s. That particular form of research was funded through a fee assessed on Bell operating companies and relied on the federal government's support. Thus, a stable flow of research funds was guaranteed for basic research within the Bell system. This basic research resulted in the discovery of a number of marvelous technologies, including but not limited to transistors (1947), photovoltaic cells (1954), lasers (1958), and the Unix operating system (1969) [6].

Following restructuring and particularly after Alcatel-Lucent took over, it began to downsize and place greater focus on applied research. During the 1980s, Bell Labs had approximately 25,000 employees, around 30,000 in 2001, and as recently as 2014 had only 700, which is definitely disturbing. In a nutshell, the overall research funding has been significantly reduced as compared to the 1980s and now we do not hear of new discoveries coming from such labs.

2.3.2   HIGH LEVEL SUGGESTIONS

It is critically important that the global ICT industry and governments focus not only on those items that they are sure will bring economic payoff in the short term. However, at the same time, they should also devote resources, both human and capital, to explore things that may revolutionize the industry and take our knowledge and understanding to the next level.

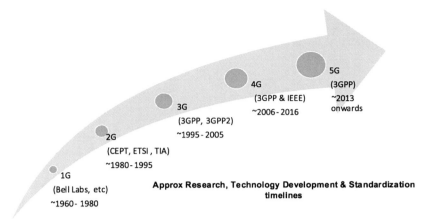

FIGURE 2.4   Telecom generation evolution.

Furthermore, the telecom sector has to look beyond the 10-year evolution cycle (Figure 2.4) and think for the long-term future as well; perhaps having a 30 to 40 (2040/2050) year ICT plan/vision from the ITU (International Telecommunications Union) would be a good starting point. In addition, for the long term sustainability of the ICT research ecosystem, the following items are vital.

- Consistent and considerable investment in basic research by the governments,
- Re-emphasis on basic research from industrial research institutions along with continued focus on applied research,
- Generous effort on reducing the R&D gap between developed and developing nations, and
- Emphasis on having a reasonable gap from one telecom generation to the next in order to allow for return-on-investment.

2.4   SUMMARY

The constant evolution in telecommunications is recognized both as a market driver and as a driver for R&D, since R&D activities are built to support consumer and business demands.

The market pressures and recent past recessions have considerably shifted the focus of companies to product development and incremental research, rather than innovating for the future and seeding technology development. This current focus on short term gains has paved the way for applied research to the furthest extent. There is much more focus on the "D" rather than on "R" in today's world. The efforts and investments in basic research have been diminishing, thus sincere and consistent hard work is required at both the government and industry levels for its revival.

Research and technological development, which is a founding block of any field of science and technology, was explained from the point of view of ICT with a primary focus on the telecom industry. The research ecosystem along with its shortcomings were also presented. It is important that governments and the ICT industry re-evaluate the R&D ecosystem and consider having both a 3 to 5-year view along with a 30 to 40-year long vision.

PROBLEMS

1. Differentiate between Basic versus Applied research and provide an example.
2. What are the benefits of Operator/Vendor joint innovation centers?
3. What is mobile receive diversity?
4. What are the key ingredients of a robust research ecosystem?

5. Why basic research is dwindling in the ICT sector?
6. Discuss in a group how to improve the standings of basic research in telecommunications and overall R&D ecosystems?

REFERENCES

1. Bush, V. 1945. Science, the Endless Frontier: A Report to the President. United States Government Printing Office, Washington, USA.
2. National Science Foundation 1953. What Is Basic Research?
3. Noll, M.A. 2003. Basic Research in Telecommunications. Annenberg School for Communication at the University of Southern California, Columbia Institute for Tele-Information at Columbia University, Guglielmo Marconi International Fellowship Foundation at Columbia University.
4. The University of Texas at El Paso. Basic vs. Applied. Research. http://cossrvfile00.utep.edu/couri/index. php?option=com_content&view=article&id=50&Itemid=243
5. Townes, C.H. 1999. *How the Laser Happened: Adventures of a Scientist.* Oxford University Press, New York, NY, USA.
6. Andersen, J.C. and Coffey, D. 2011. U.S. ICT R&D Policy Report: The United States: ICT Leader or Laggard? TIA Innovation White Paper. U.S. ICT R&D Policy Report Telecommunications Industry Association.
7. Battelle and R&D Magazine 2016. 2016 Global R&D Funding Forecast.
8. Battelle and R&D Magazine 2013. 2014 Global R&D Funding Forecast.
9. Electronic Specifier. The Top 10 Semiconductor R&D Spenders in 2015. http://www.electronicspecifier. com/around-the-industry/the-top-10-semiconductor-r-d-spenders-in-2015
10. IC Insights 2017. Research Bulletin—Intel Continues to Drive Semiconductor Industry R&D Spending.
11. IC Insights 2016. Research Bulletin—Semiconductor R&D Growth Slows in 2015.
12. Arthur D. Little 2013. Suppliers—On the Road to Redemption?
13. Nokia Corporation 2015. Nokia in 2014.
14. Ericsson 2015. Welcome to the Networked Society, Ericsson Annual Report 2014.
15. Truong, A. 2015. Huawei's R&D Spend Is Massive Even by the Standards of American Tech Giants, Quartz. http://qz.com/374039/huaweis-rd-spend-is-massive-even-by-the-standards-of-american-tech-giants/
16. Alcatel-Lucent 2014. 2014 Annual Report on Form 20-F.
17. Teleocompaper 2015. 4G Drives Increase in ZTE Results. http://www.telecompaper.com/ news/4g-drives-increase-in-zte-results--1073194
18. Quiller Media, Inc. 2014. Apple Spent Record $1.7B on Research & Development Last Quarter, $6B in Fiscal 2014. http://appleinsider.com/articles/14/10/28/apple-spent-record-17b-on-research-development-last-quarter-6b-in-fiscal-2014
19. NTT Docomo 2015. Financial Indicators. https://www.nttdocomo.co.jp/english/corporate/ir/finance/ indicator/
20. Applied Value 2015. Financial Performance & Trends in the Telecom Industry.
21. JSON.TV 2014. R&D Policy of the Largest Global Telecommunication Companies. http:// json.tv/en/ict_telecom_analytics_view/rd-policy-of-the-largest-global-telecommunication-companies-2014090101540859
22. Wikipedia. http://en.wikipedia.org/wiki/Main_Page
23. Fortune 2015. The Top Technology Companies of the Fortune 500. http://fortune.com/2015/06/13/ fortune-500-tech/
24. Forbes 2015. The World's Largest Tech Companies: Apple Beats Samsung, Microsoft, Google. http:// www.forbes.com/sites/liyanchen/2015/05/11/the-worlds-largest-tech-companies-apple-beats-samsung-microsoft-google/#14b0ca3d415a
25. AOL Tech (Engadget) 2015. Huawei Passes Microsoft as Third-Largest Mobile Phone Maker. http:// www.engadget.com/2015/07/31/huawei-microsoft-smartphone-sales/
26. Campbell, M. 2015. Apple R&D Spending Hit $8.1B in 2015, Suggests Continued Work on Massive Project. http://appleinsider.com/articles/15/10/28/apple-rd-spending-hit-81b-in-2015-suggests-continued-work-on-massive-project
27. Statista Inc. 2015. Apple Inc's Expenditure on Research and Development from 2007 to 2015 (in billion U.S. dollars). http://www.statista.com/statistics/273006/apple-expenses-for-research-and-development/
28. Ericsson. http://www.ericsson.com/

29. Huawei, Research and Development. http://www1.huawei.com/en/about-huawei/corporate-info/research-development/index.htm
30. Bell Labs, History of Bell Labs. https://www.bell-labs.com/about/history-bell-labs/
31. Alcatel-Lucent 2015. Fast Facts. http://www3.alcatel-lucent.com/wps/portal/BellLabs/AboutBellLabs/FastFacts
32. Baron, J. and Pohlmann, T. 2015. Mapping Standards to Patents Using Databases of Declared Standard-Essential Patents and Systems of Technological Classification.
33. Orange 2016. Innovation. http://www.orange-business.com/en/innovation
34. The Japan Times Ltd 2014. DoCoMo Counting on R&D to Stay Ahead. http://www.japantimes.co.jp/news/2013/05/25/business/docomo-counting-on-rd-to-stay-ahead/#.VwefzPl97IU
35. TechTarget, Smart Antenna. http://searchmobilecomputing.techtarget.com/definition/smart-antenna
36. Asif, S. 2004. Mobile Receive Diversity Technology Improves 3G Systems Capacity. *2004 IEEE Radio and Wireless Conference*, Atlanta, GA, Sep. 19–22, 2004, pp. 371–374.
37. Telecommunications Industry Association 2011. Investing in Telecom for Tomorrow's Innovations: Recommendations for Telecommunications Research and Development.
38. Finpro, Tekes, Verso-Program 2010. China Runway: Guide for ICT companies.
39. European Commission 2011. Orientations for EU ICT R&D & Innovation Beyond 2013, 10 Key Recommendations—Vision and Needs, Impacts and Instruments. Report from the Information Society Technologies Advisory Group (ISTAG).
40. Slywotzky, A. 2009. The Future of Tech—Where Have You Gone, Bell Labs? *Bloomberg Businessweek Magazine.* http://www.businessweek.com/magazine/content/09_36/b4145036681619.htm

3 Radio Frequency Spectrum

The electromagnetic spectrum is the range of all types of electromagnetic radiation that includes radio waves to gamma rays (Figure 3.1). The progression from radio waves (longest wavelength) to gamma rays (shortest wavelength[*]) is also a sequence in terms of energy from lowest to highest, that is, the energy carried by a radio wave is low while gamma rays carry high energy [1].

Mobile communication depends on the radio spectrum to carry out its tasks. The term radio spectrum refers to the frequency range from 3 kHz to 300 GHz corresponding to wavelengths ranging from 100 km to 1 mm. The exchange of information takes place by varying the amplitude, phase, and frequency of carrier radio waves. This radio spectrum is one of the most tightly regulated resources of today's world. From cell phones to broadcasting TV sets, microwave ovens to garage door openers, maritime affairs to flight tracking, virtually every wireless piece of communication depends on access to the radio frequency spectrum.

This chapter will present the frequency allocation and identification process at the global level and some potential technical and financial solutions to address the spectrum needs of 5G.

3.1 RADIO SPECTRUM AND MOBILE COMMUNICATIONS

Today's 2G/3G/4G mobile communications primarily use frequencies in the range of 700 MHz to 42 GHz. Additionally, some communication take place in the 400 MHz and 70/80 GHz range, however, the use of this set of frequencies is relatively very small (just like a needle in a haystack). The frequencies are allocated by the International Telecommunications Union Radiocommunication Sector (ITU-R) through World Radiocommunication Conferences (WRC) on both a primary and secondary basis. On a wider scale, spectrum sharing of a primary allocation with other primary and/or secondary services has not been attempted as such. For example, the 698–806 MHz band had been historically allocated by ITU-R on a primary basis for both broadcasting and mobile use, but it was only used for broadcasting and not for mobile use in the U.S. (i.e., no sharing was taking place). During 2008–09, the Federal Communications Commission or FCC auctioned this band for mobile communications while ceasing television broadcasting in the same range, thus reducing the opportunity for any sharing between the two services [3].

5G envisions the use of high capacity broadband applications and services that will require a huge amount of spectrum. Beyond excessive mobile broadband and gigabits of data rates, applications like the Internet of Things, use of wireless sensors, and so on have necessitated the search for additional spectrum. It is widely established that the world will need an additional 1000 or so MHz to meet the demands of mobile broadband by 2020. Spectrum scarcity is emerging as one of key problems for 5G and so far the world has not found a solution or solutions (as readers will see later in the chapter). The ITU through the World Radiocommuication Conference 2015 (WRC-15) only allocated 51 MHz for IMT (International Mobile Telecommunications) on a global scale, which is quite infinitesimal compared to what is required. However, the conference has identified several bands in the range of 24.25–86 GHz for studies to address this requirement. The sharing and compatibility studies of these bands are expected to be shared during WRC-19; thus we can expect to find additional spectrum for 5G and broadband in due time.

---

* One wavelength equals the distance between two successive wave crests or troughs.

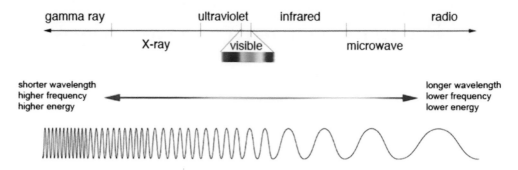

FIGURE 3.1   Electromagnetic spectrum. (From NASA 2013. Imagine the Universe. https://imagine.gsfc.nasa.gov/science/toolbox/emspectrum1.html [2].)

Furthermore, innovative techniques are needed to introduce spectrum sharing and effectively manage implications of air-interface design. And spectrum also has to be managed and regulated to avoid interference related bottlenecks.

3.2   FREQUENCY ALLOCATION AND IDENTIFICATION

The ITU-R is the body that identifies frequency bands for almost any type of wireless communications all around the world. These types include but are not limited to aviation, broadcasting, maritime, mobile communications, public protection and disaster relief, satellite services, and so on. The allocation and identification of frequencies take place at the ITU WRCs. These inter-governmental events take place every three to four years to address the frequency related needs of the world. WRC is the most significant conference related to the frequency spectrum organized by ITU with a mandate to review, and, if necessary, to revise the radio regulations which govern the use of a scarce resource, that is the frequency spectrum.

As far as mobile communications are concerned, the frequencies have to be allocated primarily for mobile communications and may have to be identified for International Mobile Telecommunications (IMT). Second, these have to be allocated for fixed wireless communications links to support backhaul in mobile networks which is termed as a fixed service in the Table of Frequency Allocations (TFA) maintained by ITU at the international level. In simple terms, frequencies have to be identified for wireless communications that take place between mobile users and cellular towers and between cellular towers (i.e., terrestrial backhaul using microwave radios). Finally, satellite based links are also used to support backhaul traffic in remote and far flung areas. Thus, some frequency bands allocated for satellite communications are used in mobile networks in the form of VSAT (Very Small Aperture Terminal), but in the whole scheme of things, their relative use is quite small compared to the frequencies that are used for communications between users and cell phone towers and in terrestrial backhaul. Table 3.1 shows various bands that have been allocated for mobile and fixed services in all three ITU regions* on a primary basis.

The challenge that should be noted is that allocation for mobile communications does not necessarily guarantee its use for IMT, and in today's world, the identification for IMT has become a necessary requirement for broadband communications. So far, only a few bands have been identified for IMT and these are shown in Table 3.2.

---

* As per Article 5 of ITU Radio Regulations: Region 1 comprises Europe, Africa, the former Soviet Union, Mongolia, and the Middle East west of the Persian Gulf, including Iraq; Region 2 includes the Americas including Greenland and some of the eastern Pacific Islands; Region 3 covers non-FSU (former Soviet Union) east of and including Iran and most of Oceania [5,6].

TABLE 3.1
Allocation for Mobile and Fixed Services

Frequency	Region 1	Region 2	Region 3
410–430 MHz	FIXED, MOBILE	FIXED, MOBILE	FIXED, MOBILE
440–470 MHz	FIXED, MOBILE	FIXED, MOBILE	FIXED, MOBILE
470–890 MHz	790–862 FIXED, MOBILE	698–806 MOBILE	FIXED, MOBILE
	862–890 FIXED, MOBILE	806–890 FIXED, MOBILE	
890–960 MHz	890–960	890–902 FIXED, MOBILE	890–960
		902–928 FIXED	
		928–960 FIXED, MOBILE	
1427–1525 MHz	FIXED, MOBILE	FIXED, MOBILE	FIXED, MOBILE
1525–1530 MHz	FIXED	–	FIXED
1668.4–1690 MHz	FIXED, MOBILE	FIXED, MOBILE	FIXED, MOBILE
1700–1710 MHz	FIXED, MOBILE	FIXED, MOBILE	FIXED, MOBILE
1710–2170 MHz	FIXED, MOBILE	FIXED, MOBILE	FIXED, MOBILE
2170–2520 MHz	FIXED, MOBILE	FIXED, MOBILE	FIXED, MOBILE
2520–2690 MHz	FIXED, MOBILE	FIXED, MOBILE	FIXED, MOBILE
2700–4800 MHz	3400–4200 FIXED	3400–3500 FIXED	3400–3500 FIXED
	4400–4800 FIXED, MOBILE	3500–4200 FIXED, MOBILE	3500–4200 FIXED, MOBILE
		4400–4800 FIXED, MOBILE	4400–4800 FIXED, MOBILE
4800–5000 MHz	FIXED, MOBILE	FIXED, MOBILE	FIXED, MOBILE
5150–5350 MHz	MOBILE	MOBILE	MOBILE
5470–5725 MHz	MOBILE	MOBILE	MOBILE
5850–8500 MHz	FIXED, MOBILE	FIXED, MOBILE	FIXED, MOBILE
10–10.45 GHz	FIXED, MOBILE	–	FIXED, MOBILE
10.5–10.68 GHz	FIXED, MOBILE	FIXED, MOBILE	FIXED, MOBILE
10.7–11.7 GHz	FIXED, MOBILE	FIXED, MOBILE	FIXED, MOBILE
11.7–14 GHz	11.7–12.5 FIXED, MOBILE	11.7–12.1 FIXED	11.7–13.25 FIXED, MOBILE
	12.75–13.25 FIXED, MOBILE	12.2–13.25 FIXED, MOBILE	
14–15.4 GHz	14.3–14.4 FIXED, MOBILE	14.4–15.35 FIXED, MOBILE	14.3–14.4 FIXED, MOBILE
	14.4–15.35 FIXED, MOBILE		14.4–15.35 FIXED, MOBILE
15.4–18.4 GHz	17.7–18.1 FIXED, MOBILE	17.7–17.8 FIXED	17.7–18.1 FIXED, MOBILE
	18.1–18.4 FIXED, MOBILE	17.8–18.1 FIXED, MOBILE	18.1–18.4 FIXED, MOBILE
		18.1–18.4 FIXED, MOBILE	
18.4–22 GHz	18.4–19.7 FIXED, MOBILE	18.4–19.7 FIXED, MOBILE	18.4–19.7 FIXED, MOBILE
	21.2–22 FIXED, MOBILE	21.2–22 FIXED, MOBILE	21.2–22 FIXED, MOBILE
22–24.75 GHz	22–23.6 FIXED, MOBILE	22–23.6 FIXED, MOBILE	22–23.6 FIXED, MOBILE
	24.25–24.75 FIXED		24.25–24.75 FIXED, MOBILE
24.75–29.9 GHz	24.75–25.25 FIXED	–	24.75–25.25 FIXED
	25.25–29.5 FIXED, MOBILE	25.25–29.5 FIXED, MOBILE	25.25–29.5 FIXED, MOBILE
29.9–34.2 GHz	31–31.3 FIXED, MOBILE	31–31.3 FIXED, MOBILE	31–31.3 FIXED, MOBILE
	31.8–33.4 FIXED	31.8–33.4 FIXED	31.8–33.4 FIXED
34.2–40 GHz	36–40 FIXED, MOBILE	36–40 FIXED, MOBILE	36–40 FIXED, MOBILE
40–43.5 GHz	40–40.5 FIXED, MOBILE	40–40.5 FIXED, MOBILE	40–40.5 FIXED, MOBILE
	40.5–43.5 FIXED	40.5–43.5 FIXED	40.5–43.5 FIXED
71–76 GHz	71–76 FIXED, MOBILE	71–76 FIXED, MOBILE	71–76 FIXED, MOBILE
81–86 GHz	81–86 FIXED, MOBILE	81–86 FIXED, MOBILE	81–86 FIXED, MOBILE

Source: ITU-R 2016. *Final Acts WRC-15 World Radiocommunication Conference 2015* [4].

Notes:

1. Only primary allocations for mobile and fixed services are shown in the Table. There could be additional primary allocations to other services in these bands as well. The primary allocations are shown in capital letters as per TFA.
2. Allocation of a certain band for a certain service to a particular Region does not mean that all the countries in the region have to comply that specific allocation. Any country can make an exception.
3. Any country with the permission of neighboring countries can allocate a frequency band to itself if it is not in TFA. Details on these exceptions, which are provided in the form of footnotes, can be found in [4].
4. Only frequencies between 400 MHz to 42.5 GHz and 71–76 GHz and 81–86 GHz which are currently under use in mobile networks and/or identified in WRC-15 are listed in the Table 3.1.

TABLE 3.2
IMT Identification Worldwide

Region 1	Region 2	Region 3
450–470 MHz	450–470 MHz	450–470 MHz
694–790	698–960	698–790
790–960		
1710–1885	1710–1885	1710–1885
1885–2025	1885–2025	1885–2025
2110–2200	2110–2200	2110–2200
2300–2400	2300–2400	2300–2400
2500–2690	2500–2690	2500–2690
3400–3600	—	3400–3600

Source: ITU-R 2016. Final Acts WRC-15 World Radiocommunication Conference 2015; GSMA 2014. The Impact of Licensed Shared Use of Spectrum. Report by Deloitte and Real Wireless [4,7].

3.3 FREQUENCY SPECTRUM NEEDS OF 5G

It is well known that the availability of new spectrum bands is a key requirement for the provision of 5G or IMT-2020 services. The ITU-R has estimated that the total global spectrum requirements for IMT will be in the range of 1340 (for lower user density settings) to 1960 MHz (for higher user density settings) for the year 2020 [8].

Spectrum allocation is required not only in air-interface, but also for backhaul and to some extent in the fronthaul. Fronthaul is the link between a pool of base band units and remote radio units (RRUs) which collectively formed the concept of C-RAN (cloud/centralized radio access network). Backhaul (first leg between RRUs and Core Network) is a major challenge for 5G, but to some extent, it can be fulfilled with wired media such as optical fiber cable and technologies such as very-high-bit-rate digital subscriber line 2, and so on. However, for the most part, the air-interface (link between wireless user/device and remote radio unit) is where the vast majority of the spectrum is required.

Almost every frequency band, particularly up to 100 GHz as stated in Tables 3.1 and 3.2, is quite extensively in use and it is well established that the sector will require an additional 1000–2000 MHz by 2020. Keeping this challenge in mind, the recently completed WRC-15 identified a few hundreds of megahertz of spectrum for mobile broadband at the regional levels. However, at the global level, WRC-15 was only able to identify 51 MHz for IMT. The spectrum was allocated to ensure harmonization at the regional and global levels. The harmonization will bring reduced prices for the production of mobile broadband equipment, and thus assist in delivering more affordable broadband to all. The key outcomes of WRC-15 in relation to IMT are as follows:

UHF (470–694/698 MHz):

- In Region 1, no allocation was made for IMT and it primarily remained assigned to broadcasting service.
- In Region 2, for the most part, all or some portions of this band were identified for IMT.
- In Region 3, the band 610–698 MHz was identified in two countries, namely Bangladesh and New Zealand, and five island nations.

L-Band 1427–1518 MHz:

- 51 MHz of L-band (1427–1452 MHz and 1492–1518 MHz) was identified for IMT worldwide.
- A total of 91 MHz of L-band (1427–1452 MHz, 1452–1492 MHz, and 1492–1518 MHz) was identified for IMT in ITU Regions 2 and 3 and in some countries of Region 1.

It can be seen that this allocation is much less than what is required. However, several bands have been identified between 24.25 GHz and 86 GHz for studies to address this requirement (Table 3.3). These sharing and compatibility studies will be completed in time for WRC-19 [4,9–13].

3.3.1 TECHNICAL SOLUTIONS

This section will briefly discuss some solutions that could potentially open doors for more frequencies for 5G. Spectrum sharing tops the list while advancements in semiconductors, antenna technology, and interference avoidance technique will provide opportunities to extract more juice from the existing allocated spectrum.

3.3.1.1 Spectrum Sharing

Spectrum sharing is defined as the collective use of a frequency band by two or more parties in a specific geographical area. Sharing can take place in both licensed and license-exempt bands [7]. For 5G, sharing may also need to be considered with incumbents such as FSS (fixed satellite service), radar, and so on. This form of sharing requires compatibility studies between broadband and nonmobile incumbents which ITU-R is planning to complete in conjunction with its preparation for WRC-19.

Spectrum sharing is a three-dimensional challenge that not only needs to allow for frequency but must also encompass time and geographical factors in providing access across multiple classes of users.

The Licensed Shared Access (LSA) mechanism allows LSA licensees to access the spectrum that has already been assigned to an incumbent. This method allows sharing based on certain rules guaranteeing some level of QoS (Quality of Service). It is different from certain cognitive approaches

---

TABLE 3.3
Potential Bands for Future IMT Services

Band (GHz)	Bandwidth (GHz)	Key Current Allocation Service
24.25–27.5	3.25	FIXED, FIXED-SATELLITE, EARTH EXPLORATION-SATELLITE, MOBILE, INTER-SATELLITE
31.8–33.4	1.6	FIXED, INTER-SATELLITE, SPACE RESEARCH, RADIONAVIGATION
37–40.5	3.5	FIXED, FIXED-SATELLITE, SPACE RESEARCH, MOBILE, MOBILE-SATELLITE
40.5–43.5	3	FIXED, FIXED-SATELLITE, BROADCASTING, BROADCASTING-SATELLITE
45.5–50.2	4.7	FIXED, FIXED-SATELLITE, MOBILE
50.4–52.6	2.2	FIXED, FIXED-SATELLITE
66–76	10	FIXED, FIXED-SATELLITE, BROADCASTING-SATELLITE, MOBILE, MOBILE-SATELLITE, RADIONAVIGATION, RADIONAVIGATION-SATELLITE
81–86	5	FIXED, FIXED-SATELLITE

Source: Americas 2016. Mobile Broadband Transformation LTE to 5G. Rysavy Research, LLC; ITU-R 2012. Final Acts WRC-12 World Radiocommunication Conference 2012; ITU-R 2008. Final Acts WRC-07 World Radiocommunication Conference 2007; ITU-R 2000. Final Acts WRC-2000 World Radiocommunication Conference 2000; ITU-R 1997. Final Acts WRC-97 World Radiocommunication Conference 1997 [9–13].

that allow access to TV white space on an unlicensed basis without any QoS guarantees. LSA allows mobile operators to use/obtain additional spectrum on a secondary basis and with guaranteed access for an agreed geographic area, time frame, and frequency range. Furthermore, licensed sharing can be horizontal, which normally involves sharing between two similar parties (like two mobile operators) whereas with vertical sharing, the frequency can be shared between different types of parties (like mobile operators and a government organization) [7,14]. All of these licensed sharing options can take place within a specified area (geographic sharing), at specific or random times (temporal sharing), and these need to be coordinated as well to avoid harmful interference [15]. The license-exempt approach, as the name suggests, allows sharing among parties without requiring a license. It enables best effort access and operations for data offloading and so on, and thus it is not well suited for carrier grade performance.

The value of a spectrum depends on the profitability of the services that have been assigned to by the respective spectrum. For example, the spectrum assigned for mobile telecommunications will yield much higher economic value than the one that has been assigned to a government entity. Sharing involves tradeoffs whereby allowing a new user in the band will likely diminish what an existing subscriber can do and capitalize on. In general terms, spectrum sharing creates costs and restricts revenues compared to the exclusive use of the same frequency band [7,15]. Many case studies are available that have determined the impact of sharing on the value of a spectrum. The studies presented in [7,15] have shown that sharing reduces the economic value of the spectrum.

3.3.1.2   Air-Interface Design

A primary question, that is, whether there will be a single air-interface or a collection of air-interfaces for 5G, still remains to be resolved. However, for frequency agility, sharing, coexistence, and scalable spectral efficiency, an effective air-interface design is a fundamental enabler and there are several reasons for this [16].

First, there are expected to be several bands for 5G distributed over a large range of frequencies so the air-interface has to be flexible enough to accommodate all such bands. These bands could be contiguous/non-contiguous and can fall anywhere from sub 1 GHz up to 100 GHz. Second, the air-interface has to deal with all the possible spectrum sharing scenarios. The sharing can take place in both licensed and unlicensed bands, with or without the involvement of 5G networks. The main challenge that it needs to manage concerning sharing is interference while also optimizing the efficiency of the spectrum. The interference can be managed in various dimensions including time intervals, orthogonal/nonorthogonal frequency resources, locations with sufficient separation, spatial, and orthogonal codes. Last but not least are the significant variations in uplink and downlink traffic ratios (that even exist today), which imply the need for a flexible air-interface design to effectively manage traffic asymmetry. Several air-interfaces (multiple access schemes) are discussed in Chapter 5 "5G Concepts."

3.3.1.3   Technological Advancements

To improve the spectral efficiency over the current LTE/LTE-Advanced systems, advancements in semiconductor design and antenna technology will be crucial. The bands that have been identified for studies for 5G/broadband services are primarily in the millimeter wave range (i.e., 24.25–86 GHz). In this particular range of frequencies, there is an additional 20–30 dB path loss as compared to 2 GHz frequencies; however, this can be compensated by using large scale phased antenna arrays. These large scale phased arrays are feasible at higher frequencies since, as the frequency gets higher, the size of the antenna decreases [17]. A similar concept of massive MIMO (Multiple Input Multiple Output) that subscribes to the use of multiple antennas both at the transmitting and receiving ends has been widely discussed in the literature to increase the capacity and spectral efficiency for 5G.

Concerning advancements in semiconductors, highly integrated Radio Frequency Integrated Circuit (RFIC) and MMIC (Monolithic Microwave Integrated Circuit) solutions are desirable to meet the size, cost, and power consumption needs of high end broadband millimeter wave radio products. A good example is the availability of MMIC for microwave radio products that operate in the 70/80 GHz range. However, challenges do exist and work is in progress for their resolution. The details on these enhancements are provided in the subsequent chapters of the book.

3.3.1.4 Spectral Efficiency

The technical measure of efficiency for a frequency spectrum is called spectral efficiency, which is measured in bits/sec/Hz. Spectral efficiency refers to the information rate that can be transmitted over a given bandwidth in a specific communication system. Numerous results are present on the Internet showing the efficiency of various mobile technologies. Simulation versus on-ground results sometimes differ to some extent as most simulations are conducted in close to ideal conditions, whereas there are practical considerations in implementing technologies in the field [9,18].

Peak and average spectral efficiencies differ due to changing radio conditions. Peak spectral efficiency is calculated using the highest throughput per sector achieved with the combination of a high order modulation scheme, low code rate, and at a high SNR (signal-to-noise ratio) in a given amount of spectrum [18,19]. In practical terms, average spectral efficiency is the better unit of measurement which considers aggregate cell throughput per sector within the assigned spectrum.

Beside modulation and coding, there are other factors that can impact the efficiency of the spectrum. These include mobile receive diversity, MIMO antenna technology, equalization, and so on. Furthermore, the spectral efficiency can be improved by using wider radio channels. LTE provides roughly 5% better spectral efficiency with a 20 MHz channel as compared to a 10 MHz channel [9]. It may be noted that the spectral efficiency of a radio technology is independent of the frequency it uses to operate, since modulation, coding, and antenna diversity remain the same at different frequencies.

Figure 3.2 shows the spectral efficiency of various 3G, LTE, and 4G (LTE-Advanced) technologies. The values are indicative and results may vary depending on the testing and deployment conditions.

3.3.2 FINANCIAL ASPECTS

Traditional approaches for valuation of a spectrum are based on estimating the opportunity costs. This method is suitable for green field and wide area network deployments. This opportunity

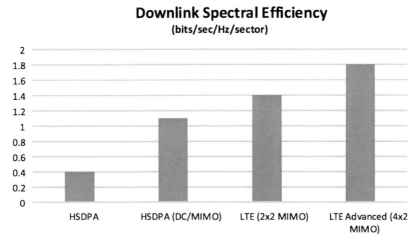

FIGURE 3.2 Spectral efficiency of radio access technologies.

cost is calculated by the savings that can be achieved by acquiring an appropriate amount of new spectrum rather than investing in new radio base station cell sites (which will use the existing assigned spectrum). In other words, demands for better service can be met either by adding newer cell sites (using the existing spectrum) or through acquisition of additional spectrum (adding more radio equipment to the existing cell sites) [16]. In some cases, operators may simultaneously need to inject capital for additional cell sites as well as for newer spectrum to stay competitive.

In macro cellular network deployments, the costs associated with radio base station sites are the largest including both capital and operational expenses. These include radio equipment, towers, masts, power, alternative power sources such as generators and batteries, nontelecom equipment such as air conditioners, and operational expenses such as installation, site rentals, and power bills. The cost of network equipment has come down drastically in the past few years due to intense competition. For example, 3G base stations have been replaced with more effective multi-standard (2G/3G/4G) nodes. The replacement of 2G base stations with multi-standard ones can cost less than USD 10,000 and these are more technologically advanced. The deployment cost of a new base station site is approximately USD 80,000–USD 100,000 (excluding network equipment) in Pakistan. With such lower prices, adding new sites makes more sense.

However, network densification with an abundance of small cells, which is expected in the 5G scenario, gets complicated. In such situations, the density of radio access points can be equal to or greater than the number of concurrent active mobile users. In denser networks, higher data rates and tougher quality of service requirements drive the need for more spectrum.

Authors in [16,20] have shown that about 2.2 times densification is needed to support the doubling of data throughput in a sparse network (base stations are much fewer in number than users). In this case, user throughput is linearly increasing with the base station density while the needed corresponding amount of spectrum is high. On the other hand, to support the doubling of data throughputs in already very dense networks, 16–20 times more densification may be needed. Contrarily, the additional amount of spectrum which is needed in such very dense networks, is relatively small, implying that the spectrum is an effective resource to handle the task at hand.

3.3.2.1 Spectrum Economic Value

The most common financial measure for spectrum is Price/Megahertz/POP (point-of-presence). It is the price paid by a buyer for one megahertz of spectrum for every individual of a country/area where the spectrum would be utilized. It should be noted that there are significant variations in price across the different bands. At the same time, there are tens of factors that play a role in determining the price of the spectrum that is utilized in telecommunication networks including, but not limited to, competition between players, timing of spectrum, financial health of the buyers, average revenue per user (ARPU), previous auctions, potential future auctions, and so on.

Figure 3.3 presents the Price/Megahertz/POP value of some of the recent auctions that took place for 1800 MHz, one of the most valuable bands for LTE. The graph is normalized for a 15 year license period. However, it is not an apples-to-apples comparison since every market has its own dynamics.

3.3.2.2 Spectrum Trading

Spectrum trading is a process that allows the licensees of a scarce resource to transfer or lease the license rights to another entity. There are different types of spectrum trading such as outright transfer, concurrent transfer, partial transfer, and leasing. With outright transfer, the rights and obligations are fully transferred from the transferor to the transferee. In the case of concurrent transfer, both transferor and transferee jointly hold the rights to use the spectrum, while with partial transfer, only certain rights are transferred to the transferee and it can be partitioned on the basis of frequency, geography or time. With leasing, the leaseholder does not own a license, but uses the spectrum by virtue of a lease contract with the license holder [21].

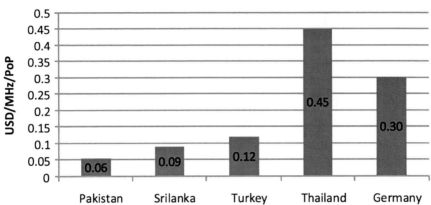

FIGURE 3.3   1800 MHz Price/MHz/POP.

A spectrum can be leased or rented out after it has been assigned, which is perhaps more economical than full scale trading in some circumstances [22]. A key reason why a service provider may prefer to rent or lease rather than trade is that leasing carries relatively lower transaction costs. Another reason is that leasing avoids payment of maintenance fees and charges, and more importantly, an operator will still have control over the spectrum. Regulators have to play a key role in all the leasing and trading exercises and manage them in an effective manner.

3.4   CONCLUSION

WRCs are the pivotal events that allocate frequencies for any technology/service that requires airwaves for its operation. WRC-15 fell short of allocating appropriate amount of frequencies for IMT/5G, however, WRC-19 is expected to fill up this gap. The chapter provided potential technical solutions and financial aspects to handle this shortage. Lastly and importantly, spectrum harmonization is needed to ensure higher economies of scale and reduce interference along borders.

PROBLEMS

1. What is the electromagnetic spectrum?
2. What is the radio spectrum?
3. What are the roles of ITU-R and WRCs in spectrum management?
4. What are the anticipated spectrum requirements for 5G?
5. What is spectrum sharing?
6. What is LSA?
7. Why is an effective air-interface design needed for 5G spectrum management?
8. What is the spectral efficiency when the aggregate sector throughput in a downlink is 1.6 Mbps in a 5 MHz channel?
9. What is the opportunity cost related to a spectrum?
10. What is the price/MHz/POP of 2 × 10 MHz of an 850 MHz spectrum that was auctioned at a price of USD 395 million in Pakistan in 2016? The population of the country at that time was 193 million?
11. What is spectrum trading?
12. Research the pros and cons of the potential 5G frequency bands listed in Table 3.3?

REFERENCES

1. Palma, C. 2016. Radio Waves to Gamma-Rays. Pennsylvania State University. https://www.e-education. psu.edu/astro801/content/l3_p4.html
2. NASA 2013. Imagine the Universe. https://imagine.gsfc.nasa.gov/science/toolbox/emspectrum1.html
3. Federal Communications Commission 2008. Auction 73: 700 MHz band – Fact Sheet. http://wireless. fcc.gov/auctions/default.htm?job=auction_factsheet&id=73
4. ITU-R 2016. *Final Acts WRC-15 World Radiocommunication Conference 2015.*
5. ITU 2012. Radio Regulations Articles Edition of 2012.
6. Wikipedia 2017. ITU Region. https://en.wikipedia.org/wiki/ITU_Region
7. GSMA 2014. The Impact of Licensed Shared Use of Spectrum. Report by Deloitte and Real Wireless.
8. ITU-R 2015. Report of the Conference Preparatory Meeting on Operational and Regulatory/Procedural Matters to the World Radiocommunication Conference 2015.
9. Americas 2016. Mobile Broadband Transformation LTE to 5G. Rysavy Research, LLC.
10. ITU-R 2012. *Final Acts WRC-12 World Radiocommunication Conference 2012.*
11. ITU-R 2008. *Final Acts WRC-07 World Radiocommunication Conference 2007.*
12. ITU-R 2000. *Final Acts WRC-2000 World Radiocommunication Conference 2000.*
13. ITU-R 1997. *Final Acts WRC-97 World Radiocommunication Conference 1997.*
14. Moore, L.K. 2013. *Spectrum Policy in the Age of Broadband: Issues for Congress.* Congressional Research Service, Washington, DC, USA.
15. Bazelon, C. and McHenry, G. 2014. Spectrum Sharing Taxonomy and Economics. The Brattle Group.
16. Li, Z. et al. 2015. Deliverable D5.4 Future Spectrum System Concept. Document Number: ICT-317669-METIS/D5.4. Mobile and Wireless Communications Enablers for the Twenty-Twenty Information Society (METIS), April.
17. Americas 2015. 5G Spectrum Recommendations.
18. Rysavy, P. 2014. Challenges and Considerations in Defining Spectrum Efficiency. *Proceedings of the IEEE*, 102(3):386–392.
19. Kim, H. 2015. Coding and Modulation Techniques for High Spectral Efficiency Transmission in 5G and Satcom. *Proceedings of the European Signal Processing Conference (EUSIPCO)*, Nice, France, August 31–September 4, 2015, pp. 2746–2750.
20. Sung, K.W. and Yang, Y. 2015. Tradeoff between Spectrum and Densification for Achieving Target User Throughput. *Proceedings of the IEEE 81st Vehicular Technology Conference (VTC Spring)*, Glasgow, Scotland, May 11–14, 2015, pp. 1–6.
21. PTA 2016. Spectrum Trading Framework (draft).
22. Jiang, W. et al. 2014. White Paper Novel Spectrum Usage Paradigms for 5G. Special Interest Group Cognitive Radio in 5G, Cognitive Networks Technical Committee, IEEE Communications Society, IEEE.

4 Standardization

Mobile telecommunications is highly governed by Standards to achieve economies of scale. For mobile communications, technical standards are documents that provide specifications primarily about technologies. These are then used to develop products (equipment) enabling such technologies. The standards are published and maintained by Standard Development Organizations (SDOs).

There are number of standard development organizations and industry forums involved in defining the specifications for telecommunications systems. The standards drive innovation, reduce implementation challenges, and address interoperability between systems and architectures.

Standardization is the second step in the development and evolution of the mobile telecommunications industry. In simpler terms, the standard development organizations like 3GPP (3rd Generation Partnership Project), IEEE (Institute of Electrical and Electronics Engineers), and so on, take the vision of ITU (International Telecommunication Union) and transform it into standards. Such SDOs have representation from non-manufacturing vendors, original equipment manufacturers, network operators, research institutions, academia, and country specific SDOs; basically, from all the players of the telecom arena. Besides SDOs, a number of commercial driven forums also exist that promote specific areas or technologies like W-iFi Alliance (wireless fidelity), Small Cell Forum, and so on. The standardization activities are currently run by developed and emerging economies and participation from developing nations is negligible. In recent times, the Service Provider (Operator) community has also significantly reduced its role and now standardization is primarily driven by the vendor/manufacturing community.

One of the main economic goals of telecom manufacturers and non-manufacturing vendors is to pitch their research (contribution) in front of SDOs, which includes their Intellectual Property Rights. If this contribution becomes part of a standard, then the vendor can expect to get considerable amount of royalty for years to come. The operators on the other hand have a relatively small patent portfolio, thus their target is not along this line. Their target, however, is to have their current technical challenges and future business requirements get incorporated into the standards. To a maximum possible extent, the industry as a whole works to address challenges and requirements in the standards: minimize the risk on current investments, reduce implementation hurdles, and reduce delay in returns on future investments.

The standardization then leads to detailed design and development of the integrated circuits, equipment/products, and devices. In a nutshell, it usually takes two to two-and-a-half years from standardization of a technology to its first full commercial launch.

This chapter will cover the following four key topics on standards/standardization:

a. Standardization processes of some key SDOs
b. 5G standardization activities
c. ITU-T guidelines for establishing SDOs in developing nations
d. Case study—Lack of Research and Standardization in OIC (Organization of Islamic Conference) member states.

4.1 KEY SDOs AND THEIR STANDARDIZATION PROCESSES

There are multiple SDOs that are involved in the standardization of telecommunications. The most prominent ones at the international level are 3GPP (3rd Generation Partnership Project), IEEE-SA (Institute of Electrical and Electronics Engineers—Standards Association), and ITU-T (ITU Telecommunication Standardization Sector).

4.1.1 3GPP

The 3GPP was formed in 1998 with the aim of producing technical specifications and technical reports for a 3G mobile system, that is, a UMTS (Universal Mobile Telecommunications System). A UMTS is based on the GSM (Global System for Mobile communication) core network and uses the Universal Terrestrial Radio Access (UTRA) air-interface in both frequency division duplex (FDD) and time division duplex (TDD) modes. The scope was later amended to include the maintenance and development of technical specifications and technical reports for GSM including GPRS (General Packet Radio Service) and EDGE (Enhanced Data rates for GSM Evolution). The 3GPP also maintains LTE and LTE-Advanced (4G) specifications [1,2].

3GPP unites seven regional/country-specific telecommunications SDOs, namely ARIB, ATIS, CCSA, ETSI, TSDSI, TTA, TTC*, known as the Organizational Partners. Their aim is to determine general policy and produce reports and specifications that define 3GPP technologies.

4.1.1.1 Standardization Process

3GPP specifications are structured as releases and the standardization work is contribution-driven. The specifications are published up to four times a year following the quarterly plenary meetings. Companies ("individual members") participate through their membership in a 3GPP Organizational Partner. Each release consists of multiple technical reports and specifications, each of which may have gone through many revisions. Often, a release provides new radio access technology and/or advancements to an existing one. The first release, called Release 99, was completed in 2001, and specified the first Universal Mobile Telecommunications System (UMTS) based 3G radio technology incorporating a CDMA air-interface. The development of releases is an on-going process and currently, Releases 15 and 16 are under progress focusing on 5G.

4.1.1.1.1 Release Development

Each release is comprised of three stages, referring to the achievement of certain milestones. The term stage is derived from the ITU-T method for categorizing specifications. The three stages are [1]:

- Stage 1 refers to the service description from a service user's point of view.
- Stage 2 is a logical analysis, devising an abstract architecture of functional elements and the information flows among them across reference points between functional entities.
- Stage 3 is the concrete implementation of the protocols appearing at physical interfaces between physical elements onto which the functional elements have been mapped.

Time frames are defined for each release and for each stage by specifying freezing dates. 3GPP Release-12 production, including stages and timelines as a sample, is shown in Figure 4.1. Once a release is frozen, only essential corrections are allowed (i.e., addition and modifications of functions are forbidden). It may be noted that detailed protocol specifications (stage 3) may not yet be complete at the time of freezing. Moreover, releases only become official when transposed into corresponding publications of the Partner Organizations (or the national/regional standards body acting as publisher for the Partner).

4.1.1.1.2 Working Structure

The 3GPP specification work is conducted in four Technical Specification Groups (TSGs):

- TSG GERAN (GSM/EDGE Radio Access Network) is responsible for the radio access part of GSM/EDGE/GERAN and is composed of three working groups.

---

* Association of Radio Industries and Businesses (ARIB) of Japan, Alliance for Telecommunications Industry Solutions (ATIS) of the U.S., China Communications Standards Association (CCSA), European Telecommunications Standards Institute (ETSI), Telecommunications Standards Development Society, India (TSDSI), Telecommunications Technology Association (TTA) of Korea, Telecommunication Technology Committee (TTC) of Japan.

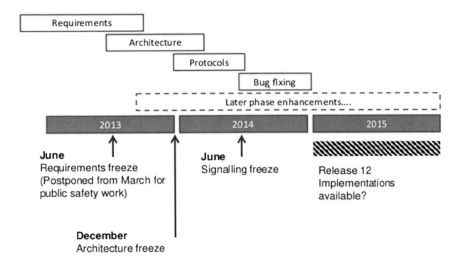

FIGURE 4.1 Release-12 production (sample). (From 3GPP Releases. Available at http://www.3gpp.org/specifications/releases [2].)

- TSG RAN (Radio Access Network) is responsible for the definition of the functions, requirements, and interfaces of the UTRA/E-UTRA (Evolved UMTS) network in FDD and TDD modes. It is also responsible for developing LTE, LTE-Advanced (4G), and 5G specifications. It is composed of five working groups.
- TSG CT (Core network and Terminals) is responsible for specifying terminal interfaces (logical and physical), terminal capabilities (such as execution environments), and the core network part of 3GPP systems. It is composed of five working groups.
- TSG SA (Service and system Aspects) is responsible for the overall architecture and service capabilities of systems based on 3GPP specifications. It is also responsible for the coordination across TSGs and is composed of six working groups.

The 3GPP structure also includes a Project Coordination Group which is the highest decision-making body. It meets formally every six months to carry out the final acceptance of 3GPP TSG work items and to approve election results and the resources committed to 3GPP.

4.1.2 IEEE-SA

The IEEE-SA, as the name suggests, is the standardization arm of IEEE. The IEEE-SA standards development process is open to IEEE-SA members and to non-members as well. However, only members can engage in a deeper level of the standardization process with additional balloting and participation opportunities. The key IEEE standards that are relevant to mobile telecommunications include, but are not limited to, WiFi (802.11 series) and Worldwide Interoperability for Microwave Access (WiMAX) (IEEE 802.16 series).

4.1.2.1 Standard Development Lifecycle

The IEEE-SA standards have a ten-year life, or in the case of trial-use standards, two years, after which they can be considered for full status or revision. Standards development is a tedious and lengthy task and requires expertise to make it a success. The IEEE standard development lifecycle consists of six stages as shown in Figure 4.2 and these are briefly discussed in this section [3].

Initiating the Project: A project gets initiated when an idea or concept has reasonable merits to be considered for standardization. This idea can be broad, such as increasing the data rates or very

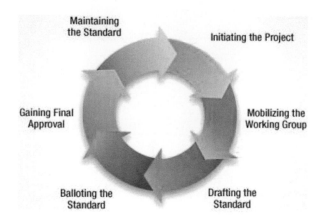

FIGURE 4.2   IEEE-SA standard development lifecycle. (From IEEE-SA. Available at http://standards.ieee. org/ [3].)

specific, like development of a new transmission technology. An idea requires a sponsor (organization) that assumes responsibility and technical oversight for it. In official terms, a standards project does not exist until a PAR (Project Approval Request) is approved. A PAR is a small, structured, legal and highly detailed document that in essence states the reason why the project exists and what it intends to do. Submission of a PAR is the first step in the standard development project.

Mobilizing the Working Group: After the approval of a PAR, a Working Group (WG) is constituted which can officially begin its work to develop the standard. The working groups are open groups where anyone with appropriate technical expertise can operate in compliance with IEEE-SA rules and working group procedures. Along the same lines, anyone can participate in projects initiated by individuals; however, for participation in corporate standards projects, IEEE-SA corporate membership is required.

Drafting the Standard: The first milestone of any working group is the completion of the first complete draft. IEEE-SA editors can assist in editing, but the overall responsibility of drafting the standard lies with the WG. The draft normally has multiple revisions before being turned into a stable standard.

Balloting the Standard: The fourth step in the process is all about getting enough votes to get it (the standard) approved. The balloting process can begin when the sponsor considers that the draft of the full standard is stable. To sponsor a ballot on standards, the sponsor either obtains an IEEE-SA membership or pays a per-ballot fee. The sponsor forms a balloting group consisting of individuals and corporations interested in the standard. Though anyone can participate by commenting on the draft, only eligible members of the balloting group can vote. Ballots usually last 30–60 days, during which balloters can approve, disapprove or abstain with or without comments. A standard will pass if at least 75% of all ballots from a balloting group are returned and if 75% of these ballots vote yes. If 30% of the balloters do not cast a vote or abstain, the ballot will fail. Each individual or corporation has one vote and can file an appeal on actions and decisions made during the process at any time. A 60-day public review process is also started simultaneously with the consensus ballot for broader acceptance and accountability. Public review comments do not have votes associated with them, but WG provides responses to all such comments.

Gaining Final Approval: The IEEE-SA Standards Board approves or disapproves standards based on the recommendation of its Standards Review Committee. This committee is responsible for ensuring that working groups have followed all procedures and guiding principles in drafting and balloting a standard. After approval, the standard is edited by an IEEE-SA editor, is reviewed by the members of the working group, and then it gets published.

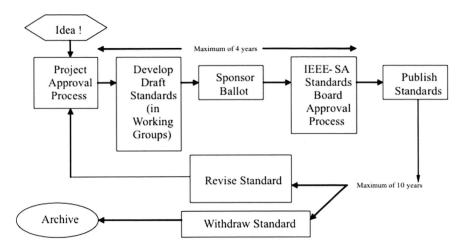

FIGURE 4.3 IEEE-SA standardization process. (From IEEE-SA. Available at http://standards.ieee.org/ [3].)

Maintaining the Standard: The standards may require technical or editorial corrections after the publication. This requirement can be addressed by issuing a corrigenda or errata sheet. The corrigenda corrects any technical errors as well as any semantic errors, requires a PAR, and needs to go through a consensus ballot. The errata sheets are separate pages issued to correct typographical or editorial errors; however, unlike the corrigenda, these require neither a PAR nor a consensus ballot. During a standard's lifetime of 10 years, working groups can develop and ballot revisions, extensions or corrections to the standard, which are appended as amendments and corrigenda. After ten years, a standard is revised or withdrawn.

In a nutshell, as shown in Figure 4.3, the development (including approval) of a standard can take a maximum of four years. Once the standard is published, it can be active for 10 years, after which either it needs to be revised or withdrawn.

4.1.3   ITU-T

The ITU-T is one of the three sectors (divisions or units) of the ITU. The main function of ITU-T is to develop international standards known as ITU-T Recommendations for Information and Communication Technologies (ICTs).

The standardization work is carried out by the technical Study Groups (SGs). Each SG is comprised of representatives (experts) of the ITU-T member organizations to develop recommendations (standards) for the various fields of ICT. Currently, there are eleven SGs studying technological innovations, operational challenges, and policy issues. From a statistical perspective, there are over 4000 recommendations in force covering a plethora of ICT topics. The recommendations are guidelines which are adopted on a voluntary basis that can be used in supply chain contracts and in technology and product development. Finally, it may be noted that ITU-T has defined separate processes for standards development and standards approval which are briefly described in this section [4].

4.1.3.1   Standards Development Process

The ITU-T standards development workflow as shown in Figure 4.4 depicts the five steps of the process. The process begins when a member organization identifies an ICT issue in need of standardization to the relevant SG. If the SG approves the idea as a study question, it later assigns it to a working party (WP). The WP starts work on the development of a new ITU-T recommendation based on the study question. The SG organizes its work in the form of study questions that drive technical studies in a particular area of ICT. A question is the basic project unit within ITU-T.

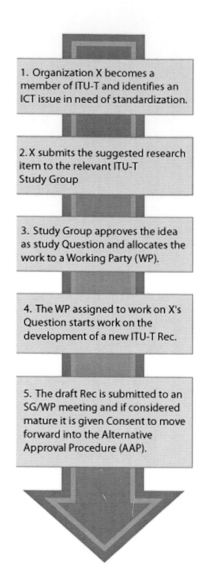

FIGURE 4.4 Standard development workflow. (From ITU-T Standards development. Available at http://www. itu.int/en/ITU-T/about/Pages/development.aspx [5].)

A study question can be published only if a number of members are committed to supporting the required work. A member's input into an SG is called a contribution. This input is normally used for suggesting new work areas and drafting recommendations or changes to existing recommendations. Once the draft recommendation from a WP is ready, it is submitted to the SG meeting and if considered mature, it is given consent to move forward into the Alternative Approval Procedure (AAP).

4.1.3.2 Standards Approval Process

The AAP was developed to shorten the standard approval timelines to meet today's industry demands. This procedure was implemented in 2001 and has cut down the lengthy process of standard approval by 80%–90%. On average, a standard can be electronically approved in two months now, while it used to approximately take four years when the vast majority of the approval process was conducted via physical meetings.

The AAP is a seven-step process as shown in Figure 4.5 addressing today's fast pace of commercialization. In the first step, after securing consent from SG/WP, the draft recommendation

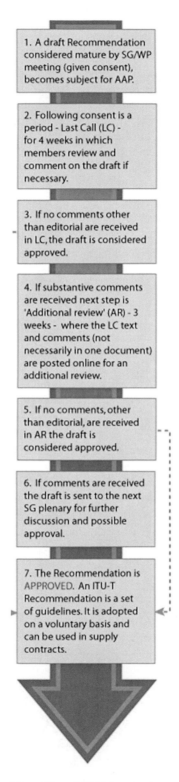

FIGURE 4.5 Standard approval workflow. (From ITU-T Standards approval. Available at http://www.itu.int/en/ITU-T/about/Pages/approval.aspx [6].)

becomes a subject for AAP. In other words, the text is mature enough to start the final review process leading to the approval of the draft recommendation. Post consent leads to the second step which signifies a period which is called the Last Call and lasts for four weeks. During this phase, the Director of ITU-T's Secretariat, the Telecommunication Standardization Bureau (TSB), announces the start of the AAP procedure by posting the draft text to the ITU-T website and asking for comments. This provides the opportunity for all members (Member States and Sector Members) to review the text and provide feedback if desired. If no comments other than editorial ones are received (step 3), then the recommendation is considered approved. However, if there are substantial comments, the SG chairman, in consultation with TSB, sets up an Additional Review (AR) process which lasts for three weeks (step 4). During AR, the last call texts and comments are posted online for review/feedback. Similar to last call, if no comments other than editorial are received in AR, then the draft is considered approved (step 5). If comments are again received, this means that there are shortcomings in the draft which need to be addressed. The draft text along with comments is then sent to the next SG plenary meeting for further discussion and potential approval (step 6). During the final step 7, the draft is approved and it becomes an ITU-T Recommendation.

4.1.4 ITU-R

The ITU-R [7] develops radio regulations and standards (Recommendations) to assure the necessary performance and quality in operating radiocommunication systems. It is responsible for global management of the radio frequency spectrum and satellite orbit resources.

The primary objective is to ensure interference free, efficient, and economical use of radiocommunication systems. This is ensured through implementation of the Radio Regulations and Regional Agreements, and the efficient and timely update of these instruments through the processes of the WRCs.

In 1999, ITU approved five radio interfaces for IMT-2000 or 3G as part of the ITU-R M.1457 Recommendation, while it added IEEE WiMAX technology in 2007. The IMT Advanced (4G) requirements were issued by ITU-R in 2008 and in 2010, LTE-Advanced and WiMAX-Advanced were officially declared as 4G standards. In a similar manner, ITU-R issued technical performance requirements for 5G radio interface(s) in 2017 and it is expected to approve the same in 2019/2020.

4.1.5 IETF

The Internet Engineering Task Force (IETF) [8] develops and promotes standards for smooth operation and evolution of the Internet. It closely cooperates with other standards bodies and conducts its tasks in WGs. It is an open standards organization, with no formal membership or membership requirements.

Similar to other standard development and approval processes, a specification undergoes a period of development and iterations. This work is conducted in WGs and by the Internet community at large. The IESG (Internet Engineering Steering Group), after the expiration of the last call period, will make its final determination as to whether or not to approve the standards action and notifies IETF appropriately [9].

If a standards action is approved, notification is sent to the RFC (Request for Comments) Editor and copied to the IETF with instructions to publish the specification as an RFC by IESG. IETF produces RFCs describing research and methods applicable to the working of the Internet and Internet-connected systems. The RFCs are submitted to communicate new concepts and information. The IETF adopts some published RFCs as Internet standards. The IETF RFCs are particularly used in the transport of voice, video, and data services over the Internet.

4.1.6 KEY REGIONAL SDOs

The regional and country-specific SDOs also play a vital role in the standardization of ICT within their regions or countries. Some prominent ones are China's CCSA, Europe's ETSI, Japan's ARIB, South Korea's TTA, and the U.S.'s Telecommunications Industry Association (TIA).

- ETSI produces globally applicable standards for ICT, including fixed, mobile, radio, converged, broadcast, and Internet technologies. It is officially recognized by the European Union as a European Standards Organization.
- The U.S. based TIA is the leading trade association representing the global ICT industry through standards development, policy initiatives, business opportunities, market intelligence, and networking events.
- The CCSA is a Chinese professional standards organization with the responsibility for developing communications technology standards.

4.1.7 KEY INDUSTRY FORUMS AND THEIR ROLES

The industry forums are usually commercially driven initiatives to speed up the promotion, development, and acceptability of a particular technology. These forums are comprised of a number of organizations including telecommunications service providers, network equipment manufacturers, device manufacturers, semiconductors vendors, and testing organizations. There are hundreds of forums in the area of telecoms and they have their own roles in the wide scheme of things. Some examples are as follows:

- The U.S. based *CDG (CDMA Development Group)* promotes CDMA/EV-DO (Evolution Data Only/Evolution Data Optimized) systems.
- The U.S. based *Metro Ethernet Forum* mission is to accelerate the worldwide adoption of carrier-class Ethernet networks and services.
- The UK based *Small Cell Forum* supports, promotes, and helps drive the wide-scale adoption of small cell technologies to improve coverage, capacity, and services delivered by mobile networks.
- The *Open Mobile Alliance (OMA)* delivers open specifications to support new and existing fixed and mobile terminals across a variety of mobile networks.

4.2 5G STANDARDIZATION

The standardization activity for 5G is currently ongoing and expected to get completed in 2019/2020. In early 2012, ITU-R initiated the program "IMT for 2020 and Beyond" that likely had set the stage for 5G research and innovation activities around the globe.

The ITU's WP 5D agreed on a roadmap during its 20th meeting held in late 2014 [10]. This roadmap includes a work plan, timelines, process, and required deliverables for the development "IMT-2020" or 5G radio interface standard(s). Based on these timelines as shown in Figure 4.6, 5G air-interface standard proposals are expected to start reaching ITU-R in late 2017 and continue until mid-2019. Later in September 2015, the ITU-R finalized its vision of the 5G mobile broadband connected society. A number of SDOs and numerous industry forums are working on the various standardization aspects (including air-interface) of 5G to meet these ITU-R targets. At the same time, the ITU-R through its working party 5D has supported the identification of additional spectrum for 5G in WRC-2015. The ITU-R vision is also instrumental in setting the agenda for the WRC-19 (during WRC-15) where deliberations on additional spectrum will take place in support of the future growth of IMT [11].

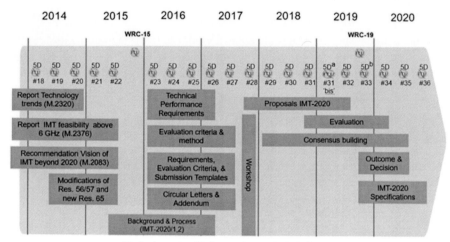

FIGURE 4.6   Detailed timelines and process for "IMT-2020" in ITU-R. (From ITU-R. Work plan, timeline, process and deliverables for the future development of IMT. http://www.itu.int/en/ITU-R/study-groups/rsg5/rwp5d/imt-2020/Documents/Antipated-Time-Schedule.pdf [10].)

3GPP is expected to accomplish its radio interface specification development task for submission to ITU-R in three releases (Rel-14, Rel-15, and Rel-16). Release-14 primarily deals with initial requirements and was frozen in June 2017. More precisely, the radio interface specification (standard) will be completed in two phases as stated by the 3GPP RAN group [12]. Release-15, which is Phase 1, the fundamental features of new radio access technology, will be formalized and is expected to freeze in H2-2018. Phase 2 (Release-16) will address all the remaining identified use cases and requirements and is expected to be completed by December 2020 in line with submission to the ITU-R [13–15].

IEEE is working on a new generation of wireless local area network standards, that is, IEEE 802.11ax known as High Efficiency WiFi to evolve as a key radio technology within 5G networks. *IETF* is also working on several building blocks including the Internet of Things, Cloud Virtualization, and RTC web (real time communications in web browsers, security, and others) for tomorrow's 5G networks.

In addition to SDOs, industry forums like NGMN (Next Generation Mobile Networks) Alliance, GSMA (GSM Association), Small Cell Forum, and others are assisting these SDOs in their respective standardization work on 5G. Additionally, there are a few large-scale government-backed industry-academic projects such as 5GPPP, METIS-2020, 5G Forum*, and others also working along the same lines.

4.3   ITU-T GUIDELINES FOR ESTABLISHING SDOS IN DEVELOPING NATIONS

ITU-T in 2014 published the first ever guidelines for establishing National Standardization Secretariats (NSS) in developing nations to coordinate standardization activities at the national level, and participate in and contribute to ITU standardization work. NSS is one of the action items of ITU's "Bridging the Standardization Gap (BSG)" initiative, which is an effort to reduce the differences between developing and developed countries [16].

---

* European Union funded/cofunded 5G Infrastructure Public-Private Partnership (5GPPP), Mobile and wireless communications Enablers for Twenty-twenty (METIS-2020), and South Korea's 5G Forum.

The standardization gap is defined as disparities between developing and developed countries in the ability to access, implement, contribute to, and influence international ICT standards, specifically ITU Recommendations. Resolution 44 (Bridging the Standardization Gap between Developing and Developed Countries) of the ITU-T World Telecommunication Standardization Assembly recognizes that the disparity in standardization between developing and developed countries includes disparities in human resources skilled in standardization and in effective participation in ITU-T activities.

4.3.1 NATIONAL STANDARDIZATION SECRETARIAT

ITU-T proposes that when a developing country sees that there is a growing awareness and use in its national environment of information and communication technologies (ICTs) and their associated international standards, there is a corresponding need to implement a national-level process to address those existing and emerging standards.

One way of addressing such functionality is to establish an NSS. An NSS could be formed under an existing government agency or any other government organization designated by the member state to represent it in ITU-T. Some of the key functions are as follows:

- Management of NSS organization structure (e.g., funding, legal authority, establishment of national advisory committees, appointment of committee chairmen, appeals processes),
- Dissemination of information from ITU-T to appropriate national stakeholders,
- Development of national strategies and policies for ICT standardization,
- Coordination of capacity building for international standardization activities, including standardization forums aimed at bridging the standardization gap, and
- Administrative functions as well as coordination activities with ITU-T.

4.3.2 EVOLUTION OF NSS

These guidelines take into account the different capability levels for standardization across the developing countries. They show how it is possible to establish an NSS at three different levels or how to grow from the NSS-General Level to the NSS-Study Group Level, and finally to the NSS-Full Sector Level. These three levels are as follows:

- General Level: this is for developing countries that have a general interest in ITU-T activities, but their involvement with any of the ITU-T study groups is minimal.
- Study Group Level: this is for developing countries that participate in some ITU-T activities and also in one or more of the ITU-T study groups.
- Full Sector Level: this is for developing countries that are deeply engaged with ITU-T activities and also actively participate in many or most of the ITU-T study groups.

Figure 4.7 shows the transition from General Level to Full Sector Level. The two key entities are RA (Responsible Agency) and NAC (National Advisory Committee). The RA in this evolution will be responsible for the higher level ITU meetings such as the ITU Plenipotentiary Conference, policy issues, and secretariat administrative functions and funding. The NAC, on the other hand, is the of all the NSS options.

Within the General Level, the NAC for ITU-T or T-NAC (National Advisory Committee for ITU-T) is the main working level advisory committee. The T-NAC is created and its leadership team is appointed by the RA, whereas participation in T-NAC is open to all interested public and private organizations. The T-NAC of a country will consider all matters of interest of that particular nation in the ITU-T. It can also create smaller ad hoc groups on an as-needed basis to address urgent and

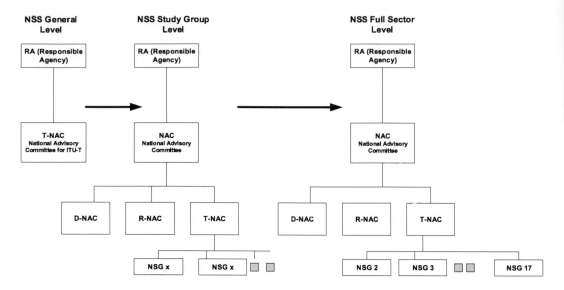

FIGURE 4.7   NSS evolution. (From Fishman, G. 2014. Guidelines on the Establishment of a National Standardization Secretariat for ITU-T—Bridging the Standardization Gap, ITU-T [16].)

specific ITU-T matters. It is further responsible for taking/defending inputs/views of the country in ITU-T WTSA (World Telecommunication Standardization Assembly), ITU-T advisory group (TSAG), and ITU-T study groups.

In the second stage, that is, the Study Group Level, the RA would create an NAC that will be an overarching organization comprised of T-NAC, R-NAC, and D-NAC (National Advisory Committee for ITU-D). The T-NAC from the previous stage would continue, however, it will report to the NAC and the R-NAC (National Advisory Committee for ITU-R) will work on the items related to the ITU-R, while the D-NAC along the same lines will work on tasks related to the ITU-D (ITU-Development Sector). During level 1, the internal ad hoc groups of the T-NAC that were formed to address specific issues will become permanent National Study Groups (NSG), participating in some ITU-T study groups and reporting to the T-NAC.

Countries that were at the Study Group Level could move to the Full Sector Level when the country's participation expands to nearly all of the ITU-T study groups. At this level, the T-NAC through its various NSGs will become part of almost all ITU-T study groups, addressing policy, network, technology, service, applications, and other issues.

4.3.3   ADDITIONAL REPORT HIGHLIGHTS

The guidelines also highlighted the importance of stable funding which is a must for NSS. In some nations, the government may be solely responsible for funding while in others, both public and private sectors can contribute. The report also suggested human resource estimates for all three options. Finally, the guidelines provided a six-step process for establishing NSS for ITU-T. This process would assist in deciding the level an NSS would be appropriate for and implementing that decision.

4.4   CASE STUDY—LACK OF RESEARCH AND STANDARDIZATION IN OIC MEMBER STATES

As illustrated in the previous sections, mobile telecommunications are highly driven by standardization. Participation and contribution in SDOs is necessary for a country or a region to have a say in the future of mobile telecommunications. The countries or regions that contribute little in the standard development process like most countries of MENAP* (Middle East, North Africa, Afghanistan, and Pakistan) correspondingly have very little or no impact on its future roadmap. Most of the OIC's 57 member states fall under the MENAP region and are collectively under discussion in this section.

There is also no meaningful role of OIC member states in the research, development, and manufacturing of mobile telecommunications. There is no single authority or SDO like ETSI that represents the OIC region at the international level. However, some country specific Telecom Ministries or Authorities do have memberships in some international SDOs. However, their memberships to these international SDOs can be best compared to an individual's health club membership that he or she attends twice a month without having any real impact and output!

In terms of R&D in mobile telecommunications:

1. The industry is driven by service providers which are solely focused on day-to-day operations and have as such no dedicated standardization or R&D functions.
2. Many service providers in OIC member states are part of the European Telecom Groups. The R&D centers of these consortiums reside in Europe, North America, and Oriental nations (non-OIC) and they do not consider it vital to have additional research centers and labs in OIC member states.
3. A home-grown telecom manufacturing community is almost non-existent. The vendor offices of international players are primarily sales offices and R&D activities hardly exist.

As a rule of thumb, in most member states, institutions for technical standards are weak [18]. Only Indonesia, Malaysia, and perhaps a couple of other nations have meaningful standardization programs out of the 57 OIC member states. Looking at this grim picture, it is not easy to find the right and balanced approach for resolving the matter. However, two possible approaches are discussed in this section as possible solutions to the challenge. These approaches are:

1. Collective Approach
2. Individual Country-Based Approach, such as NSS

4.4.1 COLLECTIVE APPROACH—COMSTECH

A science and technology platform at the international level that represents OIC member states is COMSTECH (OIC† Standing Committee on Scientific and Technological Corporation). The purpose of COMSTECH, which was established in 1981, is the cooperation and promotion of science and technology within the world of OIC [17].

There are 57 OIC member states that comprise of one quarter (1.55 billion) of the world's population. However, the share of the 57 OIC states in global R&D is only 2.1%. The COMSTECH could have been the platform for the development of standards and representation of member countries at the international SDOs for the various fields of Science and Technology (S&T), including telecom. However, COMSTECH has also not paid attention in this area and its approach is more academic in nature. The SWOT (strengths, weaknesses, opportunities, and threats) analysis of COMSTECH reveals the following sorry state of affairs [18–20]:

---

* In April 2013, the International Monetary Fund created a new analytical region called MENAP (Middle East, North Africa, Afghanistan, and Pakistan), which adds Afghanistan and Pakistan to MENA countries. The country of Israel has been excluded from the discussion as it is not part of OIC.
† OIC is the Organization of Islamic Cooperation which presents the collective voice of the Muslim world and ensures the safeguarding and protection of the interests of the Muslim world in the spirit of promoting international peace and harmony among various people of the world.

- *R&D*: As stated earlier, the OIC accounts for only 2.1% of the world total Gross Domestic Expenditure on R&D (GERD*) [20]. The contribution from the private sector is very minimal. The highest private R&D expenditure is reported for Turkey at 20% of the total R&D expenditure. In most developed countries, 60%–80% of R&D expenditure is contributed by the private sector. In South Korea, 76% of the R&D expenditure comes from private sources.
- *Standardization*: Member states' institutions for technical standards are weak and have almost no role in the development of international standards.
- *Patent portfolio*: OIC member states contributed just 1.76% to global patenting in 2012. Patent activities are mainly concentrated in a few states with Malaysia in the lead followed by Turkey, Indonesia, Iran, Egypt, Morocco, and Saudi Arabia. A single U.S./European R&D lab (Nokia Bell Labs) has more patents then the combined patent portfolio of the seven member states of OIC.

4.4.2 Country-Based Approach—Pakistan

As an example, consider the case of Pakistan where the literacy rate is very low and R&D activities are negligible (a common problem in many OIC countries). The Pakistan Standards and Quality Control Authority (PSQCA) is the country's National Standards Body. However, their focus is on development and promotion of standards and conformity assessment of non-telecom products (mainly food items, drinking water, and building materials). Another player is the Ministry of Information Technology and Telecommunication which is responsible for policy development and solving the industry's burning challenges. Perhaps the last one is the National S&T Committee on IT/Telecom headed by the Prime Minister. It is the apex decision-making body for S&T development in Pakistan.

However, the three main bodies that could have perhaps foster the development of telecom standards at the international level and/or at the MENAP level have not been specifically promulgated nor envisioned its importance for the country. The recent telecom policy that was approved in December 2015 has a clause stating that *Federal Government in collaboration with the regulator 'Pakistan Telecom Authority' and 'Frequency Allocation Board' will devise a framework for sector contribution to SDOs*. This framework and later its implementation may address some shortcomings if properly worked out. Furthermore, the contribution from the private sector is also negligible due to the same reasons pointed out in Section 4.4.1.

4.4.3 Possible Solution—Collective Approach

So, the question is—how can OIC member states promote significant improvement in the area of research and standardization. They can either collectively go through COMSTECH (or a similar platform) or through each country's specific efforts.

4.4.3.1 Why a Collective Effort?

History tells us that neither approach has worked so far; however, taking a collective approach would be better than its challenger. At the same time, there is no need to disband the existing national standard organizations which they can leverage on COMSTECH. The key reasons behind this thinking are as follows:

1. A number of OIC states have been politically and economically unstable for the last many years/decades, which continuously results in a brain drain. A common platform with states that are more stable will allow less stable states to contribute at some level.
2. In terms of funding, both rich and poor countries can contribute to a certain extent.

---

* GERD is the total intramural expenditure on R&D performed on a national territory during a given period [21].

3. Only a few states of OIC have the technological know-how on what, why, and how to contribute in regional and international SDOs. States like Malaysia, Indonesia, and Turkey are a good bit ahead of others, thus their experiences can be applied.
4. OIC member states' collective voice and contributions will have more weight and the platform that is needed is to establish an organization within COMSTECH that proactively creates regional standards and contributes to the international standardization activities already exists.

4.4.3.2 How—COMSTECH Research and Standardization Secretariat (CRSS)

The CRSS will be formed with the primary goal to become an SDO like ETSI. After developing an understanding among the members through an agreement or an MoU (Memorandum of Understanding), the first key tasks of this activity will include developing the organizational structure and functions of CRSS, developing mechanisms for funding, and getting registered at the various SDOs.

4.4.3.2.1 Organizational Structure and Functions

As shown in Figure 4.8, research, innovation, and standardization are the building blocks of the CRSS. Each block is further divided into departments to address the specific areas of mobile telecommunications. Another important aspect as shown in Figure 4.8 is the one-to-one mapping between research and standardizations blocks which is required for better coordination and problem resolution. If it is successful, the model can be replicated for other ICT fields, and later to other fields of S&T and societal implications.

Research and Innovation Block: In the research and innovation block, the following departments can be present at a minimum. The key tasks of this block are to carry out applied research, conduct simulations and proof of concept testing, produce publications, file patent applications, and conduct basic research (as applicable).

- Network Technology Research: This department can be comprised of the radio access network, transmission network, core network, and operational support systems groups.
- Device Research: A devices group can consist of subgroups that look into the various components of devices. These subgroups can include mobile software platform, radio frequency and processors, signal processing, user interface, and device management.

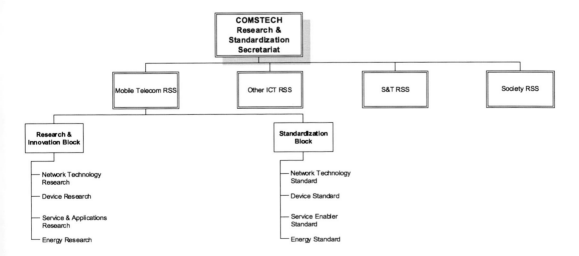

FIGURE 4.8 CRSS organizational chart.

- Services and Applications Research: This department can focus on the framework, infrastructure, and protocols that are required for the delivery of services and applications. Application development may not be the goal of the department.
- Energy Research: The energy department can focus on alternate energy and go-green initiatives.

Standards Block: In the standardization block, the following departments can be present at a minimum. The key tasks of this block are to participate in and contribute to the development of international standards, bring proposals to SDOs like ITU-T, 3GPP, and so on, develop standards for OIC member states, and create intellectual property both at country and international levels.

- Network Technology Standards: This department can participate in and contribute to technology-specific SDOs. For example,
    i. The Radio Access Network group can contribute in 3GPP GERAN and RAN TSGs and the IEEE 802 committee. It can take part in ITU-R activities and collaborate in the standardization of 5G.
    ii. The Transport Standard Development group can contribute in the transport subgroup of ITU-T Study Group 15 and the IETF.
    iii. The Core Network group can participate in 3GPP Core Network and Terminals TSG.
    iv. The Operational Support Systems group can participate in 3GPP Services and Systems Aspects TSGs and collaborate in certain communities of Tele-Management Forums.
- Device Standard: This department can focus on 3GPP Core Network and Terminals TSG, and in architecture, device management, and other WGs of the Open Mobile Alliance (OMA).
- Service Enabler Standard: This department can contribute to OMA content delivery and location WGs. It can also contribute to 3GPP Services and Systems Aspects TSG.
- Energy Standard: This department can contribute to the Smart Grid, Green and Clean Technology, and Power and Energy committees of IEEE-SA.

4.4.3.2.2 *Funding and Other Challenges*

COMSTECH needs to make a reasonable initial investment—for establishing a facility, attracting and hiring the right manpower, providing the right set of tools for research and innovation for the employees, and first-time registration/annual membership fees of SDOs such as 3GPP/ETSI, ITU, and others.

The selection of a host member state will be another major task for this endeavor. The host may need to provide 50% of the funding for establishing the facility as per certain norms of COMSTECH. Afterward, the principal source of CRSS income could be annual membership fees from each state. The annual fee will be calculated based on a country's gross domestic product (GDP). Other sources of income will include sales of standards, hosting of events and conferences and providing training. This fee will be used to cover all the above-mentioned expenses and the salaries of employees, maintenance of office(s), annual membership fees of SDOs like those of 3GPP/ETSI, ITU, and others, and capacity building. Furthermore, some ad-hoc expenses could be due to the formation of Special Task Forces which are groups of experts working together under contract to CRSS to urgently produce draft standards and other documents or to perform special studies.

The foremost ongoing expenditure is related to travel since almost all the 3GPP, ITU, and other SDOs meetings take place outside the OIC region. This travel is directly proportional to how many study groups CRSS is planning to start the process and how quickly it wants to expand its horizons.

4.4.4 FINAL REMARKS—CASE STUDY

It may be prudent for COMSTECH to make considerable contributions to the ongoing standardization activity of 5G. Thus, the creation of a COMSTECH Standardization Secretariat is of utmost importance and is urgently needed. The creation of 57 separate Standardization Secretariats is perhaps not needed when a single Standardization Secretariat can administer to the needs of low, mid, and high income, stable and less stable OIC states. The current NSSs which are present in a handful of member countries can continue to exist as needed, but their focal point at the international level should be the common OIC Standardization Secretariat, that is, CRSS, particularly for those states where the participation in international standardization activities is negligible.

The contributions can serve as a baseline for future development in the areas of research, technology development, standardization, intellectual property rights (IPR) development, and later perhaps ignite semiconductor fabrication and full-fledged product development (manufacturing). To achieve this goal, it is important that intellectual property regulations and harmonized standards development processes across OIC members are established. More importantly, it will be prudent for the CRSS to make itself a very prominent player like the ETSI and perhaps lead the standardization activity of the next generation of mobile telecommunications, that is, 6G in the next decade.

The most important element of success is the mindset change from purely operational and short-term gains to goals that lead to research and standardization, and include long term goals as well. This means involvement of domestic private businesses and government in the promotion of S&T via R&D, standardization, and patent portfolios.

In the end, CRSS must follow a checks-and-balances approach where everyone is accountable in order to become a success story.

4.5 CONCLUSION

This chapter covers the key topic of standardization from a global perspective. The standardization processes of some key SDOs, including 3GPP, IEEE-SA, and ITU-T, were discussed. The research and standardization activities of 5G were also briefly elaborated. The ITU guidelines for establishing national standardization secretariats in developing nations were presented. Finally, a case study highlighting the absence of standardization activities in OIC member states and a way forward were discussed.

PROBLEMS

1. Define standardization from the perspective of vendors and operators?
2. Briefly describe the release development process of 3GPP?
3. What are the steps of the IEEE-SA standard development lifecycle?
4. What is the role of ITU-T Study Groups?
5. What is the primary objective of ITU-R?
6. Define 3GPP activities for 5G?
7. What is NSS?
8. What are the three levels of NSS?
9. Define the role of COMSTECH?
10. Discuss, in groups, the pros and cons of having Standardization Secretariats at National versus Regional levels?
11. Develop, in groups, a case study on the formation of NSS for a non-telecom ICT industry?

REFERENCES

1. 3GPP. Available at http://www.3gpp.org/
2. 3GPP Releases. Available at http://www.3gpp.org/specifications/releases
3. IEEE-SA. Available at http://standards.ieee.org/
4. ITU-T in brief. Available at http://www.itu.int/en/ITU-T/about/Pages/default.aspx
5. ITU-T Standards development. Available at http://www.itu.int/en/ITU-T/about/Pages/development.aspx
6. ITU-T Standards approval. Available at http://www.itu.int/en/ITU-T/about/Pages/approval.aspx
7. ITU Radiocommunication Sector. http://www.itu.int/en/ITU-R/Pages/default.aspx

8. The Internet Engineering Task Force (IETF). https://www.ietf.org/
9. IETF. RFC 2026 The Internet Standards Process—Revision 3. https://datatracker.ietf.org/doc/rfc2026/?include_text=1
10. ITU-R. Work plan, timeline, process and deliverables for the future development of IMT. http://www.itu.int/en/ITU-R/study-groups/rsg5/rwp5d/imt-2020/Documents/Antipated-Time-Schedule.pdf
11. ITU-R. ITU towards "IMT for 2020 and beyond". http://www.itu.int/en/ITU-R/study-groups/rsg5/rwp5d/imt-2020/Pages/default.aspx
12. 3GPP. RAN 5G Workshop—The Start of Something. http://www.3gpp.org/news-events/3gpp-news/1734-ran_5g
13. InterDigital Inc. 5G Standardization. Mobile World Congress 2015.
14. Anritsu 2016. 5G Standardization Status in 3GPP.
15. Flore, D. 2015. RAN Workshop on 5G: Chairman Summary.
16. Fishman, G. 2014. Guidelines on the Establishment of a National Standardization Secretariat for ITU-T—Bridging the Standardization Gap, ITU-T.
17. COMSTECH. Available at http://comstech.org/
18. SESRIC 2012. Annual Economic Report on OIC countries. Statistical, Economic and Social Research and Training Centre for Islamic Countries.
19. Naim, T.S. and Sheraz, U. 2013. SWOT analysis of OIC member states. COMSTECH.
20. SESRIC 2012. Current Stance of Science and Technology in OIC Countries.
21. Organisation for Economic Co-operation and Development (OECD). https://stats.oecd.org/glossary/detail.asp?ID=1162

5 5G Concepts

Historically, the mobile technologies have transitioned from one generation to the next over a period of ten years. Figure 5.1 (reproduced from Chapter 2 with minor modification) shows this evolution and indicates the need to look beyond 4G over the next few years to define and develop the next generation of mobile networks.

This next generation, which is now commonly called the 5G (Fifth Generation) of mobile communications, has gained tremendous momentum during the last two/three years. The key difference from the previous generations is the involvement of many vertical industries (automotive, energy, finance, health, etc.) that in the past have not been very vocal.

5G, which is expected to be standardized by 2019/2020, will support a new radio access technology called 5G-NR (new radio) and an enhanced core network called NGC (Next Generation Core). The details on 5G-NR and enhanced core network are provided in Chapters 9 and 11, respectively. Some other 5G related concepts, such as massive MIMO for increasing capacity, device-to-device communications for faster connectivity, and IoT (Internet of Things) for connecting billions of devices and challenges such as Signalling Storms and Massive HetNets (Heterogeneous Networks), are also described in relevant chapters. The essence of this chapter to a certain extent is on the concepts that are applicable to both NR and NGC such as V2X (Vehicle to Everything) and network slicing and on the air interface such as cognitive radio, channel access techniques, and cloud radio access network as shown in Figure 5.2.

5.1 5G OBJECTIVES AND USAGE SCENARIOS

The ITU-R Recommendation ITU-R M.2083-0 [1], published in September 2015, defines the framework and overall objectives of the future development of International Mobile Telecommunications (IMT) for 2020 and beyond. The underlying goal is that IMT-2020 and beyond could play a significant role in serving the needs of the networked society for both developed and developing countries in the future.

This recommendation defines three usage scenarios (Figure 5.3), namely:

- enhanced Mobile Broadband (eMBB): provides higher speeds for applications such as web browsing, streaming, and video conferencing.
- Ultra-reliable and Low-latency Communications (URLLC): enables mission critical applications, industrial automation, new medical applications, and autonomous driving that require very short network traversal times.
- massive Machine Type Communications (mMTC): extends LTE IoT capabilities to support a huge number of devices with enhanced coverage and long battery life.

5.2 5G ACTIVITIES

Before jumping into the technicalities, it is worthwhile to look into the current landscape of 5G development. It is safe to say that activities started in 2013, at least a couple of years before 3GPP and ITU formally published any specific agenda on 5G.

The applied research and technological developmental activities are ongoing across the globe. Europe and Asia (China, Japan, and Korea) are, to some extent, ahead of North America. USA, however, took the lead in setting the frequency spectrum for 5G networks and applications in 2016.

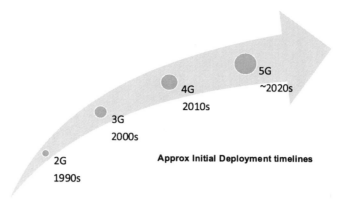

FIGURE 5.1 Mobile telecom evolution.

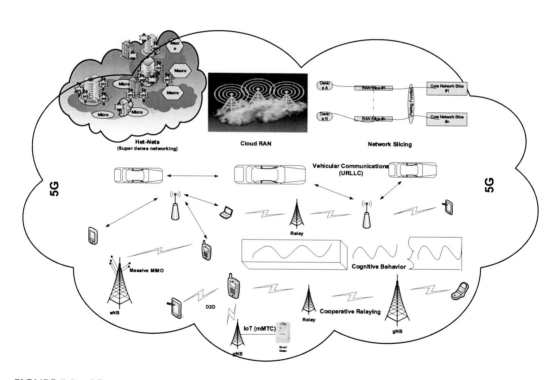

FIGURE 5.2 5G concept.

During the 2013/2014 time frame, academia was primarily involved in 5G research in U.S., while there was a governmental level support during the same time in Europe and Asia.

Details on 5G activities can be found in [2] while a summary is as follows (Table 5.1).

Earlier technology trials starting in 2014 were driven by Korean and Japanese operators in preparation for showcasing pre-standard 5G in the 2018 Winter Olympics and the 2020 Summer Olympics, respectively. WRC-15 identified several frequency bands in the range of 24–86 GHz for 5G/IMT. The ITU-R published IMT-2020 (5G) technical performance requirements in 2017. 3GPP is expected to approve 5G NR specifications (excluding NGC) in its Rel-15 by 2018. Rel-16 covering NGC is expected to get frozen in 2020.

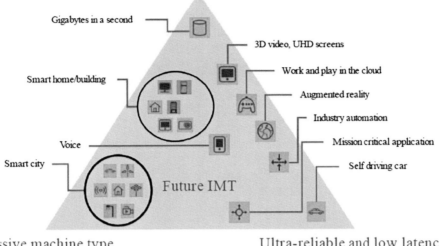

FIGURE 5.3 Usage scenarios of IMT for 2020 and beyond. (From ITU-R 2015. Recommendation ITU-R M.2083-0—IMT Vision—Framework and Overall Objectives of the Future Development of IMT for 2020 and Beyond [1].)

TABLE 5.1
Key 5G Activities

Year	Activities
2013	China formed IMT-2020 Promotion Group to promote 5G R&D by steering country's trials and other activities.
	South Korea formed 5G Forum for becoming the leading force in developing the next generation communication technology.
	University of Texas at Austin and Stanford University faculty members were awarded a National Science Foundation (NSF) grant for research on 5G wireless networks.
2014	5G-PPP (5G Infrastructure Public-Private Partnership) is a 2014 European Union initiative. 5G-PPP selected 19 projects during its Call-1, which started in the second half of 2015. 5G-PPP is part of European Union 8th Framework Project (known as Horizon 2020).
	European Union and South Korea signed a Memorandum of Understanding to collaborate on systems, standards, and radio frequency harmonization for 5G.
	NGMN (Next Generation Mobile Networks) Alliance started the work on 5G.
2015	The U.S. administration launched a $400 million Advanced Wireless Research Initiative led by the National Science Foundation (NSF).
	China National Key Project on 5G.
	3GPP's Services and Requirements Working Group started the study phase for 5G service requirements (known as SMARTER work). A few other working groups also began work.
	ITU published ITU-R Recommendation M.2083: "IMT Vision – Framework and Overall Objectives of the Future Development of IMT for 2020 and Beyond".
	Huawei conducted a joint field trial of 5G new radio access technologies with NTT DoCoMo.
2017	ITU-R published IMT-2020 technical performance requirements.

5.3 CHANNEL ACCESS METHOD/AIR INTERFACE

One of most attractive traits of mobile communications is the wireless connectivity between the device and the network. Having a perfect mobile wireless connection for voice, video, and data indulgence is the need for today and tomorrow. However, this wireless connectivity presents one of most daunting and fundamental challenges of this field.

Wireless connectivity in mobile communications is directly associated with frequency spectrum and channel access method (multiple access method). Spectrum is scarce and expensive but is a must-have to run a mobile network. The channel (atmosphere in this case) is unpredictable and beyond anyone's control. The daunting challenge is to squeeze in more bps/Hz/km² (bits per second per hertz per square kilometer), which is called system spectral efficiency. To address this challenge, beside other countless innovations, almost every generation of mobile communication has come up with a new multiple access method as an improvement over the previous one. The 2G systems use FDMA (Frequency Division Multiple Access) and TDMA (Time Division Multiple Access) techniques. The 3G systems use CDMA (Code Division Multiple Access) while LTE, WiMAX, and 4G systems employ OFDMA (Orthogonal FDMA). Thus, it can be safely said and with almost certainty that 5G will be embedded with one or more new and improved multiple access method(s) or implanted with an existing one with sufficient improvements.

Varied methods of orthogonal and nonorthogonal multiple accesses for 5G systems have been under research and investigation for at least the last five years. Nonorthogonal multiple access is more suitable for uplink since the base station can afford the multiuser detection complexity. On the other hand, for downlink, orthogonal multiple access is more suitable due to the limited processing power of the user equipment [3,4]. The focus of this section is on some of the developing channel access methods which are under consideration for 5G systems. However, before diving into the details of such potential methods, it may be productive to refresh basic understandings of channel capacity and spectral efficiency.

5.3.1 FUNDAMENTAL CONCEPTS

Air-Interface: The air-interface defines the method for transmitting/receiving information over the air between mobiles and base stations. The air interfaces of 2G, 3G, and 4G were all designed while keeping certain KPIs (Key Performance Indicators) in mind (for example, mean opinion score for voice, dropped/blocked call rates, data throughput, etc.). However, the emerging trends of IoTs, M2M (Machine to Machine), V2X, and so on are all demanding to go beyond such a static/specific approach.

Channel Capacity: Communicating messages from one location to another requires some form of pathway or medium. The communications channel is any medium (wired or wireless) over which information can be transmitted/received. Cellular communications use radio waves to carry information over the air from the user to the base station and vice versa.

Channel capacity is the tight upper bound on the rate at which information can be transmitted with an arbitrarily small error probability over a communications channel. The famous Shannon–Hartley theorem provides this channel capacity as elaborated in Equation 5.1.

$$C = B \log_2\left(1 + \frac{S}{N}\right)$$

(5.1)

where
    C is the channel capacity in bits per second
    B is the bandwidth of the channel in Hertz
    S is the average received signal power over the bandwidth, measured in watts
    N is the average noise or interference power over the bandwidth, measured in watts (or volts squared)

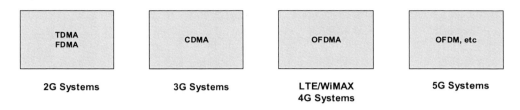

FIGURE 5.4 Channel access methods (physical layer).

In a point to point case, if R (actual bit rate in bps) ≤ C, the theorem provides the maximum rate at which information can be transmitted over a communications channel under a specified bandwidth in the presence of noise [5]. If R > C, then errorless communication is next to impossible. When there is more than one user, that is, in a multiuser case, the concept may be extended to a set of all pairs (R1, R2) such that both user 1 and user 2 can simultaneously communicate at rates R1 and R2, respectively. Under this scenario, when the bandwidth is shared, one user may communicate at a higher rate and the other at a lower rate. For example, in OFDM, this tradeoff is achieved by varying the number of subcarriers allocated to each user [6].

Channel Access Methods: A channel access method is based on multiplexing allowing sharing of a communication channel between users/devices. This form of multiplexing is based on the physical layer or layer 1 of the OSI (Open Systems Interconnection) model. A channel access method can also be based on media access control (MAC) which is the sublayer of layer 2 (Data Link Layer) of the OSI model. This section will focus on the channel access methods applicable to the physical layer.

The three multiple access techniques that are currently prevailing in mobile communications are FDMA, TDMA, and CDMA as shown in Figure 5.4. FDMA provides different frequency bands to different data streams whereas TDMA provides different time slots to different data streams. In CDMA, several message signals are transferred simultaneously over the same carrier frequency, utilizing different spreading codes.

OFDMA, which is used in 4G standards, is a form of FDMA. OFDM achieves high spectral efficiency by using orthogonal subcarriers. Orthogonality allows subcarriers' spectra to overlap, which in turn, enables transmission of more data than FDMA over the same fixed bandwidth [7].

However, OFDM does have drawbacks such as the spectrum is not localized and requires a guard band. The subcarrier spacing and symbol duration are fixed and transmission is synchronous mandating a large overhead for time alignment. These shortcomings make OFDM less attractive for some usage scenarios of 5G [3].

Spectral Efficiency: As stated in Chapter 3, the spectral efficiency refers to the information rate that can be transmitted over a given bandwidth in a specific communication system. The spectral efficiency of a mobile communications system largely depends on the choice of a multiple access method. The other factors may include the type of modulation used, error correction methods, frequency reuse factor, the number of users served, radio capability, and the percentage of time a service is active. However, spectral efficiency of a technology is largely independent of the frequency at which it operates, since modulation and coding are the same at different frequencies [8,9].

This could be measured as bit/s/Hz which is called the link spectral efficiency or bit/s/Hz per cell (site) which is system spectral efficiency. The system spectral efficiency is more practical as clarifies how efficiently an operator has deployed a specific amount of spectrum [8]. The U.S. FCC TAC (Technology Advisory Council) specifically recommends bps/Hz/km^2 as the metric for Personal Communications Systems,* which takes into account both spectral efficiency (bps/Hz) and deployment density [10].

---

* The ITU describes Personal Communications Services as a component of the IMT-2000 (3G) standard.

Today's advanced wireless technologies are essentially close to reaching the Shannon Bound, which defines the maximum/upper theoretical efficiency possible relative to noise. Thus, future gains in spectral efficiency will be limited unless a better multiple access technique is developed or sufficient improvement is made over an existing one for 5G networks.

5.3.2 MULTIPLE ACCESS/WAVEFORM

A one-size-fits-all air-interface, which has been the typical solution for the past twenty plus years, may no longer be the total solution for 5G [11]. First, since the available spectrum bands for 5G can be distributed over a large range of frequencies, including even the millimeter wave bands, the air-interface should be flexible enough so that it can operate in different frequency bands. Together with advanced radio frequency (RF) architecture and RF-related signal processing, it needs to support either flexible switching between different frequency bands or simultaneous operation in several frequency bands, including fragmented usage of certain bands. For this purpose, flexible numerology and frame structure as well as adaptive configuration are needed [12].

3GPP through its Rel-14 has endorsed OFDM-based waveform for eMBB operating up to 40 GHz in downlink and uplink. DFT-S-OFDM (Discrete Fourier transform Spread OFDM) based and CP-OFDM (Cyclic Prefix OFDM) waveforms are also supported in uplink for eMBB operating up to 40 GHz [13]. The radio characteristics for URLLC service are getting defined in Rel-15 and Rel-16 [14].

OFDM avoids interference and creates a high capacity but requires a lot of signaling and increases delay. Delay may not be suitable for URLLC and heavy signaling for mMTC types of applications [15]. Therefore, to optimize a wide variety of 5G services, a number of waveforms including OFDM were discussed in 3GPP [16]. The candidate multiple access methods can be characterized by signatures (attributes) such as use of codebooks, orthogonal/nonorthogonality mode, and the presence of an interleaver/scrambler. At the receiver, multi-user detection schemes are employed to extract the original data on a per user basis. A high-level description of these access methods is discussed in this section.

5.3.2.1 OFDM

OFDM is a multi-carrier modulation technique developed in the 1960s. The first OFDM-based standard was Digital Audio Broadcasting developed by ETSI in 1995. Since then, OFDM has been part and parcel of many telecom/broadcasting standards and its CP-OFDM form is currently used in LTE, WiMAX, and LTE-Advanced (4G) standards.

OFDM capitalizes on the use of cyclic prefixes to reduce intersymbol interference (ISI) and IFFT/FFT operations. IFFT/FFT (Inverse/Fast Fourier transform) allow combining multiple carriers at the baseband leading to OFDMA. OFDMA offers bandwidth scalability, robustness to multipaths, and effective integration with MIMO. However, aside from the benefits, OFDM suffers from high PAPR (Peak-to-Average Power Ratio) and inferior frequency localization due to the use of pulse shape filters. Details on OFDM/OFDMA can be found in [7].

To overcome such limitations, some add-ons may be added to OFDM and with these additional attributes it may become a suitable access method for 5G. Add-ons such as Weighted Overlap and Add (WOLA) described in [16,17], which replaces the rectangular pulse with a pulse with soft edges at both sides, result in much sharper sidelobe decay in the frequency domain. This decay reduces the out-of-band (OOB) leakage at the transmitter end. At the receiver, WOLA provides suppression of other (asynchronous) users' interference.

5.3.2.2 GFDM

Generalized Frequency Division Multiplexing (GFDM) is one of the nonorthogonal multi-carrier transmission methods that has been considered for 5G systems. It provides low OOB radiation and frequency localization due to variable pulse shaping filters, making it an attractive choice for IoT and cognitive radios operating in TV white spaces. Studies have shown its superiority over OFDM due to low OOB radiation and low PAPR.

As stated above, GFDM employs variable pulse shaping filters to achieve frequency localization. This localization allows the waveform to fit into narrow spectral holes eliminating interference to adjacent frequency bands [18]. An ideal pulse shape needs to attenuate very sharply both in frequency and time domains to avoid overlap with adjacent carriers/symbols to avoid intercarrier interference (ICI) and ISI. However, such pulse shapes do not exist and thus compromise has to be made with attenuation depending on the channel characteristics [19]. Such filters also affect orthogonality between the subcarriers resulting in ICI/ISI, which can be addressed by efficient detection techniques at the receiver side.

The GFDM transceiver is similar to an OFDM transceiver except it uses pulse-shaped filters for each subcarrier and a tail biting* technique. A few Rx-filter (receive filter) types come in handy for GFDM, that is, matched filter receiver, zero forcing receiver, and minimum mean square error (MMSE) receiver with varying performances.

5.3.2.3   NOMA

Today's LTE and 4G (LTE-Advanced) technologies are based on an OMA scheme, that is, OFDM. It is widely known that OFDM suffers from high PAPR, can introduce ICI due to loss in subcarrier orthogonality, and has some other impairments as well [21]. Thus, to improve spectral efficiency, a nonorthogonal scheme, namely NOMA (Non-Orthogonal Multiple Access), has been considered for 5G.

NOMA brings an additional attribute to the picture, that is, power which has not been considered to differentiate users by any currently deployed multiple access scheme. In NOMA, multiple users can transmit at the same time using the same code and frequency but with different power levels [21]. In this access method, multiple users are multiplexed in the power domain on the transmitting end and on the receiving side, SIC (Successive Interference Cancellation) can be used for multi-user signal separation [22–25].

This power sharing reduces the amount of power allocated to each user, therefore, users with high channel gains are assigned less power as compared to users with lower channel gains. The performance gain compared to OMA increases when the difference in channel gains (e.g., path loss between user terminals) is large. NOMA superposes multiple users in the power domain (forming a superposition coding) while enabling user separation at the receiving end through SIC. NOMA introduces additional complexity and delay due to the use of SIC and the performance gain is also insignificant at low SNR [26]. NOMA is suitable for both eMBB and mMTC (Massive Machine-Type Communications) types of services, but perhaps not for URLLC due to the inherent delay associated with SIC.

5.3.2.4   UFMC

The UFMC or Universal Filtered Multicarrier is a modification of the well-known waveform CP-OFDM. The term UF-OFDM (Universal Filtered OFDM) is also used synonymously with UFMC.

In CP-OFDM, symbols are separated using CP and the entire frequency band is digitally filtered as a whole. UFMC, however, applies filtering on a per sub-band (i.e., a block of subcarriers) basis and avoids use of CP [27]. The sub-band wise filtering approach was investigated since time-frequency misalignments normally occur between blocks of subcarriers (for example, sub-band wise resource allocation of different uplink users). Additionally, as the filters are broader in frequency, these become shorter in time, providing better communications in short bursts which is required for mMTC/IoT applications [28]. It may be noted that the use of zero padding instead of CP improves

---

* Tail biting is used to eliminate the need for additional guard periods that would be necessary in a conventional system in order to compensate for filtering tails and prevent overlapping of subsequent symbols [20].

spectral efficiency; however, it makes UFMC more sensitive to time misalignment as compared to the CP-OFDM waveform [29].

5.3.2.5 FBMC

Filter bank multicarrier (FBMC) is one of the potential 5G waveforms where filtering is considered at a very granular level, that is, on a per subcarrier basis. In simple terms, FBMC represents a multi-carrier system where single subcarrier signals are individually filtered with prototype filters [30]. FBMC has been proposed for cognitive radio applications [16].

For a typical multi-access system to work, the receiver (FFT) must be perfectly aligned in time with the transmitter (IFFT). During multipath propagation, the multicarrier symbols overlap at the receiver input resulting in ISI. The ISI further results in the loss of orthogonality of the carriers making demodulation harder with just the FFT. FBMC addresses this challenge by adding some additional processing to the FFT while keeping the timing and the symbol duration as it is. This additional processing together with the FFT constitutes a bank of filters [31].

The FBMC approach is different from both OFDM where filtering is applied on the entire frequency band and UFMC which filters on a sub-band basis. Thus, instead of having sinc-pulses like OFDM, the subcarriers have an appropriate shape according to the filter design and with negligible sidelobes [27]. However, the prototype filters are very narrow in frequency, necessitating rather long filter lengths (typically 3–4 times the basic multicarrier symbol length). The longer filter lengths require long ramp up and ramp down areas to address bursty data transmissions. The subcarrier filtering eases spectrum sharing and spectrum sensing, making FBMC highly applicable for cognitive radio networks [27,30,32]. FBMC also offers higher robustness against Doppler and time and frequency impairments compared to OFDM due to the use of appropriate filters. It also provides higher spectral efficiency as it does not use a cyclic prefix [30,33].

5.3.2.6 SCMA

Sparse Code Multiple Access (SCMA) is a developing nonorthogonal codebook based multiple access technique. In SCMA, coded bits are directly mapped to multi-dimensional sparse codewords. In other words, the QAM (Quadrature Amplitude Modulation) mapper and the CDMA (or Low Density Signature*) spreader are merged together to directly map incoming bits to a complex sparse vector called codeword. Each layer has a specific SCMA codebook set and a large number of layers enables massive connectivity. The layers are nonorthogonally superimposed on top of each other [35–37]. Codewords are sparse and nonorthogonal and can be detected with fewer complex detection techniques at the receiving end [38].

SCMA replaces QAM modulation and LDS (Low Density Signature) spreading with multi-dimensional codebooks enabling coding gains of multi-dimensional constellations as compared to a simple repetition of LDS. Thus, SCMA provides better spectral efficiency than CDMA/LDS due to multi-dimensional coding gains of codebooks while keeping the benefits of LDS in terms of overloading and moderate complexity of detection [35].

5.3.3 SUMMARY

To summarize, one or more multiple access methods may be prescribed for 5G networks. These methods need to provide higher spectral efficiency, lower in-band and OOB emissions, enable asynchronous access, lower overhead, lower power consumption, and reasonable complexity in the design of the transceiver [39]. Examples of waveforms supported by an OFDM-based multi-carrier approach include CP-OFDM with WOLA, UFMC, FBMC, GFDM, and so on. The nonorthogonal access methods include NOMA and so on, whereas codebook- based methods includes SCMA and so on. A category based on an interleaver/scrambler that includes methods such as RSMA (Resource

---

* A large number of chips in the sequence are equated to zero [34].

TABLE 5.2

Comparison of Multiple Access Methods

Wireless Generation	Waveform/ Multiple Access Method	Key Application	Minimum Channel Bandwidth	IFFT/FFT Operation	Cyclic Prefix	Codewords
1G	FDMA	Voice	30 kHz	No	No	No
2G	FDMA, TDMA	Voice, low speed	200 kHz	No	No	No
	CDMA	Data (around 100 kbps)	1.25 MHz			Yes
3G	CDMA	Voice, mid speed	1.25 MHz	No	No	Yes
	WCDMA	Data (approx. 10s of Mbps)	5 MHz			Yes
LTE/4G	OFDMA SC-FDMA	High Speed Data – Mobile Broadband (approx. 100s of Mbps)	1.4 to 20 MHz	Yes	Yes	No
5G	OFDM-based (orthogonal)	eMBB	100s of MHz	Yes	Yes	No
		URLLC			No	Yes
	Codebook based (non-orthogonal)	mMTC		No	Yes	No
	Interleaver/ Scrambler based			Yes (multicarrier)		

Spread Multiple Access) and so on is not discussed. Table 5.2 provides a brief comparison of different multiple access methods.

5.4 COGNITIVE RADIO

Spectrum scarcity has become a challenge due to the emergence of bandwidth hungry applications. The current practice of static spectrum assignment, where a specific frequency band is assigned to a specific organization/user for a long term and over a large geographical area, is no longer considered as a suitable option, particularly when it comes to 5G.

Cognitive Radio (CR) is one of the potential solutions to increase the effectiveness of the underutilized assigned spectrum. The concept of cognitive radio was first proposed by Joseph Mitola in his thesis and published in an article by Mitola and Gerald Q. Maguire, Jr., of the Royal Institute of Technology, Stockholm in 1999. According to them, CR extends the software radio with radio domain model-based reasoning about radio etiquette. Radio etiquette is the set of RF bands, air interfaces, protocols, and spatial and temporal patterns that moderate the use of the radio spectrum [40,41].

FCC states that a CR is one that has the ability to change its parameters based on the interaction with the environment it operates in to maximize the utilization of the radio spectrum. This interaction may involve active negotiation or communication with other spectrum users and/or passive sensing and decision making within the radio [42].

Furthermore, the primary objective of the CR is to obtain the best available spectrum through cognitive capability without interfering with the transmission of primary users. It enables the utilization of the temporarily unused licensed spectrum which is commonly known as the white space or spectrum hole [43].

This section will briefly describe the standardization efforts for CRs and regulatory aspects and highlight the key elements of CR networks.

5.4.1 STANDARDIZATION

A considerable amount of research has been conducted on such radios by various standard organizations, industry alliances, and independent researchers. The research, which is ongoing, has led to the development of multiple standards on CRs which may aid in the development of 5G standard/specification. The ETSI, IEEE, and IETF are the primary organizations that have produced standards on CRs.

5.4.1.1 IEEE Standards

A number of IEEE committees and working groups have provided specifications on CRs, dynamic spectrum sharing, and efficient utilization of TV white spaces. The key ones include IEEE 802.22 standard for Wireless Regional Area Networks (WRAN), IEEE 802.11af standard for Wireless Local Area Networks (WLAN), IEEE 802.15.4m standard for Wireless Personal Area Networks (WPAN), IEEE 802.19.1 standard for coexistence, and IEEE 1900 standards for dynamic spectrum access radio systems and networks.

IEEE 802.22: The IEEE 802.22 standard specifies air-interface and enabling of CR technologies for WRANs by enabling spectrum sharing with TV white spaces. The WRANs prescribe the use of vacant channels in the VHF and UHF bands operating in the range of 54–862 MHz that are allocated to TV broadcast services on a noninterfering basis. WRANs are suitable for less populated areas providing broadband connectivity. The standard specifies the air-interface, including the cognitive MAC and physical layer. WRANs can cover a radius of 10–30 km and go up to 100 km, supporting data rates up to 22.69 Mbps [44].

The IEEE 802.22-based systems are comprised of a data plane, management/control plane, and a cognitive plane. The data plane consists of the physical layer, the MAC, and the convergence sublayer. These layers can communicate with each other's service using Service Access Points (SAPs) that are provided with a well-defined interface or set of primitives. SAPs are added in between these layers to allow modularization of the system where different components may be disjointed and/or from different vendors. An SAP is provided with a well-defined interface or set of primitives to exchange the information providing a venue by which these different components can talk to each other. The data and control/management plane of the MAC is comprised of three sublayers including a service specific convergence sublayer, the MAC common part sublayer, and the security sublayer 1. The management/control plane consists of the Management Information Base (MIB) where SNMP (Simple Network Management Protocol) is used to communicate with the MIB database. The cognitive plane looks after spectrum sensing and geolocation. The details of these planes can be found in [44].

IEEE 802.11af: The purpose of this standard [45] is to allow WLAN operation in the TV white space spectrum (54–790 MHz) on a noninterference basis. It defines technologies and mechanisms for WSDs* (white space devices) to share the underutilized TV white space (TVWS) with the primary incumbents. The standard specifies necessary TVWS operating parameters, frequency bands, channel bands, and regulatory domains so that WSDs can operate anywhere in the world. The key difference between regulatory domains is the timescale in which WSDs are allowed by the GDB (Geolocation Database) to transmit/receive/relinquish their respective transmissions in TVWS. The standard is also known by the names of "Super Wi-Fi" or "White-Fi" and provides a common architecture assisting WSDs to satisfy multiple regulatory domains. The physical layer is based on IEEE 802.11ac standard supporting MIMO and a peak data rate of 462.7 Mbps with 6 and 7 MHz channels and 568.9 Mbps with an 8 MHz channel. The MAC employs the traditional

---

* WSDs employ cognitive capabilities to use the white space spectrum without causing harmful interference to protected/ incumbent services.

CSMA/CA (Carrier Sense Multiple Access/Collision Avoidance) protocol for transmission on a shared transmission medium.

The *IEEE P1900* initiative of the Dynamic Spectrum Access Networks Standardization Committee (DySPAN-SC) focuses on improving spectrum utilization of dynamic spectrum access radio systems and networks by enabling new techniques of dynamic spectrum access and coordination with wireless technologies for coexistence [46]. The committee has produced multiple standards through its seven working groups [47,48]. For instance:

- IEEE 1900.1 provides definitions and explains fundamental concepts of spectrum management and related topics.
- IEEE 1900.2 provides guidelines for analyzing the potential for the coexistence/interference between radio systems.
- The IEEE 1900.3 working group was dismantled; however, it was supposed to provide recommended practices and methods for evaluating the 1900.2 standard.
- IEEE 1900.4 specified the architectural building blocks to enable network-device distributed decision making in order to optimize radio resource usage in heterogeneous wireless networks.
- The IEEE 1900.5 standard defines a vendor independent set of policy-based control architectures and corresponding policy language requirements to manage CR for dynamic spectrum access applications.
- The IEEE 1900.6 standard specifies the interfaces and data structures to exchange sensing information of spectrum access systems.
- The IEEE 1900.7 prescribes a radio interface of dynamic TVWS spectrum access systems to support both fixed and mobile operations.

IEEE 802.15.4m: This is a standard [49] that defines the operation of low-rate WPAN (LR-WPAN) in TVWS. For the execution of this operation, three types of physical layer modes, namely Frequency Shift Keying (TVWS-FSK) PHY (physical), Orthogonal Frequency Division Multiplexing (TVWS-OFDM) PHY, and Narrow Band Orthogonal Frequency Division Multiplexing (NB-OFDM) PHY are defined. The TVWS Multichannel Cluster Tree personal area network (PAN) (TMCTP) technology is employed for cost effective and spectrum efficient communications. It targets on low-data-rate TVWS networking applications in sensor, smart grid/utility, and IoT networks [47].

IEEE 802.19.1: The purpose of the standard is to facilitate the family of IEEE 802 wireless standards to efficiently use TVWS. It fulfills this purpose by providing methods for coexistence among dissimilar or independently operated networks of TVBD (TV band devices) and dissimilar TVBDs [50]. The standard has specified two classes of coexistence algorithms, namely coexistence discovery algorithms and coexistence decision algorithms. A discovery algorithm can be used to detect white space objects (devices or networks) that may affect each other's performance. The decision algorithm, on the other hand, may make a decision on the channel and power allocation through negotiations [47].

5.4.1.2 Other Key Standards

IETF: IETF RFC 7545 [51] defines a Protocol to Access TV White Spaces or PAWS. PAWS is an extensible protocol, which is built on top of HTTP (Hypertext Transfer Protocol) and TLS (Transport Layer Security). It is used by a device with geolocation capability to obtain available spectrum information from a geospatial database. The standard defines a master TVWS device which can directly query the database and/or on behalf of a slave device to obtain available spectrum information while the slave TVWS device can only obtain information from the master TVWS devices [47].

ETSI: ETSI's Technical Committee Reconfigurable Radio System (TC RRS) performs the work of standardizing *Reconfiguration through Radio Applications and Cognitive Radio*. The TC has produced twenty plus standards on reconfigurable radio [52]. It has defined a Functional Architecture (FA) for the Management and Control of RRS and a Cognitive Pilot Channel (CPC) specific reports such as ETSI TR 102 682 [53] defined FA whereas CPC is described in ETSI TR 102 683 [54]. The FA and CPC are outlined in the respective technical reports, including the extent necessary to identify architectural elements (blocks and interfaces) and possible implementations for the CPC as candidates for further standardization. The RRS defines a CPC for efficient discovery of available radio accesses in a heterogeneous wireless environment. CPC is a channel which conveys the elements of necessary information facilitating the operations of CR systems.

ITU-R: The ITU-R SM.2152 [55] report, published in 2009, provides definitions of software defined radio and the CR system. The report ITU-R M.2242 [56], published in 2011, addresses aspects of CR systems specific to IMT systems. The ITU-R passed the resolution ITU-R 58–1 [57] in 2012 to conduct studies on the implementation and use of CR systems. In preparation [58] for WRC-19 and Radio Assembly 2019, ITU-R is conducting further studies for the implementation and use of CR systems.

5.4.2 Spectrum Management

From the perspective of cellular communications, radios were technology specific in the 1990s, and in the 2000s, became multi-standard radios by incorporating the SDR (software defined radio) technique. In the coming years, CR may become part of the networks. The CR is considered to be an advanced form or evolution of SDR. SDR is well known as a radio where a number of components that were traditionally implemented in hardware are instead implemented via software. A CR, as the name implies, additionally senses its environment, tracks changes, and reacts to its findings [59].

A CR-based network identifies two types of users, namely primary users and secondary users. The primary users (licensed users) use traditional wireless communication systems with static spectrum allocation and have priority in spectrum utilization within the band. Secondary Users (SU) are equipped with CRs and exploit spectrum opportunities to sustain their communication activities without interfering with primary user (PU) transmissions [60]. This requires spectrum aware operations which can consist of four steps, namely spectrum sensing, spectrum decision, spectrum sharing, and spectrum mobility as shown in Figure 5.5 [43,61].

5.4.2.1 Spectrum Sensing

The spectrum sensing function involves probing the spectrum, capturing the information, and locating an unused part of it for sharing [43,61,62]. Spectrum sensing schemes can be classified either as narrowband or wideband [63].

5.4.2.1.1 Narrowband Spectrum Sensing

Narrowband sensing implies that the frequency range is sufficiently narrow such that the channel frequency response can be considered flat [64]. Matched filter detection, energy detection, and cyclostationary feature detection are the three key methods that fall under this category [63].

Matched Filter Detection depends on the prior knowledge of the characteristics of the PUs. However, if that is not accurate, then the performance of the filter is not satisfactory. It requires CRs to be equipped with synchronization and timing devices.

Energy Detection uses energy detectors to detect the PU based on the energy of the received signals (e.g., Gaussian noise). The energy detector can be easily implemented; however, it cannot differentiate signal types. Thus, the energy detector often results in false detection and its performance is also susceptible to uncertainty in noise power.

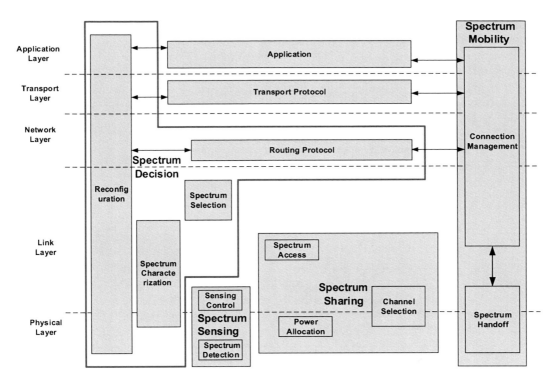

FIGURE 5.5   Reference model for spectrum management within WRAN.

Cyclostationary Feature Detection scheme determines the presence of PU signals by separating the modulated signal from the additive noise [65]. This dissimilarity is due to the fact that noise has no correlation while modulated signals are cyclostationary. The main advantage of this particular detection scheme is its sturdiness to the uncertainty in noise power. It is an effective and robust scheme; however, it is computationally complex and requires a significantly long sensing time.

5.4.2.1.2   Wideband Spectrum Sensing

Wideband spectrum sensing is intended to find more spectral opportunities over a wide frequency range. The key sensing technique in this category is cooperative detection.

Cooperative Detection refers to spectrum sensing methods where information from multiple CR users is incorporated for the detection of PUs. Cooperative detection can be implemented either in a centralized or in a distributed manner [66]. In the centralized method, the central unit (e.g., CR-based base station) receives signals from the CR users. After receiving, it then combines these signals, detects the spectrum holes, and sends this information back to the CR users. On the other hand, distributed solutions operate without a centralized unit but require exchange of observations among CR users for detection of PUs. The detection in distributed mode can take place by utilization of a neighborhood criterion. Although cooperative sensing improves detection accuracy, it increases signaling, resulting in higher latency in collecting this information due to channel contention and packet retransmissions. Furthermore, each cooperating user may have different sensing accuracies according to location. Thus, CR networks may need to consider these factors to find an optimal operating point.

5.4.2.2   Spectrum Decision

After the detection of the spectrum, the next step is to find the best available channel for the CR users. The selection of a spectrum involves a number of steps that are briefly described in this section [43,61].

Spectrum Characterization/Analysis: Spectrum analysis shall be conducted for characterization of different spectrum bands to get the best channel appropriate for the CR users' requirements. The spectrum characterization may involve detection/determination of parameters such as interference level, channel error rate, path-loss, received signal strength, the number of users that are using the spectrum, and so on. Channel capacity is one of the most important factors for spectrum characterization. Several methods [67–69] including the famous Shannon-Hartley theorem can be used for estimating the channel capacity.

Communication Features Selection: After the selection of the channel, CR users then adaptively select the appropriate modulation types, error control schemes, and upper layer protocols to meet the application requirements. These communication characteristics have to be adaptable to allow for changes in the characteristics of the spectrum.

5.4.2.3 Spectrum Sharing

Spectrum is expected to be shared between primary and secondary (cognitive) users in 5G networks. The basic goal behind spectrum sharing is that it shall not cause interference to the PUs while maintaining QoS for cognitive users. The task is performed by coordinating the channel access as well as adaptively allocating communication resources. This coordination is required to prevent multiple users from colliding in overlapping portions of the spectrum.

Various approaches have been exploited to determine the communication resources for both PUs and CRs. Each CR user has the goal of using the spectrum resources to the fullest extent. However, at the same time, they also have competing interests to maximize their own share of the spectrum resources. Thus, the activity of one CR user can impact the activities of others. Game theory provides an effective distributed spectrum sharing scheme by describing the conflict and cooperation among CR users, and hence allowing each CR user to rationally decide on the best course of action [61].

5.4.2.4 Spectrum Mobility

The need for spectrum mobility arises when current channel conditions become worse for CR users or a PU appears in the area. This activity leads to a new type of handoff in CR-based networks called spectrum handoff in which CR users transfer their connections to an unused spectrum. The mobility management protocols are required to learn in advance about the duration of a spectrum handoff so that these transitions are made in an effective fashion. Once the protocol has learned about the latency, it is imperative that it minimizes the performance degradation of ongoing communications of CR users.

5.4.3 REGULATORY ASPECTS

Spectrum sharing is essential for the success of CRs and for 5G. As defined in Chapter 3, spectrum sharing is categorized as the collective use of a frequency band by two or more parties in a specific geographical area. Sharing can take place in both licensed and license-exempt bands [70]. The value of CR will depend on how spectrum sharing is technically and economically devised. For PUs, it means that their licenses allow sharing with SUs. During sharing, a PR allows the SU(s) to use any unused spectrum on a temporary basis.

For a spectrum that is currently licensed to the PU through an auction, the SUs can acquire spectrum either statically or opportunistically. During the static mode of operation, SUs may have signed an agreement with the PU with guaranteed access for an agreed geographic area, time frame, and frequency range. In opportunistic mode, SUs can compete for channels on a noninterference basis but without a formal agreement with the incumbent [71].

5.4.3.1 Example: FCC 3.5 GHz Decision

Spectrum sharing between a government application, such as radar or a satellite, and a commercial mobile network appears to be efficient, particularly if the government use is limited to certain areas or only some portion of time [14]. In this respect, one of the first spectrum sharing attempts was made by the FCC in 2016, where it allowed the use of 150 MHz in the 3.5 GHz band for both incumbents and SUs [72,73]. The incumbents, the U.S. Department of Defense (DoD) and fixed satellite services providers, shared the spectrum with commercial users such as mobile operators. The services will be primarily unlicensed or 'lightly licensed' services, and with this approach, an operator will not be required to buy and permanently own the spectrum. The FCC envisioned a three-tiered plan or framework for the 3.5 GHz band that includes an incumbent access tier, priority access tier (acquire spectrum for up to three years through an auction process), and general authorized access tier (any user with an authorized 3.5 GHz device). The three tiers are to be coordinated through a dynamic Spectrum Access System (SAS) which is a database. The FCC has so far conditionally approved seven SAS Administrators, namely Amdocs, Inc.; Comsearch, CTIA (Cellular Telecommunications and Internet Association)—The Wireless Association, Federated Wireless, Google, Inc., Key Bridge, and Sony Electronics, Inc. The ultimate success/failure of this endeavor has not yet been determined.

5.4.3.2 Case Study: Economics (L-Band Case)

Spectrum sharing has the potential to increase the price/MHz/POP or the monetary value of the frequency spectrum. For example, the anticipated future use of the 1500 MHz spectrum can result in more money for a government for two reasons' (i) its primary allocation for IMT, that is, for mobile services in the last WRC-15 and (ii) spectrum sharing (if allowed). The mathematical factor that may need to be considered to identify the increase in price may depend on the geographical area, willingness of the current user, current use, and future prospects of the spectrum and technology.

During WRC-15, 51 MHz of L-band (1427–1452 MHz and 1492–1518 MHz) was identified for IMT worldwide. An additional 40 MHz (1452–1492 MHz) was also identified for IMT in ITU Regions 2 and 3 and in some countries of Region 1. The frequency range of 1427–1518 MHz is already used for commercial IMT services in Japan, however, in rest of the world, it is mainly used for fixed links, radars, and military purposes [74].

Band 32 (1452–1496 MHz) for supplemental downlink only, and Band 45 (1447–1467) for TDD-LTE, have been assigned in EU and China, respectively, but no deployments have been reported as such. Band 32 has been receiving a healthy reception in Europe. The UK telecom regulator Ofcom has allowed the transfer of this band from a fabless company Qualcomm, UK, to mobile operators Vodafone UK and Three UK. During 2015–16, band 32 was auctioned in Germany and France resulting in a reasonable profit for the respective governments. Bands 32 and 45, however, are not under discussion and focus is on the WRC-15 allocated frequency spectrum.

Considering the case of Pakistan, the L-band (1425–1535 MHz) has been assigned to Railways for quite some time. Pakistan Railways is currently running an analogue microwave system (i.e., as a fixed service) for its communication needs. The spectrum assigned to Railways was free of cost since it is a government entity.

A two-tiered approach as shown in Figure 5.6 can be considered to allocate this spectrum to be shared between Pakistan Railways and commercial users (mainly mobile operators). Tier-1 is comprised of the incumbent that is protected from all other users, while tier-2 is for mobile operators. Mobile operators will get this band through auction for a licensed period of 10 years, but at the cost of increased complexity. The users can be served while traveling on trains as well as in the rest of the country.

The opportunity is quite promising as 40 MHz (1452–1592 MHz) would be sufficient to operate a TDD-based 4G/5G network and/or for IoT for the four cellular operators of the country. An SAS [75], which is an automated radio spectrum coordinator for the protection of incumbents, will be required to support this approach. SAS can be maintained by each of the operators, as users of the shared spectrum resources need to be under its authoritative control.

To calculate a price tag for this futuristic opportunity is rather challenging as equipment is not available (beyond Japan) and there is as such no benchmark auction value available. If the equipment becomes available globally in the next 2–3 years (by 2020), this band will be a good choice for 5G.

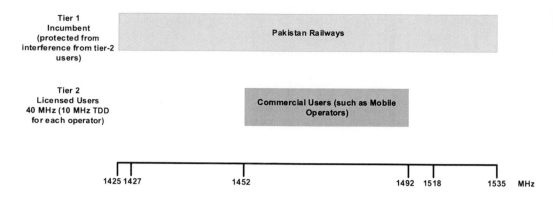

FIGURE 5.6   2-tiered approach.

At this moment, the national exchequer is not earning on this particular band, but if the approach in Figure 5.6 is adopted, it will get millions of dollars for the years to come beyond 2020.

5.5   MASSIVE CENTRALIZED RAN

Today's mobile networks primarily follow a distributed RAN architecture where a single RRU (remote radio unit) is connected to a single and specific BBU (baseband unit). The RRU (located on the pole of the tower) is connected to the BBU (placed at the base of the tower) with a point-to-point cable link based on either fiber or coax. This one-to-one RRU-BBU configuration is neither financially nor energy wise efficient.

5.5.1   CENTRALIZED RAN/CLOUD RAN

Centralized RAN and Cloud RAN approaches have changed this static configuration model. The term C-RAN was first likely introduced to the world by China Mobile during 2010 and refers to the nonstatic relationship between BBUs and RRUs. The name comes from the four "C"s, that is, Centralized processing, Clean (Green), Collaborative radio, and Cloud Radio Access Network [76].

5.5.1.1   C-RAN ARCHITECTURE

Primarily, there are two approaches to split base stations, functions between RRU and BBU within the C-RAN architecture as shown in Figure 5.7. With full centralization, the baseband (i.e., layer 1), layer 2, and layer 3 functions are located in the BBU pool. In partial centralization, the RRU integrates layer 1/radio functions, while all other higher layer functions are still in BBU. Both approaches have pros and cons. Though the full centralization technique will ease the network upgrade process, it requires very high transmission bandwidth between BBU and RRU. The partial centralization approach requires certain lower transmission bandwidths between the units, but it gives less flexibility for upgrading [76–78].

Some key elements and functionalities of this architecture are as follows:

Transport: The connectivity between BBUs and RRUs in C-RAN is provided through fronthaul which is for the most part is supported via optical fiber cable, but in some cases, wireless links are also used. Wireless links are essential since it is very difficult to extend fiber to every RRU site. It is pertinent to note that as the distances for BBUs and RRUs run to 10s of kilometers, a mix of wired (fiber, Ethernet) and wireless technology may have to be utilized. Furthermore, in some cases, repeater sites may be placed between RRU sites and pools of

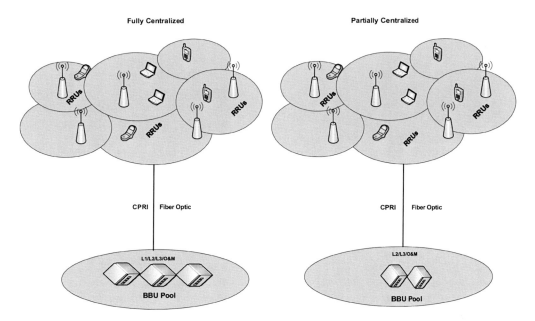

FIGURE 5.7 C-RAN architecture.

BBUs to address terrain, right-of-way, and monetary challenges. In fronthaul, BBU and RRU communicate using an analog RF signal, while backhaul which connects the base transceiver station (BTS) to the Radio Controller in 3G and to the core network in 4G is primarily Ethernet/IP based. This puts stringent requirements on fronthaul in terms of capacity and latency. Fronthaul needs to support high capacity and low latency which mandates operators to utilize fiber to the deepest level in this segment of the network. Dark fiber and wave division multiplexing are suitable for RRU-BBU connectivity, but cost is a prohibitive factor, at least in the case of the latter. Carrier Ethernet can also be applied from RRU toward the BBU pool. However, if Ethernet is chosen as the transport method for C-RAN, a CPRI (common protocol radio interface) Ethernet gateway will be needed. This gateway maps CPRI data to Ethernet packets, close to or at the interface of RRU toward the BBU pool [78].

Interface(s): Today, C-RAN is primarily supported by CPRI (common protocol radio interface). The current CPRI release 7.0 which was published in October 2015, specifies ten-line bit rate options with the lowest at 614.4 Mbps and the highest at 24.3 Gbps. It is essential for BBUs and RRUs to support at least one such bit rate [79]. CPRI protocol is sensitive to latency and the synchronization performance of a transmission system, so in a number of cases, it is limited to a distance of about 40 km [77,78,80]. Other interfaces, such as OBSAI (Open Base Station Architecture Initiative) and ORI (Open Radio equipment Interface) are also used/specified. Thus, these existing protocols may need to be revised to allow for high volume transmission over long distances [78]. China Mobile is currently working on developing a new interface, that is, Next Generation Fronthaul Interfaces (NGFI), to address the low transmission efficiency and scalability of CPRI/OBSAI interfaces [81].

Virtualization: The placement of a number of BBUs in a centralized pool while distributing RRUs according to targeted RF strategies means that operators employ virtualization technology that maps radio signals from/to one RRU to any BBU processing entity in the pool. The functions of BBUs may be realized through software instances making those virtual base stations. However, full virtualization is more of a long-term solution necessitating the use of virtualized BBUs running on commercial servers within an NFV (Network Function Virtualization) platform [77,82].

5.5.2   KEY ADVANTAGES

The key benefits of C-RAN are as follows [76,82]:

- *Energy Efficient Infrastructure*: C-RAN is an eco-friendly and energy efficient concept. The BBU pool in C-RAN is a shared resource, thus low power consumption and better load balancing can be achieved by dynamically allocating processing capability during a 24-hour period. During night time, several BBUs can either be turned off or put on low power to save energy without affecting the 24/7 service commitment.
- *Cost Savings*: The centralization allows placement of BBU pools in a few locations which will save O&M (operations and maintenance) costs. With C-RAN, transmission equipment can be shared as well, reducing CAPEX (Capital Expenditure) as well OPEX (Operational Expenditure) to some extent.
- *Improved Spectrum Utilization*: C-RAN allows implementation of joint processing and scheduling to mitigate intercell interference which improves spectral efficiency. For example, CoMP (Cooperative Multi-point Processing) technique of LTE-Advanced, which mitigates intercell interference [83], can be implemented under the C-RAN infrastructure.
- *New Business Models*: Cloud RAN may bring new as well as enhance existing business models such as base station (BS) pool resource rental system, cellular infrastructure sharing, intellectual property pooling (patent pooling), and so on.

5.5.3   KEY CHALLENGES

A number of challenges are associated with virtualization, including real time baseband processing algorithm implementation and dynamic processing capacity allocation to deal with the dynamic cell load in C-RAN [76]. The amalgamation of BBUs into a central location leads to security concerns and requires effective disaster recovery solutions.

C-RAN requires a robust transport system to address the high bandwidth and stringent jitter and latency requirements of 4G and 5G. It is pertinent to mention that carrier aggregation 4×4 MIMO has been deployed and the focus is on 8×8 MIMO, which translates to more capacity and antennas. The CPRI or some other interface may need to support high bandwidth and low latency to meet such demands.

5.5.4   MARKET STATUS/FORECAST

Before the advent of 5G, LTE and LTE-Advanced were perfect candidates for C-RAN. Features of LTE/4G such as CoMP and eICIC (Enhanced LTE Inter Cell Interference Coordination) can be supported with C-RAN. These features are greatly facilitated since signal processing of many cells can be done within a single BBU pool, which in turn reduces transmission delays. CoMP makes efficient use of the radio resources and provides processing gains due to the reduction in X2 (eNodeB to eNodeB interface) traffic (an inherent feature of C-RAN).

A mix of wired (fiber) and wireless (high capacity 70/80 GHz based microwave radios) is a key for the success of C-RAN. From both technical and business perspectives, it is difficult to connect every RRU (macro and small cell sites) with fiber. Thus, wireless links will be essential for many networks around the world. Orange and some other operators are trialing the e-band based solutions for small cell type RRU deployments.

As of today, fronthaul remains a niche market and it has yet to fully take off. The early fronthaul market was driven by Japan and South Korea and is expanding to China, Hong Kong, and the U.S. Although China Mobile pioneered the concept of C-RAN, the world has not seen many deployments from the operator until recently. According to IHS Markit Ltd. [84], in 2015, Chinese operators managed to convince their government that C-RAN is the right way to go for energy efficiency and to have a green footprint.

According to IHS, the worldwide C-RAN architecture revenue is forecast to top \$12 billion in 2020 with a compound annual growth rate (CAGR) of 19.8% from 2015 to 2020. This is primarily driven by RAN expansion in the West and the beginning of 5G rollouts in Japan and South Korea [85].

5.6 VEHICULAR COMMUNICATIONS

Vehicular communications (VC) is envisioned to improve road safety, increase efficiency in traffic flow, reduce environmental impacts, and provide additional information/services to travelers on the road. In the not too distant future, vehicles will be equipped with necessary computing, communication, and sensing capabilities and user interfaces to enable automated driving.

The research on vehicular communications was started nearly two decades ago and involved the automotive industry, telecom sector, the U.S. government, and others. The research efforts resulted in the development of the Dedicated Short Range Communications (DSRC) technology. DSRC is the incumbent technology defined in IEEE 1609 and 802.11p standards. It supports V2V (Vehicle-to-Vehicle) and V2I (Vehicle-to-Infrastructure) modes of communication, primarily focusing on enabling vehicular safety applications [86–88].

The next section will look into its evolution from simple V2V/V2I to a more meaningful V2X form, architecture, challenges, and regulatory aspects.

5.6.1 FROM V2V TO V2X

In the early stages of development and until the recent past, V2V and V2I [89–93] were envisioned as the highpoints of vehicular communications (Figure 5.8):

- V2V: Using V2V communication, a vehicle can detect the position and movement of other vehicles for short distances. The vehicles communicate with each other to support applications such as cooperative driver assistance, slow vehicle warning, and so on.
- V2I: V2I enables vehicles to communicate with fixed infrastructure along the side of the road in order to provide user communication and information services such as Internet access, mobile advertising, and so on.

As technology evolves, so do the services that run on top of those technologies. Similarly, vehicular communication is evolving from V2V/V2I to V2X to keep up with the demands of the automotive

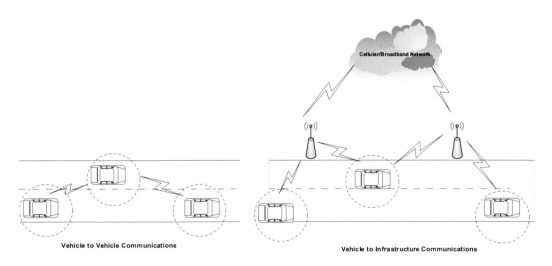

FIGURE 5.8   V2V and V2I.

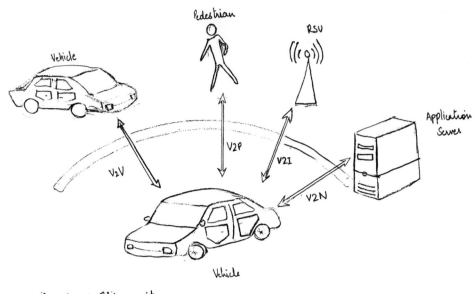

FIGURE 5.9 V2X. (From 3GPP TS 22.185 (V14.3.0) 2017. Service Requirements for V2X Services; Stage 1. Technical Specification (Release 14), Technical Specification Group Services and System Aspects, 3GPP, March [95].)

industry and to take advantage of LTE/5G developments. V2X includes C-ITS (Cooperative Intelligent Transport Systems) which is a critical component of the connected car and automated driving of the future. V2X is comprised of four forms, namely V2V, V2I, V2N (Vehicle-to-Network), and V2P (Vehicle-to-Pedestrian) as shown in Figure 5.9 [94].

V2V and V2P communications are primarily based on broadcast capability between vehicles and vehicles and road users. These two forms exchange information such as location, velocity, and direction to avoid accidents and mishaps. V2I and V2N on the other hand also involve infrastructure/network sending messages to the vehicles. V2I communication is between vehicles and roadside units (RSUs) and between vehicles and nearby traffic control devices. V2N transmission is between vehicles and application servers through cellular networks such as LTE.

5.6.1.1 Cellular V2X

Cellular V2X or C-V2X is an umbrella term for 3GPP defined V2X services [86,96]. Initially defined as LTE-based V2X, it now also encompasses forthcoming 5G technology. Cellular technology brings extensive coverage in rural, suburban, and urban areas, giving V2X services an enormous absorbable market. This will directly benefit drivers and passengers by having safer and more enjoyable travel and operators with increased usage of cellular networks along with monetary benefits [96,97].

3GPP based C-V2X offers a more future proof radio access than IEEE 802.11p because of the worldwide use of cellular technologies and durable longer term paths. It can also leverage SAE (Society of Automotive Engineers), ETIS, ISO (International Organization for Standardization), and other upper layer/transport standards that have been defined and tested by the automotive and telecom sectors. The evolution path from LTE based V2X toward 5G based V2X is developing. In the meantime, LTE will deliver C-V2X services and once 5G is available, V2X will have more capabilities and be able to offer additional services [86,98].

C-V2X defines two complementary transmission modes, namely direct communication and network-based communication [86,99].

- Direct communication is independent of the network and uses the PC5 interface (defined as a sidelink at the physical layer, between two users/devices) to address the V2V, V2I, and V2P markets. The PC5 interface was specified for 3GPP Release-12's ProSe (proximity services) feature that allows direct communications between devices in out-of-network coverage. This mode operates in Intelligent Transport Systems (ITS) bands (e.g., ITS 5.9 GHz).
- Network-based communications rely on cellular technologies such as LTE or in the future on 5G to support the V2N market. It uses the LTE Uu (between eNodeBs and users) interface to communicate between eNodeB and vehicles. It operates in the traditional mobile broadband licensed spectrum.

5.6.2 KEY STANDARDS

The IEEE 802.11p [100] is an approved amendment to the IEEE 802.11 standard and adds a vehicular communication system, that is, Wireless Access in Vehicular Environments (WAVE). It defines the enhancements to IEEE 802.11 required to support ITS applications. DSRC, defined in IEEE 802.11p and IEEE 1609 family of standards for WAVE allows high-speed data exchange between vehicles and between the vehicles and the roadside infrastructure using a licensed ITS band of 5.9 GHz (5.855–5.925 GHz).

The IEEE 1609 standards [101] define architecture, security protocols, management functions, and a standardized set of services and interfaces for enabling V2V and V2I communications. It provides two options, that is, WAVE short message and IPv6 (Internet Protocol version 6) for communication between vehicles and between vehicles and roadside units. Applications utilizing these standards in conjunction with 5.9 GHz radio can provide services to drivers, traffic operators, and RSU operators [88].

The DSRC protocol stack is based on IEEE 1609 and IEEE 802.11p standards. The latter provides physical and MAC layers while the higher layers are defined by the former. Some functions (messages and performance requirements) in the higher layers are also defined by SAE International [102]. V2V communications mainly rely on lightweight WAVE Short Message Protocol (WSMP) while TCP/IP are used in V2I and V2N modes [96].

Because LTE is a robust radio technology, it can offer a better V2X experience and feel than some existing V2X solutions. LTE using existing IEEE 1602.9 security mechanisms, SAE J2735 V2X specific messages, ITU transport technologies, and so on, can become a better V2X service provider than DSRC. It may also be noted that currently no significant efforts are in progress to evolve DSRC technology to address future needs of travelers [86]. This creates an opportunity for C-V2X to step in and take the lead.

Capitalizing on this opportunity, 3GPP has completed several studies and produced several Technical Reports [94,96,103,104]*. For example, 3GPP TR 36.786 [103] suggests that LTE operating band 47 (5.9 GHz) may be considered for V2X operations over the PC5 interface. The same report recommends frequency bands 3, 7, 8, 39, and 41 for operations over the Uu interface. Additionally, bands 3, 7, 8, 39, 41, and 47 are outlined for multi-carrier operation. 3GPP is also carrying out studies for the coexistence of different technologies in the 5.9 GHz ITS band. V2X support for vehicle platooning† as outlined in 3GPP TR 22.886 [104] can lead to shorter distances between vehicles, reduce fuel consumption, and lead to automated driving. The next step of this work, that is, 3GPP Technical Specification, is currently under development and expected to be released in time for 5G rollouts.

---

* Only some key work is referred to; details can be found at 3GPP.
† Vehicles platooning can be thought of as the carriages of a train attached with virtual strings.

5.6.3 VC Architecture

VC architecture is not expected to be developed from scratch; rather, it will be derived from the multitude of V2V and V2I and, in general, V2X projects. Vehicles, in particular high-end ones, are equipped with multiple processors and micro controllers dedicated to tasks such as fuel injection, braking, transmission, battery charging, and so on, and are also equipped with navigation systems such as GPS and sensors for velocity, direction, temperature, airbag status, rear and front cameras, and so on. The VC computing functionality is expected to be independent and responsible for running the V2X communication protocols and the supported applications [89,90,93].

Figure 5.10 compares the current protocol stack with a probable C-V2X one. The key difference between the two is at the lower layers. In the case of DSRC/WAVE, the physical layer is based on IEEE 802.11p standard, while in C-V2X, it can be based on LTE and/or 5G. The MAC is divided into lower and upper MAC supported by IEEE 802.11p and 1609.4 standards, respectively, in DSRC, while MAC is based on cellular standards in C-V2X. The upper layers, security mechanisms, and management services can be similar in the two stacks [99,102,105,106].

LTE/5G wireless transmission and medium access technologies will further enrich the VC environment as DSCR/WAVE can only support V2V and V2I based applications. The DSCR/WAVE protocol stack is split above the MAC layer between WAVE Short Message Protocol (WSMP) which is non-IP and IP. A wide variety of user cases that require low latency and high reliability can be addressed by LTE such as "do not pass warning" at highway speeds and ADAS (Automated Driving and Advanced driver assistance Systems). It is likely not feasible to address these two use cases with DSRC as these require a longer range or equivalently a reasonable driver reaction time [86].

The existing networking and transport technologies on top of physical and data link layers can be used to support data exchange between vehicles and between vehicles and roadside infrastructure/cellular networks. Routing of data can be performed using a position based forwarding strategy which is a good fit for vehicular communications. For example, the geocast method consists of broadcasting a packet in a limited geographic area around the source node. GeoCast or position-based routing assumes that every node knows its geographical position (e.g., by GPS) and maintains a locan table with the geographical positions of other nodes. The packets are forwarded in the direction of the destination based on the position of the source, destination, and the intermediate node.

The network/transport layers, in turn, support a wide range of safety and non-safety applications. Prevailing IEEE/ETSI/ISO security mechanisms along with management features can be used in support of the protocol stack [89,93].

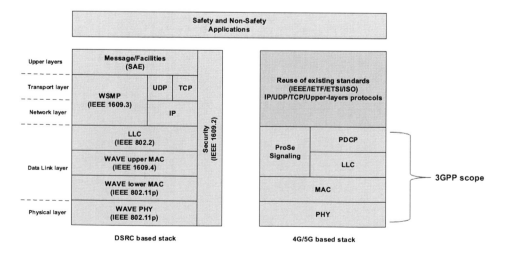

FIGURE 5.10 VC protocol stack.

5.6.4 V2X Use Cases

The primary objective of V2X is to improve transportation safety and reduce the number of accidents. The US Department of Transportation (USDoT) announced in December 2016 that it requires automakers to include V2V technologies in all new light-duty vehicles. This ruling can prevent hundreds and thousands of crashes in the U.S. Additionally, the U.S. DOT's Federal Highway Administration is also planning to issue guidance for V2I communications to improve mobility, reduce congestion, and improve safety [107]. Similar rulings can be expected on V2N and V2I in the future.

In the whole scheme of things, the screen in a vehicle can be considered as the fourth screen after TV, computer, and mobile phone, and can be used for safety as well as for infotainment. DSRC has limitations such as a shorter range and no certain evolution path. V2X by way of its anticipated design and implementation attempts to address a number of use cases including and beyond the scope of WAVE/DSRC. The following are some key use cases:

- ADAS: Advanced driver-assistance systems require high reliability and low latency message transfer at high speeds. Examples include blind spot warning, control loss warning, and so on.
- Situational Awareness: Contrary to ADAS, this requires longer wait/latency requirements such as queue warnings and hazardous road condition warnings.
- Road Travel Services: This may include providing communication support for intermodal travel that includes more than one mode of transportation.
- Auxiliary Services: This includes flexible types of communication such as route planning, infotainment, and so on.
- 3GPP defines a number of use cases in TS 22.185 such as forward collision warning, automated parking system, road safety services, and so on.

5.6.5 VC Challenges

VC brings a number of technical but also behavioral challenges (financial, privacy, legal, and organizational issues) because a number of industries are involved in the upbringing. One such challenge is related to the security of such systems. To protect VC systems from information contamination such as false warnings, and other such attacks, strong and efficient security mechanisms are required. Furthermore, security measures such as beaconing, neighbor discovery, and geocasting must be taken in order to ensure vehicle identification methods are effectively implemented. Similarly, the mobility management aspects of VC systems including location management and handover mechanisms between cellular infrastructure and RSUs demand further investigations.

A fully implemented V2X system can address some of the above-mentioned challenges. However, it may be noted that V2X also has limitations. For example, not all roads are covered with cellular infrastructure and RSUs, thus additional investment may be required. V2V is a sort of built-in feature, but for V2N/V2I, the cellular industry may have to develop business models that can justify additional investment.

5.6.6 Regulatory Aspects

The FCC assigned 75 MHz in the 5.9 GHz (5.850–5.925 GHz) band for DSRC in 2006. Ten years later, the FCC released a public notice on June 1, 2016, for potential spectrum sharing between DSRC and unlicensed WiFi. Primarily, two methods of spectrum sharing between unlicensed operations and DSRC are under investigation. The first technique is a "detect and avoid" approach where the WiFi operating in the U-NII-4 (Unlicensed National Information Infrastructure) band* evades using

---

* U-NII-4 operates at 5.850 to 5.925 GHz.

the said band when a DSRC signal is detected. The second technique looks into rechannelization where the upper 30 MHz portion remains allocated to latency sensitive safety-of-life DSRC communications while the lower 45 MHz is shared with unlicensed communications.

In August 2008, the European Commission allocated part of the 5.9 GHz band (5875–5905 MHz) for safety related aspects of ITS. Eight years later, ECC (Electronics Communications Committee) decided that CEPT (European Conference of Postal and Telecommunications Administrations) would designate the 63–64 GHz band as an alternative band for ITS applications to reduce traffic accidents/fatalities. For some futuristic ITS applications that may require 5G, the 63–64 GHz band may be appropriate.

In many other countries, investigations are underway to determine the right approach and bandwidth for V2X. In all these studies, the focus is on the 5.9 GHz band to achieve economies of scale and to have a uniform band for the operation of V2X applications.

On the opposite side of the equation, it is quite difficult to apply V2X in many countries where lanes are not marked on roads, stop signs are not available, and where a basic sense of driving is hard to find among a large number of drivers.

5.6.7   VC Evolution

The debate on VC started over two decades ago and has come a long way. The inception of WAVE/DSRC for V2V and V2I applications by IEEE along with the allocation of 5.9 GHz band in some countries was a significant milestone.

In the last few years, 3GPP has joined the bandwagon and conceptualized V2X. 3GPP Rel-14 established the foundation of V2X by setting requirements along with use cases. V2X use cases such as forward collision warning, control loss warning, emergency vehicle warning, queue warning, and so on have been envisioned in Rel-14. How these use cases can be supported through 5G are still under investigation in 3GPP.

Taking advantage of IEEE's ineffectiveness in defining a meaningful evolution path for DSRC, 3GPP has been putting requirements for low latency and higher reliability V2X applications in its Rel-15. The goal is perhaps to bring automated driving closer to reality, which requires higher reliability (up to 99.999%) and lower latency (sub 1 ms) [99]. Rel-15 is expected to be frozen in September 2018.

5.7   NETWORK SLICING

Network slicing is an end-to-end logical instance of a network with at least the following attributes [108–113]:

   a. Runs on a physical or virtual network
   b. Optimizes use of network for each intended usage scenario
   c. Uses a set of access and core network functions
   d. Is controlled and managed independently
   e. Created on demand, and
   f. Does not interfere with other functions and services on coexisting slices

Network softwarization is an emerging trend that seeks to transform a network's designing, planning, implementation, and operations through software programming. Network softwarization provides necessary flexibility and modularity to create logical networks (i.e., network slices) through NFV (Network Functions Virtualization) and SDN (Software Defined Networking) technologies [109,111].

The global standardization efforts are in the early stage and are primarily focused on vertical slicing (defined later). Industry forums such as 5GPPP (5G Infrastructure Public Private Partnership) and NGMN (Next Generation Mobile Networks) Alliance are leading the way, while SDOs like

3GPP have just started work [114]. Network slicing is identified as a key technology for 5G, but a great amount of work is needed to turn it into a successful reality.

5.7.1 E2E Slicing

Network slicing can be considered an evolution of network sharing, which is a proven business model for operators to reduce CAPEX and OPEX. Network slicing goes beyond sharing and envisions using virtualization and softwarization to improve user experience, increase network usage, and add on to operators' revenues. The slicing manifests the resolution of many complex issues including slice design, instantiation, implementation, and operations that requires new thinking. For instance, the design of a slice for a particular case requires enablement of functions in control and data planes, instantiation demands provision mechanisms over infrastructure, implementation seeks availability of all required network elements and the respective functions, while for operations slice configuration and monitoring setups are required in addition to other procedures.

Each network slice may consist of air-interface, radio access, and core networks as shown in Figure 5.11. Fronthaul in the case of C-RANs and transport area may have to be considered for some cases. Furthermore, virtualization and softwarization through NFV and SDN are also key building blocks of network slicing. The slicing is realized through network functions which provide the tailored capabilities needed to entertain a specific demand. Network functions could be slice-specific or could be used across multiple slices. The network functions could be physical (comprised of both hardware and software) or virtual (where software is decoupled from the hardware it runs on) [109].

In today's cut throat competition environment, operators are working to maximize their return on investment and utilization of their networks. E2E (end-to-end) slicing is an instrument which can achieve this goal, is making headlines, and is an important area of 5G research and development.

5.7.1.1 Slicing in RAN

A slice in the RAN relies on Radio Access Technology (RAT) and configuration of radio resources to carry what it is intended to deliver. There could be separate RATs in 5G designed to address different services. For example, one RAT could be for IoT and another perhaps for mobile broadband. RAN related configuration, which is customized to a particular slice, may include access control, load balancing, and resource scheduling [111,112]. To activate a slice, an access point/base station may allocate radio resources for the slice and enable all radio and network functions required for the operation of the slice. The slices in RAN may require slice-specific control-plane/user-plane and slice on/off operations [115].

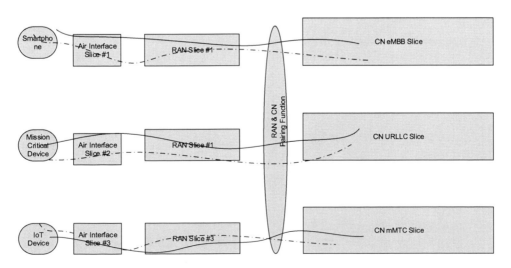

FIGURE 5.11   E2E network slice.

The slice in RAN will share radio resources such as time/frequency/space with other slices, along with the corresponding communication hardware either in a dynamic or static fashion according to the configuration rules. The radio resource sharing is performed by a central scheduler that resides in the 5G base station. If static resource sharing is used, the slice will get guaranteed resource allocation, while with dynamic resource sharing, usage is optimized. The type and amount of RAN resources required for a network slice depends on the service such as mobile broadband, IoT, autonomous driving, and their respective QoS requirements, that is, high capacity, massive connections, and ultra-low latency, respectively [111,112].

As far as C-plane (control plane) and U-plane (user plane) configurations are concerned, three alternatives are possible. Alternative 1 may allow a common C-plane across all slices and a dedicated U-plane for each slice. Alternative 2 may provide a dedicated C-plane and U-plane for each of the slices, while the third alternative would be a case with a common U-plane and a dedicated C-plane for each of the slices. Common C-plane slice functions include functions in idle mode such as paging, cell selection, and so on, while the functions in connected mode such as handover, dedicated bearer setup, and so on, can be categorized into slice-specific control plane functions [111,112,115].

5.7.1.2   Slicing in Core Network

Traditionally, the core networks have been architected as a single network serving multiple purposes and have been tailored to one or more RANs while supporting backward compatibility and interoperability. Network slicing, if implemented correctly, allows core networks to be logically separated making each core network slice operate independently while likely running on the same shared infrastructure.

A key element of 5G architecture is the separation of control and user planes' functions in the core network that allows selective choice of the U-plane functions needed for different slices and distribution of U-plane to sites closer to the devices, besides other features. The C-plane is agnostic to many U-plane functions such as physical deployments and L3 transport specifics. The control plane, as the name suggests, manages signaling messages, location information, cell selection/reselection, and so on. The C-plane can be placed in a central location, making management and operations less complex, whereas the user plane can be distributed to a number of sites. By bringing the U-plane closer to the users, the round-trip time between user and network services can be shortened. U-plane functionality can be deployed to address specific use cases. For example, an MBB (mobile broadband) service can be divided into video streaming and web browsing subservices which can be implemented by different feature sets within a network slice [112].

The slices can be defined to different support services/applications with a targeted set of radio/core network functions. Slice pairing functions are defined to pair RAN and core network (CN) slices to form end-to-end slices. Mapping among devices, RAN and CN slices, can be 1:1:1 or 1:M:N, for example, a device can have RAN slices while a RAN slice can connect to multiple CN slices [112,118].

5.7.2   SDN and NFV in Slicing

SDN and NFV are essential for the effective working of network slicing. SDN separates the C-plane and U-plane to optimize the performance of the network while NFV enables virtualization of networks.

5.7.2.1   SDN Overview

The SDN architecture defined by Open Networking Foundation (ONF) enables a common architecture to efficiently support diverse slices tailored for different services with different requirements. The SDN architecture is technology neutral, thus it can support wired, wireless, and mobile technologies. SDN consists of two key components, namely resources and controllers. Resources could be anything, but in this case, it could be a physical network function consisting of a piece of hardware and software or a virtual network function (VNF) where the software is decoupled from the hardware for the realization of a particular slice. Resources are managed through controllers which are central entities in SDN architecture. A controller maintains isolation in the control and data planes, allowing each network slice instance to be operated as a distinct and logically separate network. It provides resource isolation as well for a multitude of slices running on a common infrastructure. In addition to resources and controllers, the following key concepts are critical for the implementation of SDN based networks [109,116]:

- *Virtualization* is a function of a controller to abstract the underlying resources it manages. Network virtualization assists in the creation of isolated virtual networks that are dissociated from the underlying physical network and run on top of it.
- *Orchestration* by definition brings disparate things into a coherent whole. Within slicing, it is defined as the responsibility of the controller to dispatch resources to address the demands of the client in an optimal manner.
- *Recursion*, which could be hierarchical or federated, allows the controller to use resources from a lower level controller and provides services to a higher level controller (hierarchical scheme) while federation works on an equal level.

SDN provides extensive control plane functions for enabling network slicing, however, to efficiently manage the lifecycle of slices and their constituent resources, NFV is needed.

5.7.2.2 NFV Overview

ETSI NFV standard GS (Group Specification) NFV 002 [117] defines the lifecycle management of network services. A network service according to ETSI GS NFV 003 [118] is a composition of network functions and is defined by its functional and behavioral specifications. Thus, the concept of lifecycle management can be reused for network slicing [112].

NFV envisions the implementation of network functions as software-only entities (virtualization) that run over the NFV Infrastructure (NFVI). Virtualization means that the Network Function (NF) and part of the network infrastructure are implemented in software, that is, a decoupling of software from hardware. This decoupling allows the independent evolution of each. Along the same lines, an end-to-end network service (voice, data, IoT, etc.) can be described by an NF Forwarding Graph* of interconnected NFs and end points. To facilitate virtualization, the NFV reference architectural framework defined by ETSI can be used. This framework enables dynamic construction and management of VNF instances. It also manages the relationships between VNFs related to data, control, and other attributes.

5.7.2.3 SDN + NFV for Slicing

To take advantage of the benefits of both SDN and NFV for network slicing, an appropriate cooperation between the two is required. However, integration of both entities into a common framework is not an easy job.

Consider a slice in LTE that may consist of a radio access component eNB, and core network components such as Mobility Management Entity (MME), Serving Gateway (SGW), Packet Gateway (PGW), and so on, to provide broadband to a population of mobile users. The eNB is a PNF (physical network function) while other entities could be either PNF or VNF. These PNFs and VNFs, which are resources as far as SDN is concerned, can be made part of the end-to-end network slice as defined as shown in Figure 5.12.

These VNFs of an LTE slice can interact with ETSI NFV MANO (Management and Orchestration) [119] framework (part of the NFV reference architectural framework defined in [117]). Figure 5.13

---

* A graph of logical links connecting NF nodes for the purpose of describing traffic flow between these network functions is an NF forwarding graph [118].

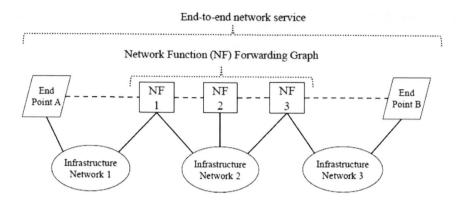

FIGURE 5.12 E2E network service. (From ETSI GS NFV 002 (V1.1.1) 2013. Network Functions Virtualisation (NFV); Architectural Framework, ETSI, October [117].)

shows a potential architecture which manages both virtualized and nonvirtualized network functions using NFV and SDN for 3GPP technologies. For the nonvirtualized components or PNFs, NFV MANO may not be used and thus a separate lifecycle management is required. PNFs are not under the control of MANO, but are also shared between numerous slices, particularly the PNF part of the radio access network. 3GPP currently has a study item on the management of such nonvirtualized

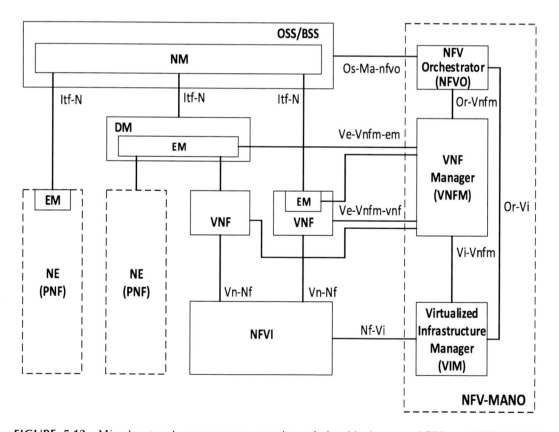

FIGURE 5.13 Mixed network management mapping relationship between 3GPP and NFV-MANO architectural framework. (From 3GPP TR 32.842 (V13.1.0) 2015. Telecommunication Management; Study on Network Management of Virtualized Networks. Technical Report (Release 13), Technical Specification Group Services and System Aspects, 3GPP, December [120].)

parts [120]. The same report, that is, TR 32.842, also clarifies the relationship between the 3GPP management framework and NFV-MANO framework.

3GPP defined VNFs will still require some underlying hardware. These hardware resources are expected to be managed independently from the virtualized entities required for the 3GPP system. The NFV objective is that the VNFs are procured independently from the hardware resources and applied to partially virtualized (e.g., via SDN) or completely virtualized systems. The term partial versus complete partial virtualization is defined in ETSI GS NFV-INF 001 [121]. According to the said specification, a large NF or an NE (network element) can be broken down into a number of constituent NFs. If all such constituents are implemented as VNFs, then the virtualization of such a large-scale NF or NE can be considered complete. If not all but only some constituent NFs are implemented as VNFs, then virtualization of such a large-scale NF is said to be partial.

5.7.3 Benefits/Challenges/Future

Virtualization and softwarization bring a number of challenges and opportunities. Some examples of network slicing include a slice serving a utility company (requesting a water tanker, etc.), a slice for a Mobile Virtual Network Operator (MVNO), a slice for streaming video service, and so on.

Today, one of the difficult challenges faced by operators is the clogging of networks by certain devices, impacting the service delivery to other users. Network slicing may assist in containing such rouge devices to particular slice(s) while keeping other slices unaffected. A misbehaving sensor, for example in a network slice, will not impact a critical public safety service running on another slice.

Network slices are deployed over a common underlying infrastructure which has a finite number of resources. This implementation has two challenges, namely isolation and resource management. Without proper isolation, slices may not be able to perform adequately. However, if slices are assigned dedicated resources, these may lead to over-provisioning. Resource management mechanisms are needed to strike a balance for the implementation of dedicated and shared resources.

For close to perfect collaboration between SDN and NFV, it would be mandatory to formalize interfaces through which either could query or invoke the services of the other. Along the same lines, interfaces between 3GPP and NFV MANO also require research and investigation for smooth operation. For the success of this collaboration, it is important that the three different disciplines understand each other's concepts and terminology and resolve the differences for a successful outcome [122].

For the success of slicing, new business models are required that necessitates innovative partnerships between several players including but not limited to traditional service providers, OTT (over-the-top) service providers, utility companies, media houses, and so on, and business friendly regulatory frameworks.

In a nutshell, present day techniques provide a number of services over one network; 5G network slicing can set up an optimized network environment for every service (*at least this is the expectation*) [123].

5.8 5G POLICY

The implementation 5G is getting closer and closer and becoming a reality day by day. Research is materializing, standardization is ongoing, and ITU is all set to endorse the IMT-2020 (5G) standard by 2019/2020.

Billions are getting invested, but only from certain economies and key players of the telecom sector. For example, South Korea through its Ministry of Education, Science, and Technology (MEST) is investing $1.5 billion. Chinese manufacturers Huawei and ZTE have already spent at least USD 500 million, European Union and European ICT industry are investing €1.4 billion in the joint 5G Infrastructure Public-Private Partnership program, and the U.S. through its National Science Foundation (NSF) is pouring $400 million into the Advanced Wireless Research Initiative.

Countries such as Bhutan, Nepal, Pakistan, and similar developing economies have not joined the bandwagon and are watching from the sidelines. They are mainly waiting for the finished product. In some developing economies, the telecom industry may conduct early technology trials.

5G will encompass a lot more than its predecessors. ITU has forecast three usage scenarios, that is, enhanced Mobile Broadband (eMBB), URLLC, and massive Machine Type Communications (mMTC) for 5G. These scenarios involve various vertical markets/industries which were minimally attached to the technology in the past, such as automotive, energy, and other industries, but now are expected to get much more involved because they have a lot more at stake.

At this moment, when the mobile industry is heavily focused on developing 5G, it is equally important that governments/regulators also become active in developing long term and solid national 5G mobile policies. This section attempts to cover the key aspects that may be incorporated in such 5G policies.

5.8.1 TARGETS

Every country has its own peculiarity, culture, and economic standing, thus a one size fits all policy is not going to work. However, a 5G policy has to achieve certain targets and address certain challenges which are perhaps common for very many economies. These common targets are as follows:

  i. Provide reliable eMBB, URLLC, and mMTC services.
  ii. Connect many things and many industries on the ICT grid and make these more efficient and intelligent.
  iii. Create many more jobs within ICT and in other sectors/industries.
  iv. Provide sustainable RoI (return on investment).

5.8.2 HOW TO ACHIEVE TARGETS?

The profit margins are getting smaller day by day in the telecom arena. This reality makes it more difficult to achieve the above-mentioned targets. The efforts that are required to accomplish the goals also vary with the financial standing and technological status of the countries. For example, the U.S. FCC policy is probably not going to be appropriate for India and Pakistan and vice versa.

The four targets are interrelated, for example, jobs cannot be created without having a solid RoI. If the telecom operators, vendors, and other stakeholders are making money, then they will enter into new business areas, which will bring innovation, create jobs, and stir economic growth/activity. One of the key enablers to achieve these targets is a strong national 5G policy.

5.8.3 AREAS/CHALLENGES IN REALIZING TARGETS

5.8.3.1 Tax

The mobile communications industry is heavily taxed in many developing and emerging economies. The taxes are applicable to both the mobile industry and mobile consumers. In many such economies, tax payments from mobile operators and consumers provide a significant contribution to public revenues—it is considered a cash cow. The mobile industry and consumers also pay a number of industry-specific/mobile-specific taxes and fees, making its contributions to the national exchequer higher than that of many other industries [124–126].

The consumers pay VAT (value added tax), customs duty on handsets, and mobile-specific taxes. The mobile-specific taxes may include a special tax on usage (such as tax on data usage), luxury taxes on handsets, and SIM (Subscriber Identification Module) activation taxes. Mobile broadband costs represent over a third of the average annual income of the poorest of the population in many developing countries [125]. Additionally, it may or may not include charges that operators pass on to consumers to indirectly cover some of their portion of taxes. Contrary to this unbalanced practice of taxation in developing and emerging markets, the taxation on mobile consumers in many developed economies is attributable to a VAT (general sales tax) and is applied uniformly on all goods and services. Overall, the taxes on mobile consumers reflect a sizeable portion (20% or higher) of the cost of mobile ownership* in many emerging and developing economies.

The mobile industry is subject to corporate tax, customs duty on equipment, revenue-based contributions, and annual regulatory and annual spectrum fees. Revenue-based contributions such as a Universal Service Fund (USF) (for development of telecommunication services primarily in less developed areas) are not represented as taxes, but the charges are just like taxes. In addition to USF, mobile operators also pay a certain percentage of their revenues for incubating R&D (for example, in Pakistan) and in the form of a revenue share tax (for instance, in Thailand). Mobile operators are also subject to higher corporation taxes in some countries. For example, in Cameroon and Tunisia, operators pay 39% and 35% as corporation taxes, respectively, against standard rates of 25% [126].

In a nutshell, the mobile industry and mobile consumers are subject to higher taxation as compared to other industries in many developing and emerging economies as briefly reflected in this section. For the betterment of both, a well-balanced taxation structure based on principles of equality in comparison to other industries/sectors is needed.

5.8.3.2 Right-of-Way (RoW)

Telecom licensees normally have the right to get access to any public RoW or private RoW in order to install, establish, and/or maintain its telecommunication equipment/system. The RoW has become a source of income for governments as well as for the owners of private properties.

RoW is a challenging and complex topic as it involves almost every person and every organization that owns a piece of land, since everyone is interested in getting a share of the pie. It will become more troublesome with the abundance of small cells in future networks such as 5G. In countries such as Pakistan, where there is no uniform/consistent mechanism for: obtaining permits to acquire RoW, charging fee and resolving disputes, the situation could get worse.

In many countries, mobile operators face complex and restrictive regulations on infrastructure deployment. Authorizations are granted at the municipal level for each cell/tower site and this further complicates the installation of antennas and small cells. In the anticipation that there will be an abundance of small cells in the coming years, municipalities are changing their outside plant codes. For example [127], the city council of Lancaster, Pennsylvania, U.S., changed its zoning rules to prevent the deployment of small cells on public RoW.

There is no easy resolution to the challenge of RoW, however, remedies such as one-window operation, less complex legislation, consistency in the applicable fees, and effective dispute resolution mechanism can be very useful.

5.8.3.3 Connected Things

The IoT (Internet of Things) brings the benefits of reducing costs, increasing revenue, and increasing efficiency. IoT will become massive and will bring a number of policy/regulatory challenges. Technology is less of an issue when it comes to IoT. In countries where corruption is high particularly in government setups, adoption of IoT will face many bottlenecks as it may directly hit their under-the-table source of income. For example, connecting electricity power meters with the mobile communications grid is a challenge because it involves two separate sectors with different sets of goals. Mobile technology is purely a commercial endeavor (with minimal corruption), whereas electricity is a basic essential facility provided by the state with serious corruption problems (at least in many developing countries). The use of mobile technology will make it difficult to commit corruption, at least when it comes to meter readings.

---

* Total cost of mobile ownership includes handset and connection costs as well as VAT and mobile-specific taxes on data/call/SMS.

TABLE 5.3
5G Policy—Key Targets, Approach, and Benefits

Approach	Benefits
Well-balanced taxation structure based on principles of equality in comparison to other industries/sectors	Innovation
	Job growth (target #3)
	Higher revenue (target #4)
	Less corruption
Right-of-way one window operation	Speedy deployment of IMT services (target # 1)
Enablement of vertical markets	IoT will be a success story (target #2)
Harmonization between ICT and municipalities	Meaningful management of RoW
Availability of spectrum, regional/global harmonization, and justified cost for spectrum auctions	Speedy deployment of eMBB, mMTC, URLLC services (target # 1)
	Higher revenue (target #4)

Privacy is a concern for all. The embedded IoT sensors can become watchdogs as well as spies by sending messages that are necessary and at the same time can act as invaders of one's privacy. The risky part is that IoT consumers are surrendering their privacy slowly and gradually without realizing how their personal information is being collected, who is using it, and how it is being used.

For the true success of mMTC, harmonization between ICT, relevant sector/industries, and data safeguarding measures are needed more than ever.

5.8.3.4 Spectrum

Last but not least, spectrum related factors such as the availability of the spectrum and regional/global harmonization and its consideration by governments as a cash cow needs to be rethought.

Around 2000 MHz is needed as estimated by ITU-R by 2020 for IMT [128] and more will definitely be needed when 5G is in full swing in the next decade. ITU-R needs to take urgent steps to ensure spectrum harmonization and its availability for the next 5 to 10 years.

Governments and regulators may need to do some rethinking when it comes to spectrum auctions. The world has seen multiple examples where the operators have not been able to recover the money they have invested in a spectrum both in developed (e.g., United Kingdom, etc.,) and developing (e.g., Pakistan, etc.) economies.

5.8.4 Summary

A summary of the section is shown in Table 5.3.

This section provided a brief overview of a policy for the implementation of 5G. Countries can use it as a guideline for developing their own detailed and comprehensive subject policy.

5.9 5G TIMELINES

The world is moving at a fast pace toward enabling 5G. Some rich economies have invested millions/billions of dollars to maintain/improve their ICT standings, enhance their economies, and create jobs. On the standardization side, ITU has published 5G technical performance criteria and 3GPP is expected to release Phase-1 and Phase-2 5G specifications in 2018 and 2019, respectively, as shown in Figure 5.14.

The first live demonstration of pre-standard 5G at a larger scale is expected to be held at the 2018 Winter Olympics in South Korea. The 2020 Summer Olympics in Japan will perhaps serve as the test case for 5G. ITU is also expected to approve the 5G standard in 2019/2020 and from 2020/2021 onwards, commercial deployments will be in full swing.

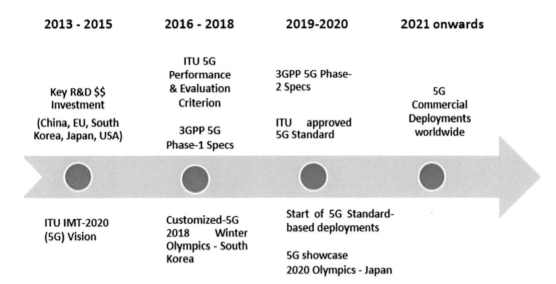

FIGURE 5.14 5G mobile technology evolution timelines.

5.10 CONCLUSION

5G is expected to be a user-centric concept instead of being operator-centric as in 3G or the service-centric concept as seen for 4G. Some key concepts of IMT-2020/5G were discussed and others will be described in the relevant chapters.

The 3GPP, from its Release-8 on, has introduced some functionality which is related to either a better handling of multi-standard systems, and/or to self-x capabilities. Even if the word cognitive is rarely used in the 3GPP specifications, some features like multi-standard radio, self-organizing networks, and so on have inherently introduced some level of cognitive capability in LTE. Regarding self-x functionalities, both ETSI RRS and 3GPP have addressed multi-RAT aspects. The difference is that ETSI RRS does not consider only 3GPP RATs, but all possible RATs.

5G is not just about new RAN or the latest technologies but is about future networks. The practical implementation of future networks will be THE daunting challenge.

PROBLEMS

1. What is 5G?
2. What are the usage scenarios defined by ITU for 5G?
3. What is air-interface and how might it be different in 5G?
4. Define the channel access method?
5. What is a communications channel and give some examples?
6. Define channel capacity?
7. What is the maximum data rate that can be transmitted over a 20 MHz LTE channel operating at an SNR of 30 dB?
8. Define spectral efficiency?
9. Is OFDM fit for URLLC and mMTC types of applications?
10. What is NOMA and how it is different from other multiple access methods?
11. Define GFDM, UFMC, and FBMC and their prospective applications?
12. What is a cognitive radio as defined by the FCC?
13. What is white space?
14. What are the key standards that describe cognitive radios?
15. What are narrowband and wideband spectrum sensing?

16. Explain the differences between centralized and distributed cooperative detection?
17. What is spectrum sharing?
18. Describe the potential of a WRC-15 identified frequency band (for IMT) for spectrum sharing?
19. Describe Cloud RAN?
20. What is massive C-RAN?
21. Define fronthaul?
22. What is the role of CPRI in C-RAN?
23. What are V2V and V2I communications?
24. Define V2X?
25. Define the two communication modes of C-V2X?
26. What is DSRC and can LTE replace it?
27. What is geocast and why it used in vehicular communications?
28. Describe network slicing?
29. Define network softwarization?
30. Define end-to-end network slicing?
31. Define the roles of SDN and NFV in network slicing?
32. Describe the key targets for your country's future (5G) Policy?

REFERENCES

1. ITU-R 2015. Recommendation ITU-R M.2083-0—IMT Vision—Framework and Overall Objectives of the Future Development of IMT for 2020 and Beyond.
2. 5G Americas 2016. Global Organizations Forge New Frontier of 5G—5G Americas 5G Global Update.
3. Zhu, P. 2014. 5G Enabling Technologies An Unified Adaptive Software Defined Air Interface. Keynote Presentation. *IEEE 25th Annual International Symposium on Personal, Indoor and Mobile Radio Communications (PIMRC)*, Washington, DC, USA, September 2–5, 2014.
4. Imran, M.A., Al-Imari, M. and Tafazolli, R. 2012. Low Density Spreading Multiple Access. *Journal of Infromation Technology Software Engineering*, 2(4):1–2. http://eprints.gla.ac.uk/136363/1/136363.pdf
5. Gokhale, A.A. 2004. *Introduction to Telecommunications*, 2nd edn. Thomson Delmar Learning, Inc., NY, USA.
6. Salehi, M. and Proakis, J.G. 1994. *Communications Systems Engineering*. Prentice Hall, New Jersey, USA.
7. Asif, S.Z. 2011. *Next Generation Mobile Communications Ecosystem: Technology Management for Mobile Communications*. Wiley Inc., UK.
8. Rysavy, P. 2014. Challenges and Considerations in Defining Spectrum Efficiency. *Proceedings of the IEEE*, 102(3):1–7.
9. Wikipedia. http://en.wikipedia.org/wiki/Main_Page
10. Federal Communications Commission Technological Advisory Council Sharing Work Group 2011. Spectrum Efficiency Metrics, September.
11. 4G Americas 2015. 5G Technology Evolution Recommendations.
12. European Commission/Seventh Framework Program 2015. Document Number: ICT-317669-METIS/D5.4 Future Spectrum System Concept. Mobile and Wireless Communications Enablers for the Twenty-Twenty Information Society (METIS).
13. 3GPP TR 38.804 (V14.0.0) 2017. Study on New Radio Access Technology; Radio Interface Protocol Aspects. Technical Report (Release 14), Technical Specification Group Radio Access Network, 3GPP, March.
14. 5G Americas 2017. LTE to 5G: Cellular and Broadband Innovation. Rysavy Research, LLC.
15. 5G Americas 2016. Mobile Broadband Transformation LTE to 5G. Rysavy Research, LLC.
16. 3GPP TSG-RAN WG1 #84b, R1-162199 2016. Waveform Candidates, Qualcomm Inc., April 11th–15th, Busan, Korea.
17. Nagapuhpa, K.P. and Chitra Kiran, N. 2017. Studying Applicability Feasibility of OFDM in Upcoming 5G Network. *International Journal of Advanced Computer Science and Applications*, 8(1):216–226.
18. Krone, S. et al. 2012. Bit Error Rate Performance of Generalized Frequency Division Multiplexing. *2012 IEEE Vehicular Technology Conference (VTC Fall)*, Quebec City, Canada, 3–6 Sept. 2012, pp. 1–5.
19. Du, J. 2008. *Licentiate Thesis: Pulse Shape Adaptation and Channel Estimation in Generalized Frequency Division Multiplexing Systems*. Royal Institute of Technology (KTH), Electronics and Computer Systems, Stockholm, Sweden.

20. Krone, S. et al. 2012. Generalized Frequency Division Multiplexing: A Flexible Multi-Carrier Modulation Scheme for 5th Generation Cellular Networks. *IEEE Transactions on Communications*, 62(9):3045–3061.

21. Ding, Z., Peng, M. and Poor, V.H. 2014. Cooperative Non-Orthogonal Multiple Access in 5G Systems. Cornell University, arXiv:1410.5846v1 [cs.IT].

22. Saito, Y. et al. 2013. System-Level Performance Evaluation of Downlink Non-Orthogonal Multiple Access (NOMA). *2013 IEEE 24th International Symposium on Personal Indoor and Mobile Radio Communications (PIMRC)*, London, UK, 11 Sept. 2013, pp. 611–615.

23. Cover, T. and Thomas, J. 1991. *Elements of Information Theory*, 6th edn. Wiley and Sons, New York.

24. Saito, Y. et al. 2013. Non-Orthogonal Multiple Access (NOMA) for Cellular Future Radio Access. *2013 IEEE 77th Vehicular Technology Conference (VTC Spring)*, Dresden, Germany, 2–5 June 2013, pp. 1–5.

25. Saito, Y. et al. 2013. Concept and Practical Considerations of Non-orthogonal Multiple Access (NOMA) for Future Radio Access. *2013 International Symposium on Intelligent Signal Processing and Communications Systems (ISPACS)*, Naha, Japan, 12–15 Nov. 2013, pp. 770–774.

26. Ding, Z., Yang, Z., Fan, P. and Poor, H.V. 2014. On the Performance of Non-Orthogonal Multiple Access in 5G Systems with Randomly Deployed Users. arXiv:1406.1516v1 [cs.IT].

27. Wild, T. et al. 2013. Universal-Filtered Multi-Carrier Technique for Wireless Systems Beyond LTE. *2013 IEEE Globecom Workshops (GC Wkshps)*, Atlanta, GA, USA, December 9–13, 2013, pp. 223–228.

28. 5GNOW 2015. D3.3—Final 5GNOW Transceiver and frame structure concept, version 1.

29. Wild, T., Schaich, F. and Chen, Y. 2014. Waveform Contenders for 5G—Suitability for Short Packet and Low Latency Transmissions. *IEEE 79th Vehicular Technology Conference (VTC Spring)*, Seoul, South Korea, May 18–21, 2014.

30. European Commission/Seventh Framework Program 2015. Document Number: ICT-317669-METIS/D2.4 Proposed Solutions for New Radio Access. Mobile and Wireless Communications Enablers for the Twenty-Twenty Information Society (METIS).

31. European Commission/Seventh Framework Program 2010. INFSO-ICT-211887 FBMC Physical Layer: A Primer. PHYDYAS (Physical Layer For Dynamic Spectrum Access and Cognitive Radio) project.

32. Wild, T. and Schaich, F. 2014. Waveform Contenders for 5G—OFDM vs. FBMC vs. UFMC. *6th IEEE International Symposium on Communications, Control and Signal Processing (ISCCSP)*, Athens, Greece, May 21–23, 2014.

33. Tensubam, B.V. and Singh, S. 2014. A Review on FBMC: An Efficient Multicarrier Modulation System. *International Journal of Computer Applications*, 98(17):6–9.

34. Razavi, R. et al. 2012. On Receiver Design for Uplink Low Density Signature OFDM (LDS-OFDM). *IEEE Transactions on Communications*, 60(11):3499–3508.

35. Taherzadeh, M., Nikopour, H., Bayesteh, A. and Baligh, H. 2014. SCMA Codebook Design. *IEEE 80th Vehicular Technology Conference (VTC Fall)*, Vancouver, BC, Canada, September 14–17, 2014.

36. Nikopour, H. et al. 2014. SCMA for Downlink Multiple Access of 5G Wireless Networks. *IEEE Global Communications Conference (GLOBECOM)*, Austin, TX, USA, December 8–12.

37. Nikopour, H. et al. 2014. Uplink Contention Based SCMA for 5G Radio Access. *2014 IEEE Globecom Workshops (GC Wkshps)*, Austin, TX, USA, December 8–12.

38. Huawei 2015. 5G: New Air Interface and Radio Access Virtualization.

39. Qualcomm Technologies Inc. 2015. 5G Waveform & Multiple Access Techniques.

40. Joseph, M. III 1999. Cognitive Radio Model-Based Competence for Software Radios. Thesis (submitted to the Royal Institute of Technology in partial fulfillment of the requirements for the Licentiate of Technology degree).

41. Joseph, M. III and Gerald, Q.M. Jr. 1999. Cognitive Radio: Making Software Radios More Personal. *IEEE Personal Communications*, 6(4):13–18.

42. Federal Communications Commission 2003. FCC 03-322 Notice of Proposed Rule Making and Order: Facilitating Opportunities for Flexible, Efficient, and Reliable Spectrum Use Employing Cognitive Radio Technologies. ET Dcoket No. 03-108. Washington DC, USA.

43. Ian, F.A., Won-Yeol, L., Mehmet, C.V. and Shantidev, M. 2006. NeXt Generation/Dynamic Spectrum Access/Cognitive Radio Wireless Networks: A Survey. *Elsevier Computer Networks*, 50(2006):2127–2159.

44. IEEE Std 802.22-2011 2011. IEEE Standard for Information Technology—Telecommunications and Information Exchange between Systems Wireless Regional Area Networks (WRAN)—Specific requirements; Part 22: Cognitive Wireless RAN Medium Access Control (MAC) and Physical Layer (PHY) Specifications: Policies and Procedures for Operation in the TV Bands IEEE, July.

45. IEEE Std 802.11af 2013. IEEE Standard for Information technology—Telecommunications and Information Exchange between Systems Local and Metropolitan Area Networks—Specific Requirements Part 11: Wireless LAN Medium Access Control (MAC) and Physical Layer (PHY) Specifications, Amendment 5: Television White Spaces (TVWS) Operation, IEEE, December.

46. IEEE Standards Association. IEEE 1900.1 Working Group on "Definitions and Concepts for Dynamic Spectrum Access: Terminology Relating to Emerging Wireless Networks, System Functionality, and Spectrum Management". http://grouper.ieee.org/groups/dyspan/1/index.htm

47. Zhou, H. et al. 2017. *Dynamic Sharing of Wireless Spectrum.* Springer International Publishing AG, Cham, Switzerland.

48. Khattab, A. and Bayoumi, M.A. 2015. An Overview of IEEE Standardization Efforts for Cognitive Radio Networks. *IEEE International Symposium on Circuits and Systems (ISCAS)*, Lisbon, Portugal, May 24–27, 2015.

49. IEEE Std 802.15.4m 2014. IEEE Standard for Information Local and Metropolitan Area Networks—Part 15.4: Low-Rate Wireless Personal Area Networks (LR-WPANs)—Amendment 6: TV White Space between 54 MHz and 862 MHz Physical Layer, IEEE, April.

50. IEEE Std 802.19.1 2014. 802.19.1-2014—IEEE Standard for Information Technology—Telecommunications and Information Exchange between Systems—Local and Metropolitan Area Networks—Specific Requirements—Part 19: TV White Space Coexistence Methods, IEEE, May.

51. RFC 7545 2015. Protocol to Access White-Space (PAWS) Databases, IETF, May.

52. ETSI. Reconfigurable Radio. http://www.etsi.org/technologies-clusters/technologies/reconfigurable-radio

53. ETSI TR 102 682 V1.1.1 2009. Reconfigurable Radio Systems (RRS); Functional Architecture (FA) for the Management and Control of Reconfigurable Radio Systems. Technical Report. ETSI, July.

54. ETSI TR 102 683 V1.1.1 2009. Reconfigurable Radio Systems (RRS); Cognitive Pilot Channel (CPC). Technical Report. ETSI, September.

55. ITU-R 2009. Report ITU-R SM.2152 Definitions of Software Defined Radio (SDR) and Cognitive Radio System (CRS), September.

56. ITU-R 2011. Report ITU-R M.2242 Cognitive Radio Systems Specific for International Mobile Telecommunications Systems, November.

57. ITU-R 2012–2015. Resolution ITU-R 58-1 Studies on the Implementation and Use of Cognitive Radio Systems.

58. ITU Radiocommunication Bureau 2017. Preparations for CPM19-2, RA-19 and WRC-19 (including outcome of RA-15 and WRC-15).

59. Friedrich, K.J. 2005. Software-Defined Radio-Basics and Evolution to Cognitive Radio. *EURASIP Journal on Wireless Communications and Networking*, 3:275–283.

60. Matteo, C., Francesca, C. and Eylem, E. 2010. Routing in Cognitive Radio Networks: Challenges and Solutions. *Elsevier Ad Hoc Networks*, 9(3):228–248, doi:10.1016/j.adhoc.2010.06.009.

61. Ian, F.A., Won-Yeol, L. and Kaushik, R.C. 2009. Spectrum Management in Cognitive Radio Ad Hoc Networks. *IEEE Network*, 23(4):6–12.

62. Ammar, M. et al. 2014. Survey: A Comparison of Spectrum Sensing Techniques in Cognitive Radio. *IIE International Conference Image Processing. Computers and Industrial Engineering (ICICIE'2014)*, Kuala Lumpur, Malaysia, January 15–16, 2014, pp. 65–69.

63. Vakil, M. and Nagamani, K. 2017. Cognitive Radio Spectrum Sensing—A Survey. *International Journal of Recent Trends in Engineering & Research (IJRTER)*, 3(2):45–50.

64. Sun, H. et al. 2013. Wireless Spectrum Sensing for Cognitive Radio Networks: A Survey. *IEEE Wireless Communications*, 20(2):74–81.

65. Patil, H., Patil, A.J. and Bhirud, S.G. 2015. Multichannel Cooperative Sensing in Cognitive Radio: A Literature Review. *International Journal of Advanced Research in Computer and Communication Engineering*, 4(5), 425–429.

66. Letaief, K.B. and Zhang, W. 2009. Cooperative Communications for Cognitive Radio Networks. *Proceedings of the IEEE*, 97(5):878–893.

67. Ganesan, G. and Li, Y. 2005. Cooperative Spectrum Sensing in Cognitive Radio Networks. *First IEEE International Symposium on New Frontiers in Dynamic Spectrum Access Networks, (DySPAN 2005)*, November 8–11, 2005.

68. Tang, H. 2005. Some Physical Layer Issues of Wide-Band Cognitive Radio System. *First IEEE International Symposium on New Frontiers in Dynamic Spectrum Access Networks, (DySPAN 2005)*, November 8–11, 2005.

69. Ramchandran, K. and Wild, B. 2005. Detection Primary Receivers for Cognitive Radio Applications. *First IEEE International Symposium on New Frontiers in Dynamic Spectrum Access Networks, (DySPAN 2005)*, November 8–11, 2005.

70. GSMA 2014. The Impact of Licensed Shared Use of Spectrum. Report by Deloitte and Real Wireless.
71. Hossain, E., Niyato, D. and Kim, D.I. 2013. Evolution and Future Trends of Research in Cognitive Radio: A Contemporary Survey. *Wireless Communications And Mobile Computing*, 15: 1530–1564 doi:10.1002/wcm.2443, John Wiley & Sons, Ltd.
72. Dano, M. 2016. CTIA Wants to Oversee Spectrum Sharing in the 3.5 GHz Band. Questex LLC (FierceWireless). http://www.fiercewireless.com/wireless/ctia-wants-to-oversee-spectrum-sharing-3-5-ghz-band
73. Lowenstein, M. 2016. 3.5 GHz Spectrum: An Opportunity for the U.S. to Lead in Wireless Innovation. Gerben Law Firm, PLLC (Tech.pinions). https://techpinions.com/3-5-ghz-spectrum-an-opportunity-for-the-u-s-to-lead-in-wireless-innovation/46514
74. Plum 2015. Global Momentum and Economic Impact of the 1.4/1.5 GHz Band for IMT. A Report for GSMA.
75. Schaubach, K. 2017. The Technology Behind Spectrum Sharing: The Spectrum Access System. CBRS Alliance. https://www.cbrsalliance.org/single-post/2017/06/02/The-Technology-Behind-Spectrum-Sharing-The-Spectrum-Access-System
76. China Mobile Research Institute 2011. C-RAN The Road Towards Green RAN.
77. Webb, R. 2017. New Frontiers in C-RAN Fronthaul: to CPRI, or to not CPRI? IHS Market.
78. Yan, Y. et al. 2015. Cloud RAN for Mobile Networks—A Technology Overview. *IEEE Communications Surveys & Tutorials*, 17(1):405–426.
79. Ericsson, A.B., Huawei Technologies Co. Ltd, NEC Corporation, Alcatel Lucent, and Nokia Networks 2015. CPRI Specification V7.0—Common Public Radio Interface (CPRI); Interface Specification.
80. Wang, X. et al. 2016. Energy-Efficient Virtual Base Station Formation in Optical-Access-Enabled Cloud-RAN. *IEEE Journal on Selected Areas in Communications*, 34(5):113–1139.
81. China Mobile Research Institute, Alcatel-Lucent, Nokia Networks, ZTE Corporation, Broadcom Corporation, Intel China Research Center 2015. White Paper of Next Generation Fronthaul Interface Version 1.0.
82. Wu, J., Zhang, Z., Hong, Y. and Wen, Y. 2015. Cloud Radio Access Network (C-RAN): A Primer. *IEEE Network*, 29(1):35–41.
83. Bendlin, R. et al. 2012. *Embracing LTE-A with KeyStone SoCs*. Texas Instruments Inc, Dallas, TX, USA.
84. Alleven, M. 2016. IHS: China Fuels $5B C-RAN Market. Questex LLC (FierceWireless). http://www.fiercewireless.com/tech/ihs-china-fuels-5b-c-ran-market
85. Teral, S. 2016. Research Note—China Now Fueling a $5 Billion Centralized RAN Market. https://technology.ihs.com/580929/research-note-china-now-fueling-a-5-billion-centralized-ran-market
86. 5G Americas 2016. V2X Cellular Solutions.
87. ITS Standards Advisory 2003. Dedicated Short Range Communications (DSRC). Advisory No. 3. United Stated Department of Transportation. https://web.archive.org/web/20130216162616/http://www.standards.its.dot.gov/Documents/advisories/dsrc_advisory.htm
88. ITS Standard Program 2009. ITS Standards Fact Sheet: IEEE 1609—Family of Standards for Wireless Access in Vehicular Environments (WAVE). United Stated Department of Transportation. https://www.standards.its.dot.gov/factsheets/factsheet/80
89. Imad, J., Nader, M. and Liren, Z. 2010. Inter-Vehicular Communication Systems, Protocols and Middleware. *IEEE Fifth International Conference on Networking, Architecture and Storage*, Macau, China, 15–17 July 2010, pp. 282–287.
90. Vaishali, D.K. and Pradhan, S.N. 2012. V2V Communication Survey—(Wireless Technology). *International Journal Computer Technology & Applications*, 3(1):370–373.
91. Mar, B., Walter, J.F. and Lars, W. 2003. Mobile Internet Access in FleetNet. *Kommunikation in Verteilten Systemen*, 107–118.
92. Andreas, F. et al. 2004. FleetNet: Bringing Car-to-Car Communication Into the Real World. *11th World Congress on ITS*, Nagoya, Aichi, Japan, 18–22 October 2004.
93. Panos, P. et al. 2009. Vehicular Communication Systems: Enabling Technologies, Applications, and Future Outlook on Intelligent Transportation. *IEEE Communications Magazine*, 47(11):84–95.
94. 3GPP TR 22.885 (V14.0.0) 2015. Study on LTE Support for Vehicle to Everything (V2X) Services. Technical Report (Release 14), Technical Specification Group Services and System Aspects, 3GPP, December.
95. 3GPP TS 22.185 (V14.3.0) 2017. Service Requirements for V2X Services; Stage 1. Technical Specification (Release 14), Technical Specification Group Services and System Aspects, 3GPP, March.
96. 3GPP TSG-RAN Meeting #73, RP-161894 2016. LTE-based V2X Services, September 19–22, New Orleans, LA, USA.
97. Flore, D. 2016. Initial Cellular V2X Standard Completion. 3GPP. http://www.3gpp.org/news-events/3gpp-news/1798-v2x_r14

98. 5G Automotive Association 2016. The Case for Cellular V2X for Safety and Cooperative Driving.
99. Qualcomm Technologies, Inc. 2016. The Path to 5G: Cellular Vehicle-to-Everything (C-V2X).
100. IEEE Std 802.11p-2010 2010. IEEE Standard for Information Technology—Telecommunications and Information Exchange between Systems—Local and Metropolitan Area Networks—Specific Requirements—Part 11: Wireless LAN Medium Access Control (MAC) and Physical Layer (PHY) Specifications—Amendment 6: Wireless Access in Vehicular Environments. IEEE, June.
101. IEEE Standards Association 2017. 1609 WG—Dedicated Short Range Communication Working Group. https://standards.ieee.org/develop/wg/1609_WG.html
102. Lansford, J. 2016. *The Connected Car: From the Highway to the Cloud.* Qualcomm Technologies, Inc, San Diego, CA, USA.
103. 3GPP TR 36.786 (V14.0.0) 2017. Vehicle-to-Everything (V2X) Services Based on LTE; User Equipment (UE) Radio Transmission and Reception. Technical Report (Release 14), Technical Specification Group Radio Access Network, 3GPP, March.
104. 3GPP TR 22.886 (V15.1.0) 2017. Study on Enhancement of 3GPP Support for 5G V2X Services. Technical Report (Release 15), Technical Specification Group Services and System Aspects, 3GPP, March.
105. RF Wireless World 2012. IEEE 802.11p tutorial-802.11p, WAVE, DSRC Protocol Stack. http://www.rfwireless-world.com/Tutorials/802-11p-WAVE-tutorial.html
106. Miucic, R. and Al-Stouhi, S. 2016. Absolute Localization via DSRC Signal Strength. *IEEE 84th Vehicular Technology Conference (VTC-Fall)*, Montreal, QC, Canada, September 18–21, 2016.
107. National Highway Traffic Safety Administration—NHTSA 2016. U.S. DOT Advances Deployment of Connected Vehicle Technology to Prevent Hundreds of Thousands of Crashes. https://www.nhtsa.gov/press-releases/us-dot-advances-deployment-connected-vehicle-technology-prevent-hundreds-thousands
108. 5G-PPP 2016. View on 5G Architecture. 5G PPP Architecture Working Group, July.
109. Lorca, J. et al. 2017. Network Slicing for 5G with SDN/NFV: Concepts, Architectures and Challenges. *IEEE Communications Magazine*, 55(5):80–87.
110. Zhou, C. et al. 2016. On End to End Network Slicing for 5G Communication Systems. *Transactions On Emerging Telecommunications Technologies*, DOI: 10.1002/ett.3058.
111. Taleb, T. et al. 2017. End-to-End Network Slicing for 5G Mobile Networks. *Journal of Information Processing*, 25:153–163.
112. 5G Americas 2016. Network Slicing for 5G Networks & Services.
113. Ericsson 2014. Network Functions Virtualization and Software Management.
114. 3GPP TR 23.799 (V2.0.0) 2016. Study on Architecture for Next Generation System. Technical Report (Release 14), Technical Specification Group Services and System Aspects, 3GPP, November.
115. Li, Q., Wu, G., Papathanassiou, A. and Mukherjee, U. 2016. *An End-to-End Network Slicing Framework for 5G Wireless Communication Systems.* Cornell University Library, Ithaca, NY, USA. arXiv:1608.00572v1.
116. Open Network Foundation 2016. TR-526 Applying SDN Architecture to 5G Slicing. Issue 1.
117. ETSI GS NFV 002 (V1.1.1) 2013. Network Functions Virtualisation (NFV); Architectural Framework, ETSI, October.
118. ETSI GS NFV 003 (V1.1.1) 2013. Network Functions Virtualisation (NFV); Terminology for Main Concepts in N FV, ETSI, October.
119. ETSI GS NFV-MAN 001 (V1.1.1) 2014. Network Functions Virtualisation (NFV); Management and Orchestration, ETSI, December.
120. 3GPP TR 32.842 (V13.1.0) 2015. Telecommunication Management; Study on Network Management of Virtualized Networks. Technical Report (Release 13), Technical Specification Group Services and System Aspects, 3GPP, December.
121. ETSI GS NFV-INF 001 (V1.1.1) 2015. Network Functions Virtualisation (NFV); Infrastructure Overview, ETSI, January.
122. Open Network Foundation 2015. TR-518 Relationship of SDN and NFV. Issue 1.
123. Alleven, M. 2017. SK, Deutsche Telekom, Ericsson Demo Network Slicing for 5G Roaming. Questex LLC (FierceWireless). http://www.fiercewireless.com/tech/sk-deutsche-telekom-ericsson-demo-network-slicing-for-5g-roaming
124. GSMA 2016. Country Overview: Pakistan—A Digital Future.
125. GSMA 2016. Digital Inclusion and Mobile Sector Taxation 2016.
126. GSMA 2015. Digital Inclusion and Mobile Sector Taxation 2015.
127. Gibbs, C. 2017. Public Policy Will Lay the Foundation for 5G and Beyond. Questex LLC (FierceWireless). http://www.fiercewireless.com/wireless/public-policy-will-lay-foundation-for-5g-and-beyond
128. ITU-R 2015. Report of the Conference Preparatory Meeting on Operational and Regulatory/Procedural Matters to the World Radiocommunication Conference 2015.

6 Semiconductors in Mobile Telecommunications

The mobile telecommunications industry heavily relies on the advancements in semiconductors and it is also one of the driving factors behind it. According to IHS Technology, the worldwide semiconductor market grew by 9.2% in 2014 which is the highest annual growth rate since the 33% boom of 2010 [1]. On a similar note, the global semiconductor market is estimated to grow by a compound annual growth rate of 7.67% from 2017 to 2024 [2]. It is now widely accepted that consumer products like smart phones and tablets account for more than 50% of the demand for semiconductors [3].

Today, mobile application processors operate at 5%–10% of a typical laptop's computing power, however, this gap is rapidly decreasing due to the exponential increase in the use smart phones and connected devices for mobile gaming, mobile videos, and so on. Similarly, the energy consumption of smart phones is also lower than that of laptops by a factor of 10–30 times. The nonstop increase in performance requirements, a rapid shift from traditional handsets to smart phones and connected devices, and an upsurge of the IoT has been reducing this energy consumption gap and posing a number of challenges for the semiconductor industry. Beside handheld devices, semiconductors are also used in the design of base station modules, optical nodes, and routers, which are the essential elements of mobile wireless networks.

This chapter will cover the following topics, highlighting their applicability to mobile and wireless communications and starting with a brief background on semiconductors which should be helpful for readers.

- Background
- RF, Analog/Mixed-Signal (RFAMS) and Millimeter Wave (MMW) Circuits
- RF and MMW Examples—Mobile and Network Components
- Micro Electro-Mechanical Systems
- Internet of Things—the next growth engine for semiconductors
- Pakistan's Semiconductor Industry Perspective

6.1 BACKGROUND

A semiconductor is a material that has electrical conductivity between that of a conductor, such as copper, and that of an insulator, such as glass. Semiconductors form the foundation of modern electronics, including transistors, solar cells, digital and analog integrated circuits, and many other items. Semiconductors such as silicon, germanium, and compounds of gallium are the most widely used in electronics including telecommunications equipment [4].

Silicon belongs to group 14 of the periodic table and has four valence electrons. Silicon is the eighth most common element in the universe by mass, however, it can rarely be found in pure free element form. It is most widely found in dusts and sands, which are various forms of silicon dioxide (silica) or silicates. As per estimates, more than 90% of the Earth's crust is composed of silicate minerals, making silicon the second most abundant element in the Earth's crust (about 28% by mass) after oxygen. With wide scale abundance and great chemical properties, silicon is considered to be an excellent choice for integrated circuits.

Germanium also belong to group 14 and like silicon reacts and forms complexes with oxygen in nature. It ranks near fiftieth in relative abundance of the elements in the Earth's crust. Gallium is in group 13 of the periodic table and is mainly used in electronics [4].

An integrated circuit is a set of electronic circuits on a small plate (chip) of semiconductor material. Integrated circuits can be classified into analog (such as power management circuits), digital (such as microprocessors), and mixed signal (both analog and digital on the same chip such as analog/digital converters). IC development is a lengthy process, from semiconductor material development to designing, fabrication, testing, assembly, and then packaging [4].

Semiconductor device fabrication, which is the costliest part of the entire chain of events, consists of hundreds of sequential process steps to create the ICs [5]. Without going into detail, the entire manufacturing process takes six to eight weeks and is performed (technology node) in highly specialized facilities called fabs (foundries). The semiconductor manufacturing process is measured in micrometers (μm) or nanometers (nm), which stood at 10 nm (10 nanometers) during 2017–2018.

Historically, the semiconductor companies controlled the entire production process from design to manufacture. However, complexity, cost, and faster time to market trends have shifted the priorities and the chip companies are becoming leaner and more efficient. Thus, now we have IDMs (Integrated Device Manufacturers), fabless semiconductor firms, semiconductor foundries, and intellectual property core developers. A brief description of these follows:

1. *IDMs* design, manufacture, package, and sell, that is, run the entire process on their own. Companies like IBM, Samsung, and a few others fall under this category.
2. Second are *fabless semiconductor firms* that design and sell hardware semiconductor devices/chips, however, they outsource the fabrication to specialized manufacturers, that is, semiconductor foundries. Qualcomm, Broadcom, and so on are fabless companies.
3. The *semiconductor foundry* (or simply fabs) are fabrication plants where ICs are manufactured; this is the third category. Fabs fabricate the designs of other companies, such as fabless semiconductor companies. Foundries are mainly found in China, France, Germany, Ireland, Japan, Malaysia, Singapore, Taiwan, and the U.S.
4. The fourth category consists of *semiconductor intellectual property core developers, designers, and licensors.* An IP (intellectual property) core or IP block is a reusable unit of chip layout design that is the intellectual property of one party and can be licensed to another party. An IP (intellectual property) core is a block of logic or data that is used in making a field programmable gate array (FPGA) or application-specific integrated circuit (ASIC) for a product. ARM Holdings, PLC, and Cadence Design Systems are the key firms in this category [4,6].

6.2 RF, ANALOG/MIXED-SIGNAL (AMS), AND MMW CIRCUITS

Radio frequency and analog/mixed-signal (RF and AMS or simply RFAMS) circuits are critical enablers for existing wireless and wireline communications and are expected to continue to be for the future IoT. These RF and AMS circuits address the 0.4–30 GHz market that primarily includes mobile cellular, broadband, and wireless LANs. Furthermore, communications products that support applications such as radar imaging, automotive radar (24/77 GHz), point to point radios (70/80 GHz), defense, and security all have functionalities that are enabled by RF, AMS, and MtM (More-than-Moore*) circuits. It may be noted that 4G phones and tablets now have a much higher RF and AMS semiconductor content as compared to only 5% of the market a few years ago [3,7]. According to [8], the famous iPad has more than 19 RF and AMS front-end components.

Silicon is the most common semiconductor used in the development of a multitude of RF and AMS components. Compound semiconductors like GaAs (gallium arsenide) for power amplifiers, and alloys like SiGe (silicon-germanium) for low-noise amplifiers (LNA) and mixers are also common. Power

---

* MtM describe technology features that do not fit the miniaturization trends as implied by Moore's Law.

amplifiers amplify strong signals from the electronic circuitry while LNAs amplify very weak signals from the antenna in wireless systems. Both of these components are critical to the success of wireless communication systems. The costs associated with SiGe processing are similar to those of silicon CMOS (Complementary Metal Oxides Semiconductor) manufacturing but are lower than GaAs [4]. Regardless, the selection of a semiconductor includes a number of performance metrics such as wafer (thin slice/substrate of semiconductor material) cost, level of integration, volume, and time-to-market.

The four key technology device subgroups that build RF and AMS integrated circuits according to International Technology Roadmap for Semiconductors include [3,9].

6.2.1 RF CMOS

Most of the semiconductor value in cell phones is derived from CMOS [10]. CMOS is a technology used in the creation of digital and analog integrated circuits. CMOS uses complementary and symmetrical pairs of p-type and n-type metal oxide semiconductor field effect transistors (MOSFETs) for logic functions [4]. MOSFET transistors are used for amplifying or switching electronic signals. The primary driver for CMOS is mobile transceivers in the 0.4 GHz–30 GHz frequency range.

6.2.2 GROUP IV SILICON BIPOLAR AND BiCMOS

As the name suggests, BiCMOS is an integrated circuit integrating a bipolar junction transistor (BJT) and the CMOS transistor in a single integrated circuit. Bipolar transistors provide high speed and gain that are critical for high frequency analog sections, while CMOS exceeds in constructing low-power digital logic gates. Si/SiGe hetero junction bipolar transistors are the key bipolar transistors for analog and mixed signal applications. High voltage NPN and PNP* bipolar transistors are used in applications operating below 5 GHz. These include cellular and WLAN power amplifiers and operational power amplifiers. The key applications of high speed NPN transistors include analog to digital converters, 60 GHz WLAN, and 40/100 Gb/s Ethernet products, whereas high PNP circuits are used in operational amplifiers.

6.2.3 GROUPS III-V COMPOUND SEMICONDUCTORS

The compound semiconductors consist of elements from groups III and V used in both bipolar and FET (field effect transistors). Silicon is primarily used for RF, microwave, and mixed signal applications because of low cost and better integration density capabilities. The drivers for groups III-IV devices are based on applications that require high yield (manufacturability), higher efficiency and dynamic range, and lower noise performance such as power amplifiers used in communication applications. The key compound semiconductors that can achieve such targets are gallium arsenide (GaAs) and gallium nitride (GaN).

The groups III-V based MOSFETs have better mobility as compared to those that are silicon based and which have fueled more than 40 years of R&D. Current efforts on groups III-V MOSFETs are going beyond GaAs and can be subdivided into those based on arsenides, phosphides, nitrides, and antimonides. These compounds have a wide range of band gaps† and carrier mobility, so they are suited to wireless systems.

GaAs and GaN are compound semiconductors with higher electron mobility than silicon making these more suitable at higher frequencies. GaAs and GaN are the key semiconductor compounds for many RF/microwave applications such as switches and power amplifiers [11].

---

* BJT is formed by two P-N junctions for the amplification of electric current. An NPN transistor is formed by sandwiching a thin region of P-type between two regions of N-type material and vice versa for a PNP transistor.
† A band gap or energy gap refers to the energy difference between the top of the valence band and the bottom of the conduction band in insulators and semiconductors [4].

GaAs tends to be two generations behind Si in wafer size, whereas InP (indium phosphide) and SiC (silicon carbide) substrates are one generation behind GaAs. It is imperative for the compound semiconductor industry to keep up the substrate size with silicon advances in order to gain benefits from the achievements in manufacturing processing. GaN relies on SiC as a host substrate which opens up the possibility of fabricating GaN circuits in a silicon foundry as well as the heterogeneous integration of GaN amplifiers with Si CMOS control circuitry. GaN-on-SiC offers better performance and higher reliability for power amplifiers.

GaN has power densities four to five times greater than silicon embedded LDMOS (Laterally Diffused MOSFET). Thus, with these power levels, GaN HEMT (High-Electron-Mobility Transistor) device capacitances are also much lower, providing the ability to increase instantaneous bandwidths without losing efficiency which is the key for LTE/4G and future 5G systems. LDMOS on the other hand is still the key device for RF power amplifiers in base stations of wireless communications systems. Additionally, GaN-on-SiC, which is the mainstay of GaN technology, is effective for remote radio heads, multi-standard and multi-frequency base stations, pico and femto cells, and for larger bandwidths where Si LDMOSFETs are not suitable [9,12].

6.2.4 PASSIVE ON-CHIP DEVICES

Passive elements including both distributed and lumped are imperative for analog and RF systems. Distributed passive elements include transmission lines, waveguides and antennas, and lumped elements consisting of inductors, transformers, linear and variable capacitors (varacators), and resistors. The lumped passive elements are primarily used in applications in low frequency analog (from DC to 0.4 GHz) and RF (0.4–30 GHz). The distributed passive elements are more frequently used for microwave and millimeter wave applications [13].

6.3 EXAMPLES (RFAMS)

This section will briefly explain the use of various RFAMS circuits in some key network equipment and components such as transceivers, power amplifiers, and solar cells.

6.3.1 TRANSCEIVER

A transceiver is a device that transmits and receives various forms of signals using a common circuitry. Cellular networks and devices employ transceivers that provide transmission and reception simultaneously but on separate frequencies, that is in full duplex mode.

The transceivers in base stations and handhelds are embedded with ARM (Advanced RISC* Machine) processors. ARM is primarily a 32-bit processor architecture widely present in the world's mobile devices. ARM, a UK based firm, develops instruction sets and processor core architectures, but does not manufacture processors. The architecture describes the rules for how the microprocessor will behave, but without constraining or specifying how it will be built. The microprocessor architecture defines the processor's instruction set, the programmer's model, and how the processor interfaces with its closest memory resources. The low power characteristics, high data bandwidth, and media processing capability coupled with a high-performance core make the ARM micro-architecture particularly suited for networking, wireless, and other consumer applications [14–16].

6.3.1.1 Material and Technology for Transceivers

CMOS is the technology of choice for mobile transceivers operating in the 0.4–30 GHz frequency range using silicon. Bipolar and BiCMOS have been the key technologies for wireless transceivers

---

* Reduced Instruction Set Computer.

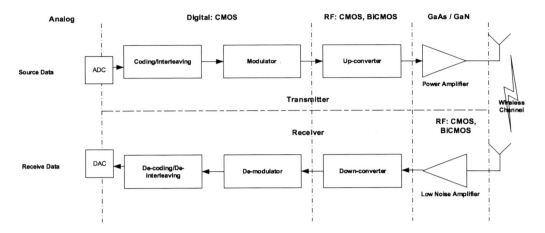

FIGURE 6.1 RF transceiver generic block diagram. (From Richard, M.S. 2005. SiGe BiCMOS RF ICs and components for high speed wireless data networks. Dissertation submitted to the Faculty of the Virginia Polytechnic Institute and State University; Razavi, B. 1997. *RF Electronics.* Prentice-Hall, Inc, NJ, USA [17,18].)

in the 0.4 GHz to 5 GHz range, however, the current inclination is toward RFCMOS also using silicon [9]. RFCMOS is a low cost, high volume digital process technology that uses less power and produces less heat compared to other processing technologies. Thus, it allows more functions, such as the RF and baseband components, to be integrated into a single chip.

Figure 6.1 shows the block diagram of a generic RF transceiver (transmitter and receiver) tagged by IC technology and circuit type (RF, analog or digital). The digital components are implemented in CMOS while the RF areas in the transmitter and receiver can be implemented with different technology options including CMOS. Although the use of multiple technologies is not cost effective, it has traditionally been used to meet the stringent performance requirements of wireless communication standards. SoC* on the contrary offers a low cost and low power alternative where all the functions are integrated into a single package/chip using a single technology.

6.3.1.2 LTE/4G Transceiver

LTE and LTE-Advanced (4G) transceivers have to meet more stringent requirements than their predecessors. These include multi-band (40 plus frequency bands), multi-mode (FDD, TDD), carrier aggregation (in 4G), and so on. A CMOS based transceiver for LTE can consist of dual receivers for antenna diversity, digital to analog and analog to digital converters, D4G for LTE/3G/2G, and D3G† for 3G/2G standards [19], an ARM7 microprocessor for sequence/hardware control, and an API (application programming interface) for faster radio platform development time. In most cases, the receivers (in LTE transceivers) do not have SAW (surface acoustic wave) filters and external LNAs (low-noise amplifiers), which reduces the component count and footprint [20,21].

6.3.2 RF POWER AMPLIFIERS

An RF power amplifier is a type of electronic amplifier that is used to convert a low power radio frequency signal into a signal of significant power able to be transmitted over greater distances. Power amplifiers are normally designed for specific applications or standards (e.g. point-to-point microwave radios, LTE, etc.) [22].

In cellular base stations operating in the range of 400 MHz to 3.5 GHz requiring 100/200 watts, the power amplifiers are typically connected in parallel providing the final amplification to the data

---

* System on chip provides the option of combining multiple technologies to form a complete (subsystem). SoC is a single IC connected inside a single package dedicated to a specific application.
† DigRF specification defines the interface between one or more baseband ICs and RFICs (Radio Frequency Integrated Circuits) in a single terminal. D3G and D4G are open standard digital interfaces.

signal in order to achieve the desired output power. Silicon LDMOS transistors are the de facto standard for such Power Amplifiers (PAs) because of maturity and low cost. GaN continues to make inroads as it has four to five times higher power densities than LDMOS. GaN may be advantageous over LDMOS in certain classes of high efficiency PA architectures such as Doherty. Circuits designed for Doherty architecture may not perform well in an input signal envelop tracking PA architecture. Additionally, a nonstop increase in operating frequency ranges and modulation schemes which are expected to be aggravated with 5G also need to be taken into consideration.

In typical portable communication sets, Si is used as power management chip whereas GaAs HBTs (heterojunction bipolar transistor—a bipolar transistor consists of at least two semiconductors) are used for the power amplifier chips. Contrary to base stations, the PA chips in portable units typically need to provide 1–4 watts of RF power to the antenna. However, the challenge for PA vendors is associated with end-of-life battery voltage in portable communication units. PAs not only need to accommodate 4 V to 5 V chargers, but also operate at lower voltages such as 2.4 V.

Thus, for both base stations and mobile phones, the above-mentioned figures-of-merit need to be understood to enhance the efficiency of power amplifiers.

6.3.2.1 Material and Technology for PAs

Groups III-V compound semiconductors GaAs and GaN are particularly used in power amplifiers, an important component of base stations and handsets. GaAs and GaN can be used to make BJTs, FETs (field effect transistor), and HEMTs (high electron mobility transistor) types of transistors.

GaAs based monolithic microwave integrated circuits (MMICs) are widely used in smartphones, tablets, and WiFi devices. Switches and amplifiers that are based on such MMICs are designed for operation at the low voltages and currents, normally available from batteries. GaN based MMICs have higher voltage capabilities making these an attractive choice for power amplifiers in base stations [11].

GaN transistors have been around for over a decade and were initially developed for IED (improvised explosive device) jammers. GaN has a high power density, that is, it has the ability to dissipate heat from a small package making it highly effective. GaN, due to its high electron mobility, can amplify signals well into the upper gigahertz ranges. However, GaN enabled equipment is costly since the manufacturing processes are expensive.

6.3.2.2 PA Performance Metric—PAE

PAs in cellular phones are normally powered up directly from the battery making system implementation easier, but this leads to nonlinearity and inefficiency in terms (e.g., shorten battery life, reduced talk time, etc.). However, there is a tradeoff between linearity and efficiency. To meet the required linearity, the operating transmitting power is backed off from the power amplifier's compression point which reduces efficiency. According to Texas Instruments [23], a major IC developer, the RF power section of a mobile phone during transmission can consume up to 65% of the overall power budget due to the PA's intrinsic inefficiencies.

PAE is the most common metric of performance for RF power amplifiers today. It is the ratio of the RF power output, P_{out}, less input power, P_{in}, to the total DC power, P_{dc}, consumed expressed as a percentage as shown in Equation 6.1. If the PA is directly connected to the battery, it translates into $P_{dc} = V_{batt} * I_{batt}$, and if it is powered by a supply regulator, $P_{dc} = V_o * I_o$.

$$\text{PAE} = \frac{(P_{out} - P_{in})}{P_{dc}} * 100\% \qquad (6.1)$$

To gain higher efficiencies, P_{dc} needs to be reduced and one way of achieving this is through the use of DC-DC supply regulators* [24]. These regulators lower the output voltage at lower RF transmitter power levels which in turn reduces I_o (current drawn by PA), thus improving PAE.

---

* A DC-to-DC converter/regulator is an electronic circuit that converts a source of direct current (DC) from one voltage level to another [4].

6.3.3 SOLAR CELLS

A solar cell is a detector and a convertor that converts sunlight into an electric current with a minimum amount of loss [25]. Solar cells were initially developed for powering satellites in space, but now are a common part of cell sites' roll out for reducing operational costs and following a green agenda through the use of photovoltaic concentrator systems.

Photovoltaics is a defining technology that directly takes advantage of our planet's ultimate source of power, the sun. Solar cells directly exposed to sunlight produce electricity without any known harmful effects to the environment. These can generate power for many years (perhaps 20 years) with minimal maintenance and operational costs [25–28].

6.3.3.1 Material and Technology for Solar Cells

One of the most effective ways to generate electricity from solar radiation is the use of multi-junction solar cells, which are made up of III-V compound semiconductors. The development of such multi-junction or monolithic tandem solar cells* was started in 1985. The first concrete development on GaInP and GaAs-based multi-junction cells was published and patented by J. Olson and coworkers at the National Renewable Energy Laboratory [29,30].

GaAs is more commonly used in multi-junction photovoltaic cells for concentrated photovoltaics [4]. A Concentrating Photovoltaic (CPV) system converts light energy into electrical energy in the same way that conventional photovoltaic technology performs. However, it uses an advanced optical system to focus a large amount of sunlight on to each solar photovoltaic cell for maximum efficiency [4,31]. Thus, by focusing sunlight onto smaller areas, the concentrator is able to use fewer solar cells than traditional photovoltaic power (i.e., traditional solar cells made of silicon) [4,25].

A concentrator has a high power generating capacity with a capacity of approximately 1–200 megawatts. A concentrator has movable parts to accommodate changes in the sun's direction, making these easy to manufacture and maintain, thus these are well suited for a large-scale PV power station [26]. Groups III-V compound semiconductors are the most efficient materials for photovoltaic concentrator systems at present. The key reason is the flexibility that allows combining a range of materials from binary to quaternary compounds with a corresponding flexibility of band gap engineering. Interestingly, as these compounds can retain band gaps due to high absorption coefficients, these also tend to radiate light efficiently. The most commonly used substrates are GaAs and InP and these have ideal band gaps for solar conversion [27].

GaAs has the optimum band gap (energy gap) for a single semiconductor material [32]. Higher efficiencies can only be achieved by dividing the solar spectrum[†] into several parts, that is, a high band gap semiconductor like GaAs is used to absorb the short wavelength radiation while the long wavelength part is transmitted to a second semiconductor with a lower band gap energy such as Ge. In such a configuration, transmission and thermalization losses of hot carriers[‡] can be minimized [25].

A high end and high efficiency solar cell utilizes a triple-junction structure to maximize efficiency. Concentrating the sunlight by a factor of between 300× to 1000× onto a small cell area allows the use of highly efficient but expensive triple-junction solar cells based on Group III-V semiconductors [33]. For example, the general structure of a GaInP/GaInAs/Ge based triple-junction cell is shown in Figure 6.2, which is advantageous as compared to silicon, even with an InP radiation hard material.

---

* Multi-junction solar cells or tandem cells are solar cells containing several p-n junctions. These cells are made up of different of semiconductor materials. Each junction is exposed to a different wavelength of light, thus improving the cell's sunlight to electrical energy conversion efficiency [4].

† The sun radiates sunlight through electromagnetic waves spreading over a range of wavelengths (290–2500 nm) and that is known as the solar spectrum [34].

‡ Hot carriers imply either holes or electrons that have gained very high kinetic energy after being accelerated by a strong electric field in areas of high field intensities within a semiconductor (especially metal oxide semiconductor [MOS]) device. These carriers can get injected and trapped because of their high kinetic energy in areas where they should not, resulting in a space charge that causes the device to degrade or become unstable. This device degradation due to hot carrier injection is called "hot carrier effects" [35].

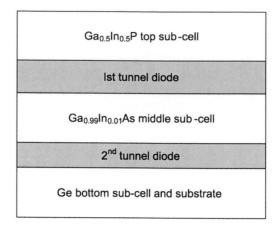

FIGURE 6.2 III-V triple-junction cell. (From Frank D. 2006. High-Efficiency Solar Cells from III-V Compound Semiconductors. *Wiley InterScience: Physica Status Solidi (c)* 3(3):373–379 doi:10.1002/pssc.200564172 [25].)

This sample is comprised of Ge (0.7 eV) as the bottom subcell formed on a Ge substrate with $Ga_{0.99}In_{0.01}As$ (1.4 eV) as the middle subcell, while a lattice matched* $Ga_{0.49}In_{0.51}P$ (1.9 eV) is the top subcell. The tunnel diodes are used to form an electrical series connection of the subcells [25].

6.3.3.2 Performance of Multi-Junction Solar Cells

The efficiency of multi-junction solar cells can be improved primarily by increasing the amount of light collected by each cell that is turned into carriers and increasing the collection of light generated carriers by each p-n junction [26].

The CPV standard IEC 62108 called "Concentrator photovoltaic (CPV) modules and assemblies—Design qualification and type approval" was issued by the International Electrotechnical Commission (IEC) in 2007 [36]. This most prominent test standard specifies minimum design requirements and types of approval procedures for CPV modules and assemblies, suitable for long term operation (~25 years) in general open-air climates.

The standard defines a number of tests including electrical insulation, thermal cycling test, hail impact, and so on, to capture the performance of the CPV. The details can be found in the standard [36]. As an example, a thermal cycling test is vital to test the reliability of concentrator solar cells whereas the focus of the electrical performance test is to find power degradation (not on the absolute power output) by comparing the CPV test module pre- and poststress relative power. The relative power is defined as the maximum power output of the sample under test divided by the maximum power output of the control sample, measured under similar test conditions.

The sample relative power P_r can be calculated from Equation 6.2.

$$P_r = \frac{P_m}{P_{mc}} * 100 \qquad (6.2)$$

where

P_r is the sample's relative power, in percentage %
P_m is the test sample's maximum power, in W

---

* Lattice matched is a structure consisting of ultra-thin layers of single crystal semiconductors (typically Groups III-V) of different chemical compositions. Matching between two different semiconductor materials allows a region of band gap change to be formed in a material without introducing a change in crystal structure [4,37].

P_{mc} is the control sample's maximum power measured at similar conditions as P_m in W. The metric, relative power degradation, is used to determine the viability of concentrator solar cells after a certain number of thermal cycles. The relative power degradation P_{rd} is described in Equation 6.3:

$$P_{rd} = \frac{P_{ri} - P_{rf}}{P_{ri}} * 100 \tag{6.3}$$

where P_{rf} and P_{ri} are the relative powers measured after and before the given test, respectively.

P_{rd} should be less than 13% for outdoor measurements and 8% for indoor. The difference of 5% is provided to account for the larger uncertainty in outdoor environments.

6.3.4 MEMS

The interest in MEMS is growing due to increasing demand for motion enabled products. The primary reason for this demand is the use of MEMS devices in consumer electronic related applications. In smart phones and tablet computers, iSuppli forecasts that the MEMS market will grow by approximately 30% compounded annually [38]. MEMS enabled products can also be found in automobiles and medical instruments [39].

MEMS by definition are miniaturized devices and structures (i.e., mechanical and electro-mechanical elements) that are fabricated using techniques similar to those used for integrated circuits. A MEMS device typically consists of a micro-electro-mechanical sensor element for receiving/sensing information from the environment and an actuator element for responding to decisions from the control system to change the environment, which is packaged together with IC. The control system can either be a fixed mechanical or electronic system, software-based system (such as printer driver), person or some other form of input [40–43].

6.3.4.1 MEMS in Mobile

One of the largest segments of growth in MEMS manufacturing is mobile internet devices. This is due to the ability of MEMS which allows the integration of a number micro-components on a single chip for both sensing and controlling the environment by using, for the most part, existing IC manufacturing processes [44].

Mobile Internet devices contain MEMS devices such as accelerometers, gyroscopes, and microphones, and also have radios (WiFi and cellular) that have application needs for RF MEMS devices, including resonators, varactors, and switches [43]. Some examples are as follows:

- 3-axis MEMS accelerometers and gyroscopes in Nintendo Wii, Apple iPhone, and so on.
- Digital MEMS microphones with built-in analog-to-digital converter circuits [4].
- Use of single chip technology called CMOS MEMS in accelerometers, smart sensors (pressure, voice), and RF related MEMS* (variators, switches) [45,46].
- RF MEMS switches, switched capacitors, and varactors are applied in software defined radios, reconfigurable antennas, tunable band-pass filters, and so on [4,46].

6.4 IOT—NEXT GROWTH ENGINE FOR SEMICONDUCTORS

The IoT is seen as the next killer application by many ICT sector pundits. IoT, as defined by ITU in Recommendation ITU-T Y.2060 [47], is a global infrastructure for the information society, enabling advanced services by interconnecting (physical and virtual) things based on existing and evolving interoperable information and communication technologies.

---

* RF MEMS devices have movable microparts that are capable of reconfiguring the RF characteristics of a device.

It is commonly said these days that IoT is the next growth engine or driver for the semiconductor industry. It will drive demands not only for sensors, but also for actuators and security chips. Any device/object that needs connectivity without or with minimal human interaction can become part of the IoT ecosystem. Hence, every industry from manufacturing, real estate, transportation, automotive, logistics, supply chain, and healthcare, to telecommunication, utilities, and finance can depend on the IoT in the foreseeable future [48,49].

6.4.1 TECHNOLOGY/PROCESS

A semiconductor is considered as the building block for the infrastructure components within the IoT ecosystem. These components sense, collect, transmit/receive, and process signals and data for IoT applications [49]. Essentially, an IoT device senses something first, then collects data on what it senses, and communicates this data through an Internet connection to another device or location where it is processed [50].

The majority of IoT devices are battery powered, thus need to operate without any maintenance or replacement for years and also have to consume minimal power. In other cases, power will be provided through external sources to the IoT infrastructure. In both cases, energy efficiency is absolutely essential for the success of IoT. In addition, a tremendous amount of personal data and organizational confidential information can be exchanged, particularly via wearables. Thus, power and security embedded integrated circuits will be required for the smooth operation of the IoT [50].

The advanced manufacturing techniques such as MEMS discussed earlier and NEMS (nanoelectromechanical systems) combine electronics and mechanical components at micro and nano scales. This integration allows the functionality of sensors, actuators, and integrated circuits into small form-factors, which makes them suitable for use in a plethora of IoT applications. Integrated circuits in use in IoT do not require complex integration and can use more than one chip. However, IoT applications ideally require a single chip at an acceptable form-factor and with very low power consumption which has driven the growth for SoC-type designs. The SoC is comprised of multiple embedded cores, an embedded GPU (graphic processing unit), and integrated wireless connectivity in a single package [49]. In a nutshell, the basic building blocks of an IoT include microcontrollers, sensors, actuators, power management functions, embedded memory, and connectivity [51].

6.4.2 IOT'S FUTURE

As more THINGs become connected, so does the demand for semiconductors. Demand for chips and ICs containing processors, MEMS, and RFAMS components will rise accordingly. If the supply is as needed, eventually economies of scale will be achieved resulting in lower costs. For example, MEMS components dominate a variety of consumer products and this prevalence has occurred due to a decline in the average selling price coupled with an increased demand [51].

Many IoT standards are in development and use cases are continuously evolving. Semiconductor companies need to keep an eye on those standards and technologies to develop a comprehensive strategy. A number of varying estimates and forecasts are available, all of them pointing to a significant increase in the number of connected devices in the next few years. Multiple estimates indicate that there will be around 30 billion connected devices by 2020 and according to International Data Corporation (IDC) the worldwide IoT market will stand at $1.7 trillion in 2020 [52].

6.5 PERSPECTIVE—SEMICONDUCTOR BUSINESS IN PAKISTAN

Today, the existence of a semiconductor sector in many developing countries including Pakistan is negligible. Only a handful of activities (as explained later) in academia and industry can be seen in this regard. However, as witnessed from this chapter and from a global perspective, it is crucial to have semiconductor facilities and firms to remain competitive in the world and to increase productivity in the country.

The hardware development of electronic devices and circuits remains a challenge for the developing nations such as Pakistan. The prohibiting factors are lack of overall political stability, socio-economic disparity, shortage of experienced manpower, and the high cost of establishing foundries and fabless companies.

The question becomes how to instigate any one or more forms of semiconductor activities in the country.

6.5.1 Status of Semiconductor Industry

The existing semiconductor industry is quite small, consisting of only a few players. A number of attempts have been made since the 1980s, but most of them were unsuccessful for one reason or other. The key reasons were that the size of the local market was not large enough to attract local and foreign companies and the availability of the latest and cheapest semiconductor devices from China/Taiwan. Due to these and above-mentioned factors, it can be safely said that no one does IC level design or IC fabrication in the country. Table 6.1 shows the dismal picture of the semiconductor sector in Pakistan [53].

6.5.2 Semiconductor Development Process

The semiconductor development from design on paper all the way to final IC packaging is a complex, expensive, and time-consuming process (Figure 6.3). This section will briefly look into this process [4,55–58].

Chip design has come a long way from the time when the first semiconductor chips were made literally by hand. Now, EDA (electronic design automation) software tools are used to do the work. Regardless, the goal of the design process is to create a working blueprint for a new chip and get it ready for manufacturing.

Once a system design is finalized, designers use hardware description languages describe this design in hundreds/thousands of lines of software codes. This step is followed by physical design, where several substeps are used to build up a succession of layers of semiconductor materials and geometries to produce thousands of electronic devices (transistors) at tiny sizes, which together

---

TABLE 6.1
Semiconductor Sector of Pakistan

Semiconductor Area	Status in Pakistan
Integrated device manufacturers	None
Fabless firms	None
Fabs (foundries)	None
Semiconductor corporate R&D labs	None
Class (100) clean[a] room supportive of proof of concept	Handful (if any)
ISO 14000+ standardized semiconductor design facilities	Handful (if any)
Non-QA (quality assurance) semiconductor design facilities (labs) for fabless solutions	Very few (primarily academia)
National policy on semiconductor sector development	None

Source: Clean Air Technology, Inc. What is a Cleanroom? http://www.cleanairtechnology.com/cleanroom-classifications-class.php [54].

[a] A cleanroom is any given controlled space where conditions are imposed to reduce particulate contamination and control other environmental parameters such as temperature, humidity, and pressure.

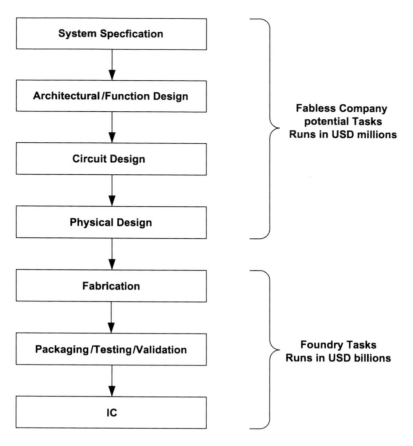

FIGURE 6.3 IC design flow (along with cost).

function as integrated circuits. The processing of semiconductor wafers to produce ICs involves a good deal of chemistry and physics.

A quick way to design new chips involves the use of reused circuitry. Large size chips include a fair amount of reused, borrowed, licensed or recycled circuitry (not physically recycled silicon), and recycled design ideas. This is called design reuse or IP reuse as described earlier. Such companies license the partial chip designs to others and collect a royalty. The next step after physical design is wafer fabrication which takes place in foundries. Turning semiconductor material into ICs requires an absolute absence of contaminants and this process is also handled by fabs. Estimates put the cost of building a new fab at over one billion US dollars with values as high as $3–4 billion not being uncommon. The cleanroom, the central part of a fab, operates under vacuum, with elemental, molecular, and other particulate contaminants rigorously controlled. The finished ICs (products) are then later stored, distributed, and get sold either by IDMs or by fabless companies.

6.5.3 POTENTIAL WAY FORWARD

The key steps in this process as discussed above include software design, physical design, manufacturing of integrated circuits, and finally, sales and distribution.

According to Fast Market Research, the Pakistan consumer electronic market is projected to grow at a CAGR of 9.1% over 2016–2020 [59] and the vast majority of such products have been/will be shipped from China. This figure makes the case of a foundry in the country unattractive, as that easily requires $1 billion of investment with no ROI in the near future. Moreover, the competition from China, the U.S., and others will be extremely difficult to beat and lack of experienced manpower will be another challenge [60]. However, the option of setting up of fabless companies is still viable where the cost of initial investment is in the order of a few million dollars.

6.5.3.1 Pakistan Fabless Company

There are number of successful and unsuccessful cases of fabless companies. The cost of such companies runs into millions of dollars (from single to triple digits). However, it should be kept in mind that the ROI is not positive before 4–7 years [55,56]. To have this in Pakistan, the following are some of the basic steps that need to take place.

- *Market*: Find the niche market at the right time. For example, GreenPeak, the Netherlands based fabless company, picked chip designing for low power ZigBee communications.
- *Manpower*: The country needs to get manpower from abroad by offering attractive packages. It would be prudent to get technical support from China and the U.S., but not in terms of funding.
- *Funding*: This is one of the most crucial elements of the puzzle. As Pakistan is a developing country, there are few choices, and government support is essential. Government may also reduce taxes and provide subsidies as well. Perhaps the funding should be secured from a mix of Ignite—National Technology Fund (government entity), and bank loans. There is, as such, no presence of venture capital firms in the country.
- *Goals*: Only realistic targets should be kept in mind, noting that only hard work and dedication can bring ROI and jobs in a few years to provide stable continuity in the program.

Finally, it will be prudent for the government to hire a mature consulting firm for truly understanding the potential of the semiconductor industry in the country.

6.6 SUMMARY

Semiconductors and ICs are vital pieces for the mobile telecommunications sector. The ongoing development in semiconductors will enrich the experience of mobile users. For example, an MIT (Massachusetts Institute of Technology) team of researchers claims to have designed the shortest-gate working transistors, yet they were built using Group III-V channels. They had experimentally proven that InAs channels outperform silicon at small device dimensions. This and similar pioneering works will augment the progress of CMOS-compatible, Group III-V-based technology research and development worldwide [61].

The GaAs product market is comparatively very large as compared to GaN, but the GaN market is growing. Continuous improvements are reducing GaN production costs at a reasonable pace. GaAs will continue to dominate the microwave radio market with MMICs and power amplifiers for cell phones. However, as GaN costs come down, GaAs could start to lose its market share [62].

MEMS will continue to receive attraction for motion enabled components in mobile Internet devices. IoT is most likely the next growth engine for semiconductors. However, the fragmented standardization work for IoT will likely impact its adoption in the mobile telecommunications to some extent. Finally, to be part of the semiconductor world, it is important for Pakistan to perhaps kick off with fabless firm(s) to gain momentum and later move to more costly endeavors.

PROBLEMS

1. Describe the four types of semiconductor firms?
2. What are the key RF applications that are developed with CMOS and bipolar technologies?
3. What are the key III-V compound semiconductors for wireless systems?
4. What are ARM processors and how important are these to the mobile industry?
5. Determine the power added efficiency of an LM123 RF amplifier when Vo = 3.4 V, Io = 400 mA, Vin = 4.2 V, and Iin = 100 mA. The PA is powered by a power amplifier regulator?

6. Why is GaN gaining importance as a substitute material for power amplifiers?
7. What is a solar cell?
8. What are photovoltaic concentrator systems?
9. Why are III-V multi-junction solar cells better than single layer ones?
10. What is an electrical performance test as defined in the standard IEC62018?
11. What is the relative power degradation of a solar cell whose relative powers before and after are 9.45 watts and 10 watts, respectively, after 5 years with 800 thermal cycles?
12. What are MEMS and what is their usage for the mobile communications industry?
13. What is IoT as defined by ITU?
14. What are basic components of IoT?
15. Discuss in a group how to create a successful fabless company in a developing nation.

REFERENCES

1. Bauer, H., Grawert, F. and Schink, S. 2012. *Semiconductors for Wireless Communications: Growth Engine of the Industry*. McKinsey & Company, New York, USA.
2. IHS Markit 2015. Global Semiconductor Market Achieves Strong, Broad-Based Growth in 2014. http://press.ihs.com/press-release/technology/global-semiconductor-market-achieves-strong-broad-based-growth-2014-accordi
3. International Technology Roadmap for Semiconductors 2013. Radio Frequency and Analog/Mixed-Signal Technologies Summary.
4. Wikipedia. http://en.wikipedia.org/wiki/Main_Page
5. Hollauer, C. 1975. Modeling of Thermal Oxidation and Stress Effects Semiconductor Fabrication Processes. *Dissertation*. http://www.iue.tuwien.ac.at/phd/hollauer/
6. Tech Target (Bitpipe). http://www.bitpipe.com/tlist/IP-Core.html
7. STMicroelectronics. BiCMOS. http://www.st.com/content/st_com/en/about/innovation---technology/BiCMOS.html
8. International Technology Roadmap for Semiconductors 2.0 2015. Outside System Connectivity.
9. International Technology Roadmap for Semiconductors 2011. Radio Frequency and Analog/Mixed-Signal Technologies for Communications.
10. Lee, T.H. 2007. Key Note: The History and Future of RF CMOS: From Oxymoron to Mainstream. *XXV IEEE International Conference on Computer Design, Lake Tahoe*, California, USA, October 7–10, 2007.
11. Brown, J. 2016. What's the difference between GaN and GaAs?
12. International Technology Roadmap for Semiconductors 2009. Radio Frequency and Analog/Mixed-Signal Technologies for Communications.
13. Bahl, I.J. 2014. *Control Components Using Si, GaAs, and GaN Technologies*. Artech House, Norwood, MA, USA.
14. Asif, S.Z. 2011. *Next Generation Mobile Communications Ecosystem: Technology Management for Mobile Communications*. Wiley Inc., UK.
15. Georgescu, D. 2003. Evolution of Mobile Processors. *2003 IEEE Conference on Communications, Computers and Signal Processing,* vol. 2, Victoria, Canada, August 28–30, 2003, pp. 638–641.
16. Khoushanfar, F. et al. 2000. Processors for Mobile Applications. *2000 IEEE International Conference on Computer Design*, Austin, USA, September 17–20, 2000, pp. 603–608.
17. Richard, M.S. 2005. SiGe BiCMOS RF ICs and Components for High Speed Wireless Data Networks. Dissertation submitted to the Faculty of the Virginia Polytechnic Institute and State University.
18. Razavi, B. 1997. *RF Electronics*. Prentice-Hall, Inc, NJ, USA.
19. MIPI Alliance, Inc. DigRF(SM) Specifications. http://mipi.org/specifications/digrfsm-specifications
20. Ganseh, K. and Claudio, R. 2012. Design: Multimedia, Multiband Transceiver Technology Delivers LTE. http://www.eetimes.com/General/PrintView/4235909
21. Rahman, M. et al. 2012. SAW-less Transceiver for 4G/3G/2G Cellular Standards. *Fujitsu Scientific & Technical Journal* 48(1):60–68.
22. Arrow Electronics, Inc. (RichardsonRFPD) RF & MW Power Amplifier. http://www.richardsonrfpd.com/Pages/Product-End-Category.aspx?productCategory=10042
23. Texas Instruments 2011. LM3203, LM3204, LM305 Optimizing RF Power Amplifier System Efficiency Using DC-DC Converters (Literature Number: SNVA593).

24. Ripley, D.S. 2016. Patent Application Title: Shared Integrated DC-DC Supply Regulator. http://www.patentsencyclopedia.com/app/20160094254

25. Frank, D. 2006. High-Efficiency Solar Cells from III-V Compound Semiconductors. *Wiley InterScience: Physica Status Solidi (c)* 3(3):373–379 doi:10.1002/pssc.200564172.

26. Lin, G.J. et al. 2013. III-V Multi-junction solar cells. In: book edited by S.L. Pyshkin and M.B. John, *Optoelectronics—Advanced Materials and Devices*. InTech, New York, pp. 445–471. http://www.intechopen.com/books/optoelectronics-advanced-materials-and-devices/iii-v-multi-junction-solar-cells.

27. Connolly, J.P. and Mencaraglia, D. 2013. III-V Solar Cells. http://arxiv.org/ftp/arxiv/papers/1301/1301.1278.pdf

28. Green, M.A. 2005. *Third Generation Photovoltics*. Springer-Verlag, Germany.

29. Olson, J.M. 1987. US Patent 4,667,059 Current and Lattice Matched Tandem Solar Cell.

30. Olson, J.M. et al. 1985. GaInP2/GaAs: A Current Lattice-Matched Tandem-Cell with a High Theoretical Efficiency. *18th IEEE Photovoltaic Specialists Conference*, Las Vegas, Nevada, USA, 21–25 October, pp. 552–555.

31. CPV Consortium. http://cpvconsortium.org/

32. Green, M.A. 1982. *Solar Cells: Operating Principles, Technology, and System Applications*. Prentice-Hall, USA.

33. Philipps, S.P. and Bett, A.W. 2016. Current Status of Concentrator Photovoltaic Technology (CPV). Fraunhofer Institute for Solar Energy Systems ISE/National Renewable Energy Laboratory. Version 1.2., February.

34. Australian Glass and Glazing Association 2012. AGGA Technical Fact Sheet Solar Spectrum. file:///C:/Users/director.wireless/Downloads/AGGA%20Technical%20Fact%20Sheet%20-%20Solar%20Spectrum%20-%20February%202012.pdf

35. EESemi.com. http://www.siliconfareast.com/hotcarriers.htm

36. International Electrotechnical Commission 2007. IEC 62108 Concentrator photovoltaic (CPV) modules and assemblies—Design qualification and type approval. Edition 1.0, December.

37. Semiconductor Glossary. http://www.semi1source.com/glossary/default.asp?searchterm=lattice+matched+structure

38. International Electronics Manufacturing Initiative 2011. Technology Roadmap, January.

39. International Technology Roadmap for Semiconductors 2013. Micro-Electro-Mechanical Systems (MEMS) summary.

40. Waldner, J.-B. 2008. *Nanocomputers and Swarm Intelligence*. ISTE John Wiley & Sons, London, p. 205. ISBN 1-84821-009-4.

41. MEMSnet. What Is MEMS Technology? https://www.memsnet.org/about/what-is.html

42. Madou, M.J. 1997. *Fundamentals of Microfabrication*. CRC Press, Boca Raton, FL, USA.

43. SmallTech Consulting. What Is MEMS? http://www.smalltechconsulting.com/What_is_MEMS.shtml

44. International Technology Roadmap for Semiconductors 2011. Micro-Electro-Mechanical Systems (MEMS).

45. Semiconductor Manufacturing International Corporation. CMOS MEMS. http://www.smics.com/eng/foundry/technology/cmos_mems.php

46. Jacopo, I. 2011. An Overview of RF MEMS Technologies and Applications. *MEMS Journal*. http://www.memsjournal.com/2011/05/an-overview-of-rf-mems-technologies-and-applications.html

47. ITU-T, Recommendation ITU-T Y.2060 2012. Overview of the Internet of Things.

48. Yeo, K.S. et al. 2014. Internet of Things: Trends, Challenges and Applications. *International Symposium on Integrated Circuits*, 10–12 December, pp. 568–571.

49. Acker, O. et al. 2015. *The Internet of Things: The Next Growth Engine for the Semiconductor Industry*. PricewaterhouseCoopers AG, May.

50. Lattice Semiconductor 2015. Energy Efficiency: The Common Denominator in the Internet of Things, March.

51. Altis 2014. IoT, a Key Enabler of Current and Future World's Semiconductor Industry Growth. Altis Insights, October.

52. International Data Corporation 2015. Explosive Internet of Things Spending to Reach $1.7 Trillion in 2020. http://www.idc.com/getdoc.jsp?containerId=prUS25658015

53. Syed, A.S. 2012. Semiconductor Radiation Engineering-Global Applications & Trends, Centre for Emerging Sciences, Engineering & Technology, presentation.

54. Clean Air Technology, Inc.. What Is a Cleanroom? http://www.cleanairtechnology.com/cleanroom-classifications-class.php

55. Olofsson, A. 2012. A Lean Fabless Semiconductor Startup Model, Adapteva. http://www.adapteva.com/white-papers/a-lean-fabless-semiconductor-startup-model/

56. EntrepreNL 2012. How to Start a Semiconductor Company: The Story of GreenPeak. http://entreprenl.wordpress.com/2012/09/27/how-to-start-a-semiconductor-company-the-story-of-greenpeak/

57. Turley, J. 2003. How Chips Are Designed, Informit. http://www.informit.com/articles/article.aspx?p=31679

58. Investopedia 2016. The Industry Handbook: The Semiconductor Industry. http://www.investopedia.com/features/industryhandbook/semiconductor.asp

59. New Market Research Report: Pakistan Consumer Electronics Report Q3 2016. https://www.clickpress.com/releases/Detailed/747268005cp.shtml

60. Pakistan Affairs 2014. Thread: Semiconductor showdown: TSMC, Intel, Samsung, Global Foundries, IBM, SMIC, and UMC. http://www.pakistanaffairs.pk/threads/66042-Semiconductor-showdown-TSMC-Intel-Samsung-Global-Foundries-IBM-SMIC-and-UMC

61. Mike, C. 2012. MIT Researchers make Smallest Working III-V Transistors Yet. http://www.semiconductor-today.com/news_items/2012/DEC/MIT_241212.html

62. MSE Supplies 2016. What Are the Differences between GaN And GaAs RF Power Amplifiers? https://www.msesupplies.com/blogs/news/what-are-the-differences-between-gan-and-gaas-rf-power-amplifiers

7 Product Development

The finalization of standards and development of semiconductor devices (integrated circuits) leads to product development. Product development means creation of products with new or different characteristics that can offer new or additional benefits to the customer. Product development may involve modification of an existing product or its presentation or the formulation of an entirely new product that satisfies a newly defined customer want or market niche [1].

The term product in this chapter refers to final developed telecom equipment (excluding mobile phones). This chapter will look into the characteristics of such products but not the actual process. The details on the product development process can be found in [2]. The telecom products could be active, such as base stations, small cells, mobile switching centers, and so on, as well as passive, such as cell towers, power backup equipment, cables, and so on.

The discussion in this chapter is primarily focused on the features and specifications of the finished telecom products. Each finished telecom product can consist of hundreds of components and ICs. Network equipment, such as radio base stations, radio controllers, small cells, microwave radios, routers, gateways, optical nodes, core network nodes, and so on is considered to be finished telecom products in this chapter. It is certain that neither each component nor each product can be described in a single chapter or even in a solo book. Thus, the purpose of this chapter is to review a handful of such products including existing multi-standard base stations, small cells, and SSGN-MME(Serving GPRS [General Packet Radio Service] Support Node—Mobility Management Entity), which will continue to be in use for the foreseeable future, and to get a peek into 5G base stations and small cells.

7.1 MULTI-STANDARD, MULTI-MODE, MULTI-RAT BASE STATIONS

The subject base stations combine multiple access technologies such as 2G, 3G, and 4G onto a single platform. The key enabler of these base stations is SDR. SDR with reconfigurable baseband and radio units allows scalability and ease of upgrade of base stations. These base stations provide OPEX (operating expense) efficiencies such as integrated backhaul, reduced power consumption and space, and site rental costs. The key benefit of these base stations is their ability to support multiple radio technologies within a given spectrum band. Spectrum refarming has become an important driver as operators move from 2G to 3G to 4G, taking advantage of unused spectrum assets. The 900 MHz GSM transition to 900 MHz UMTS is one of the great examples of how such base stations have been used [3].

3GPP defines the concept of the minimum RF characteristics of E-UTRA, UTRA, and GSM/EDGE Multi-Standard Radio (MSR) Base Stations (BS) in Technical Specification 37.104 [4]. The 3GPP defines MSR-BS as a "Base Station characterized by the ability of its receiver and transmitter to process two or more carriers in common active RF components simultaneously in a declared RF bandwidth, where at least one carrier is of a different RAT (radio access technology) than the other carrier(s)".

7.1.1 BASE STATION ARCHITECTURE

2G and 3G base stations consist of two key network elements, namely Base Transceiver Station (NodeB, for 3G UMTS) and Base Station Controller (Radio Network Controller for 3G). The LTE and 4G (LTE-Advanced) have incorporated most of the functions of the controller into the base transceiver station eNodeB (enhanced Node B).

The recent innovation of supporting multiple radio access techniques via a single base station made headlines in the 2010s. This technique allows operators to support at least two technologies out of GSM, CDMA/EV-DO, WCDMA, HSPA (High Speed Packet Access), LTE, and LTE-Advanced in a single box. This section will provide some insights into such base station products.

7.1.1.1 Base Transceiver Station

The key components of base transceiver stations (or Node Bs) include system module, RF unit (or RRU for distributed architecture) and required software (Figure 7.1) [5,6].

7.1.1.1.1 System Module

The system module manages baseband processing, power distribution, and the system clock. It also provides common public radio interfaces (CPRIs) and electrical or optical cables to communicate with the RF unit and RRU. The key components of system module include:

- *Clock board* for control and management of baseband unit and providing the system clock.
- *Baseband unit* (BBU) for baseband processing of various radio access technologies. This processes the physical layer protocols and frame protocol as specified in the relevant standards.
- *Alarm Board* provides site alarm monitoring interfaces.
- *Transport Module* provides E1/T1,* optical, and electrical transmission interfaces.
- *Power Module* provides measurement and protection of input over-voltage, over-current protection, and under-voltage and load power management.
- *Fabric Switch Board* provides baseband optical interface between BBU and RRU and processes the IQ (in-phase and quadrature) signal.

A typical UMTS enabled system module supports simultaneous downlink (High Speed Downlink Packet Access [HSDPA]) and uplink (High Speed Uplink Packet Access [HSUPA]) operations. In the case of LTE, a dedicated BBU could have a capacity of 3×20 MHz with MIMO enabling 500 Mbps in downlink and 200 Mbps in uplink.

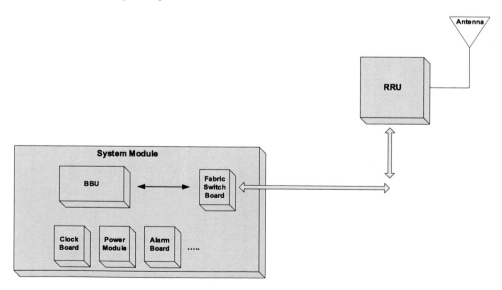

FIGURE 7.1 Distributed base transceiver station architecture. (Note: antenna not part of BTS.)

---

* E1/T1: digital data transmission formats; E1: line data rate 2 Mbps; T1: 1.544 Mbps.

TABLE 7.1

Technical Specifications of a Multi-Standard BTS

Attributes	Macro Base Transceiver Station	Micro Base Transceiver Station
Physical dimensions (H*W*D)	900 mm × 600 mm × 450 mm	650 * 320 * 480 mm
	1600 mm × 600 mm × 600 mm	
Capacity	8 GSM TRXs or 2 UMTS Carriers, +One	6 GSM TRXs or 4 UMTS carriers or One
	20 MHz LTE carrier (TRX: Transceiver)	20 MHz LTE carrier
Operating frequencies	850/900/1800/1900/2100/AWS	850/900/1800/1900/2100/AWS
	(Advanced Wireless Services)	
Power consumption (Watts)	12 W, 20 W, 40 W, 80 W per carrier	12 W, 15 W, 20 W, 40 W, 80 W per carrier
Power supply	−48 V DC, +24 V DC	48 V DC, +24 V DC
	110 V AC, 220 V AC	110 V AC, 220 V AC
Operating temperature	−40 C to +55 C	−40 C to +55 C
Baseband unit	1, 2	1
Radio unit per baseband unit	3	1
RRUs per radio technology	3, 6	1
Mean time between failures (MTBF)	≥120,000 hours	≥120,000 hours

7.1.1.1.2   RF Unit

A base station accommodates multiple RF modules to support different radio technologies. An RF module is a multi-carrier, multi-standard radio transceiver unit for processing radio frequency signals. It could have three or more independent branches to transmit and receive signals (i.e., sectorization). RF and system modules can be housed in the same rack / cabinet. When RF module is placed on the tower mast it is called as RRU (Remote Radio Unit). The RRU is the radio frequency part of a distributed base station and is installed near antennas (on tower masts). The RRU modulates, demodulates, combines, and divides baseband and RF signals similar to an RF unit.

7.1.1.1.3   Specifications

A high level technical specification of an indoor and outdoor base transceiver station is shown in Table 7.1. These specifications are by no means following a standard and are merely depicting some typical attributes.

7.1.1.2   Controller

The base station controller or radio network controller is an essential component of 2G and 3G systems, respectively. The main function of the controller is to manage the radio access network and radio channels. A multi-standard controller serves a number of BTSs/NodeBs within a geographical area. Such IP supporting controllers have a modular design that enhances resource utilization and system reliability. The functions of GSM and UMTS are integrated, effectively addressing the trend of multi-radio access technology convergence in the mobile networks.

7.1.1.2.1   Architecture

A controller can be housed in a standard 19″ rack/cabinet comprised of GSM service boards and UMTS service boards in separate subracks/modules/shelves. Each subrack can be of height 12 U (rack unit*). The smallest multi-standard controller could have two subracks, one for each radio technology (in this case, GSM and UMTS). The controller could support multiple subracks (as shown in Figure 7.2) reaching a higher capacity of around 40–60 Gbps in the IP plane and 128 kbps × 128 kbps data

---

* Each rack unit or U (or RU) is 1.752 inches (44.50 mm) tall.

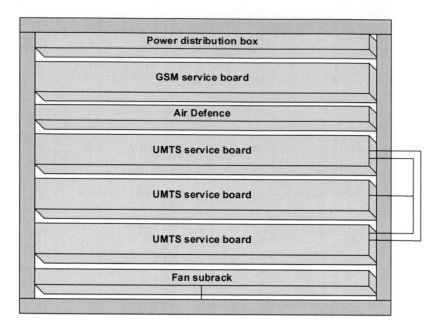

FIGURE 7.2   Multi-standard controller.

switching on the TDM plane. The boards are installed on the front and rear sides of the backplane, which is positioned in the center of the subrack. The IP interface board can be shared between the GSM and UMTS networks allowing simultaneous transmission of both GSM and UMTS data [7–9].

Additionally, each unit within the controller uses common system software. The software is usually based on a layered architecture where each layer is dedicated to its own functions and also provides services to other layers. Moreover, the implementation and topology of each layer is isolated from other layers.

7.1.1.2.2   Key Functions

The *Radio Resource Management Module (RRM)* performs unified management and intelligent scheduling of the radio resources for GSM and UMTS networks. RRM manages channel allocations, that is, they manage the number of traffic channels and signaling channels that can be used in the RAN simultaneously. To achieve this task, it handles admission control, scheduling, and load balancing. Admission control and scheduling requests are handled as these arrive, whereas load balancing is a continuous process. Furthermore, power control is needed as UMTS is an interference-limited system, that is, in order to achieve high capacity (more radio resources), interference needs to be reduced. Finally, handover management is required for ensuring that the user is connected to the strongest cell all of the time, that is, getting the best service without interruption.

The *Operation and Management Unit (OMU)* is a necessary element in the operations of the BSCs/RNCs. It provides both local maintenance and remote maintenance which helps operators to achieve cost efficiency. The OMUs could use Man Machine Language to provide operations, maintenance, and configuration functions, or the GFU to support the same. Java-based client applications can run on Microsoft Windows and on Linux Red Hat platforms.

The OMU could adopt the server-client approach where the O&M board of the controller works as the server while service board can works as clients. The LMT (local maintenance terminal) can be is used for local maintenance. There is also a centralized Operations Support System for the management of the entire operator network, which can be used for remote maintenance as shown in Figure 7.3. The OMU supports functions such as security management, fault management, alarm management, equipment management, and software management. In the event of a fault, the OMU

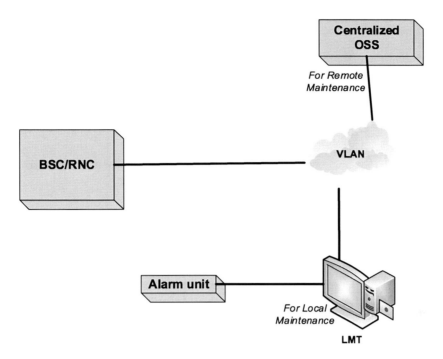

FIGURE 7.3   OMU of the controller.

automatically activates appropriate recovery and diagnostics procedures within the controller. It can also provide post-processing support for measurement and statistics tasks.

7.1.1.2.3   Specifications

Some key technical specifications of the controller are shown in Table 7.2, while the logical interfaces are shown in Table 7.3.

7.2   SMALL CELLS

According to Small Cell Forum, "a small cell (metro cell) is an umbrella term for low-powered radio access nodes that operate in licensed and unlicensed spectrum that have a range of 10 m to several hundred meters. These contrast with a typical mobile macrocell which might have a range of up to several tens of kilometers." The term covers femtocells, picocells, microcells, and metrocells [11].

A femtocell is a low-power, short range, self-contained base station. These were initially developed for indoor use in residential homes, but now they encompass higher capacity units for enterprise, rural, and metropolitan areas. A picocell is a low power compact base station used in enterprise or public indoor areas. A microcell, on the other hand, is a short-range base station aimed at enhancing coverage for both indoor and outdoor users where macro coverage is insufficient. When installed indoors, it provides coverage and capacity in areas above the scope of a picocell. Such small cells and traditional macro cells, if deployed in a network, together form a HetNet. A HetNet or (Heterogeneous Network) is a mobile network which consists of macrocells, small cells, and in some cases, WiFi access points, and different radio access technologies which work hand in hand to provide coverage with handoff capabilities between them.

7.2.1   Femtocells' Architecture

Small cells and in particular femtocells are getting a lot of industry attention. Femtocells are very small base stations and thus can be placed in a customer's residence. A femtocell network consists

TABLE 7.2

Technical Specifications of Controllers

Specifications[a]

	2G	3G
Traffic volume	6500 Erlangs[b]	16,750 Erlangs
Total PS throughput (uplink + downlink)	N/A	1000 Mbps
# of TRXs (transceivers)	600	N/A
# of NodeBs supported	N/A	500
# of BTS sites supported	400	N/A
TCH (traffic channel) for GPRS/EDGE	5000	N/A
Voice Traffic Handling Capacity		
Mean holding time		80–100 seconds
MS/UE (mobile station/user terminal) originating calls ratio		60%–70%
MS/UE terminating calls ratio		40%–50%
Handover ratio		30%–40%
Location updates		2–3 per call

Source: Wikipedia. http://en.wikipedia.org/wiki/Main_Page [10].
Note:
[a] This represents a generic configuration and one of the lowest network controller configurations.
[b] The Erlang (symbol E) is a dimensionless unit that is used in telephony as a measure of offered load or carried load on service providing elements such as telephone circuits or telephone switching equipment.
PS = Packet Switched; TRX = Cell (sector)

of two sections, namely radio access and core network. The access network consists of femtocells which are connected to a digital subscriber line (DSL)/Cable modem, which, in turn, is hooked to an ISP's (Internet Service Provider) IP backhaul/transmission network. Thus, the traffic would be routed through the home's ISP connection and would reduce backhaul costs since the mobile phone traffic is routed through an independent IP network. The IP network is connected to the femto gateway (part of the femto core network) which is connected to the core network of the operator as shown in Figure 7.4 [11,12].

TABLE 7.3

Logical Interfaces

Logical Interfaces	2G	3G
Abis	Between BTS and BSC	
A	Between BSC and circuit switched network	
Gb	Between BSC and packet core network	
Iub		Between RNC and NodeB
Iur		Between two neighboring RNCs
Iurg	Between a 2G BSC and 3G RNC	
Iu-CS		Between RNC and circuit switched core network
Iu-PS		Between RNC and packet core network
O&M		Proprietary management interface between network management system (NMS) and RNC

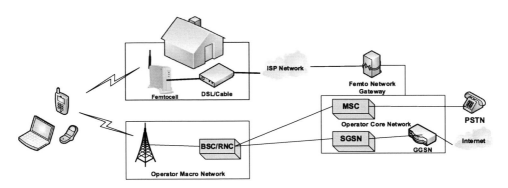

FIGURE 7.4 Femtocells' network.

The early femtocell designs supported up to 4 simultaneous active users and were targeted for residential use. With the passage of time and advancements in technology, femtocell products have evolved to deliver longer range and higher capacity, delivering service to small to large enterprises and public spaces while being part of a single coordinated operator network.

7.2.2 FEMTOCELL PRODUCTS

Femtocells have been developed for all the major radio access technologies (except for WiMAX) which are briefly discussed in this section. The development efforts on CDMA/WiMAX femtocells are almost nonexistent and vendors are heavily focusing on the rest of the existing radio access technologies. 5G will heavily rely on small cells and present a sizeable business opportunity.

7.2.2.1 2G GSM/GPRS Femtocells

The 2G femtocells are designed to support GSM, GPRS, and EDGE radio standards. These femtocells provide subscribers not only voice and messaging services, but also mobile data to complement the existing wireless broadband. The 3G subscribers are also able to use these femtocells with 2G/3G dual mode mobile devices. The dual technology is not mirrored in 3G femtocells and thus excludes the 2G subscriber base.

These small cells conform to the 3GPP specifications in terms of connectivity, security, services, and operations. The small cell is connected to a GSM core network using IPsec (IP security), thus ensuring security between the small cell and the operator. A typical 2G femtocell product specification [13] is shown in Table 7.4.

7.2.2.2 3G UMTS Femtocells

The 3G UMTS femtocells are used to extend WCDMA coverage and HSPA capacity in residences. The end user simply provides power and a broadband connection and a femtocell comes into service without any additional user intervention. Some UMTS femtocells also provide application programming interfaces that enable mobile service providers to leverage unique network capabilities, such as location and presence, to further enable new, innovative applications. A typical 3G UMTS femtocell product specification [14–16] is shown in Table 7.5.

7.2.2.3 3G CDMA2000 Femtocells

It can be safely said that the very first femtocells were based on CDMA2000 technologies. These femtocells support both CDMA2000 1X and CDMA2000 1xEV-DO standards. The characteristics of CDMA2000 based femtocells [17] are shown in Table 7.6.

7.2.2.4 LTE Femtocell

Unlike 3G technologies, the standardization process for LTE femtocells was started along with rest of the LTE standardization. LTE femtocells or HeNB (Home Enhanced Node B) can be connected

TABLE 7.4
2G GSM Femtocell Specifications

Attributes	Description
Radio standard	3GPP release 6—GSM, GPRS, EDGE
Frequency bands	GSM 1800, GSM 900
Antennas	Normally a single, either an 800 MHz or 1900 MHz antenna
	An optional connector for external antenna (e.g., GPS)
Timeslots	Total 8
	• 7 for voice for voice or data or for mix voice and data traffic
	• 1 for dedicated signalling
Security	GSM air encryption A5/0, A5/1, A5/3
	WAN via IPSec tunnelling
Voice calls	10–15 simultaneous voice calls
Speech support	Full rate, half rate, AMR (adaptive multi-rate) audio codec
Data support	GPRS CS1–4 (coding scheme) 8–20 kbps, EDGE MCS1–9 (modulation and coding scheme) 118.4–296 kbps
O&M support	3GPP based O&M
Power supply	External AC/DC adaptor (100–240 VAC @ 50/600 Hz up to 12 VDC 2A)
Physical dimension	H: 120–200 mm W: 130–220 mm D: 30–70 mm
Weight	400–700 grams
Operating temperature	0°C to +45°C

TABLE 7.5
3G UMTS Femtocell Specifications

Attributes	Description
Radio standard	WCDMA Rel-99, HSPA
Frequency bands	1900 MHz, 2100 MHz
Antennas	Omni
Data rates	3GPP Release 6 HSDPA at 14.4 Mbps and HSUPA at 5.76 Mbps
Users' support	4–8 simultaneous voice and data sessions
Security	USIM interface (Universal Subscriber Identity Module)
	IPsec, AES (advanced encryption standard), SHA (secure algorithm), DES (data encryption standard)
Sensitivity	−105 to −115 dBm
Network interface	10/100 Base-T RJ45
Power supply	12 V DC, 100–240 V AC (External AC/DC adaptor)
Power output	20 mW per carrier (maximum)
Weight	300–700 gm (without power supply)
Operating temperature	0°C to +45°C
Physical dimensions	H: 120–200 mm W: 130–220 mm D: 30–40 mm

to the EPC (Enhanced Core Network) in three different ways as specified in 3GPP TR 23.830 [18]. [18] also defines another important element of the architecture, that is, the HeNB Gateway. Its placement decides which architectural model is followed by the operator. The femto architecture also consists of an SeGW (security gateway) which could be a separate physical entity or colocated. It provides security functions for the backhaul link connecting the femto access point and core network.

TABLE 7.6
CDMA2000/EV-DO Femtocell Specifications

Attributes	Description
Radio standard	CDMA2000 1X, EV-DO (revisions 0, A, B)
Frequency bands	800 MHz, 1900 MHz
Antennas	Omni
Data rates	4.915 Mbps (downlink) and 1.8 Mbps (uplink) per carrier
Users' support	8 simultaneous voice and data sessions
Network interfaces	10/100 base-T RJ45
Power supply	100 V–240 V AC
Transmit power	20 mW per carrier
Weight	<500 grams
Operating temperature	0°C to +45°C
Physical dimension	H: 180–200 mm W: 130–150 mm D: 30–40 mm

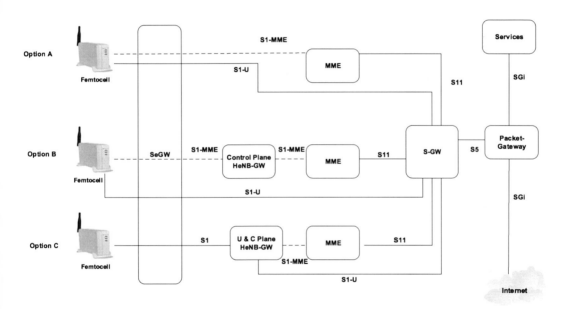

FIGURE 7.5   LTE femtocells' architectural options.

If the architecture lacks an HeNB-GW, S1 interfaces connect LTE femtocells straight to the Mobility Management Entity (MME) and Serving Gateway (S-GW). In the second option, the HeNB-GW only aggregates CP (control plane) traffic from multiple HeNBs and sends it to the MME. In the third option, HeNB-BW aggregates CP traffic from femtocells and dispatches it to the MME and also takes the UP traffic and sends to the S-GW as shown in Figure 7.5 [19,20].

The characteristics of LTE femtocells are shown in Table 7.7.

7.3   SSGN-MME NODE

SGSN and MME are the two key network elements of the packet switched core network. SSGN (Serving GPRS Support Node) as a control (signaling) plane product, is responsible for delivery, packet routing, and mobility management of data packets in GSM and UMTS networks. The MME, on the

TABLE 7.7
LTE Femtocell Specifications

Attributes	Description
Radio standard	LTE/LTE-advanced (release 8+)
Frequency bands	1800, 2600 MHz
Bandwidth	5/10/15/20 MHz
Antennas	Omni or directional
Data rates	150 Mbps (downlink) and 50 Mbps (uplink) per 20 MHz carrier
MIMO	2×2
Security	IPSec
Synchronization	IEEE 1588v2, GPS
Network interfaces	10/100 base-T RJ45
Power supply	100 V–240 V AC
Transmit power	13–26 dBm per TX channel
Weight	400 grams to 2 kg
Operating temperature	$-5°C$ to $+45°C$
Physical dimension	H: 170–250 mm W: 130–200 mm D: 30–50 mm

Source: Airspan 2016. AirVelocity & AirDensity; Sercom FDD-LTE Small Cell. http://www.ser-comm.com/contpage.aspx?langid=1&type=prod3&L1id=2&L2id=1&L3id=1&Pro did=63; BTI Wireless 2015 Small Cell Solutions. http://www.btiwireless.com/products/small-cells/ [21–23].

other hand, more or less performs the same functions, but in LTE. The GPRS/UMTS supporting the SGSN product has been upgraded to a combined SGSN-MME mobility server to support four 3GPP access technologies, namely GSM, WCDMA, LTE, and LTE-Advanced [24–26]. Some details on hardware and software components are as follows:

7.3.1 HARDWARE COMPONENT

This node or network element is based on Advanced Telecommunications Computing Architecture (i.e., ATCA). ATCA are a series of PICMG (PCI Industrial Computer Manufacturers Group) specifications, designed to provide an open, multi-vendor architecture, addressing the requirements for the next generation of carrier grade communications equipment [27].

The physical structure of the node houses in the standard 19-inch cabinet consists of subracks that contain physical boards. The board area of a subrack has a certain number of dedicated slots both at the front and rear sides. Boards can be inserted from both the front side and the rear side of the subrack. The key physical boards and their functions are as follows:

- Base Switch Unit is used for switching control plane packets in the same subrack or between subracks.
- Fabric Unit is used for switching media plane (quality of service) packets between subracks.
- OMU Unit is responsible for operation and maintenance and is placed at the front board with 1 + 1 active/standby redundancy.
- Control Plane Unit is responsible for processing the service (signaling and routing management functions) on the control plane.
- Packet Forward Interface is responsible for processing the service (packet forwarding) on the user plane.
- Multi-Protocol Interface can be configured for IP connection with Fast Ethernet/Gigabit Ethernet (FE/GE) electrical ports or GE optical ports at the back board.

TABLE 7.8
SGSN-MME Specs

Attributes	Description
Cabinet dimensions	2200 mm * 600 mm* 800 mm (HxWxD)
Full configuration	3 cabinets with 3 subracks in each cabinet
Total weight	100–120 kg (empty cabinet)
	300–400 (3 full subracks in a cabinet)
Capacity (full configuration)	• 2G/3G: 10–13 million simultaneously
	• LTE: 12–16 million simultaneously
	• eNodeB: 25000–50,000
	• Serving gateway (LTE): 3000–4000
Data throughput	Gb over IP (SGSN and BSC): 3–7 Gbps
	Iu over IP (SGSN and RNC): 25–40 Gbps
Concurrent bearers activated by a UE	11
Signalling indices	64 kbps and 2 Mbps links
Interface indices	FE/GE/E1/T1/STM-1/STM-4
Lowest clock accuracy	$\pm 4 \times 10^{-7}$
Power input	−60 V to −40 V DC
Power consumption	1500–3500 W (one sub rack)
Temperature	−40°C to +65°C
Mean time between failures (MTBF)	300,000–1,100,000 hours
System availability	≥99.999%

7.3.2 SOFTWARE COMPONENT

The node SGSN-MME can use a modular and hierarchical software structure where the functional modules of the software are distributed in different types of boards. The components include Linux based operating system, middleware technology such as DOPRA/TULIP (Distributed Object-oriented Programmable Real-time Architecture/Telecom Universal Integrated Platform), which is applied to the operating system and applications (such as service processing, protocol processing, etc.), making upper-layer service software irrelevant to the lower-layer operating system and applications.

7.3.3 SPECIFICATION

Table 7.8 shows a typical configuration of an SGSN-MME node.

7.4 5G BASE STATIONS AND SMALL CELLS

5G standards are under development so the exact base station specifications are still unknown. However, much is known, such as that these base stations will support millimeter wave frequencies, C-RAN, one or more waveforms, carrier aggregation, IoT, and at least 2×2 MIMO [28–30]. Additionally, like LTE and LTE-Advanced, the base station will directly communicate with the core network and there will be no separate entity, that is, radio controller.

5G base station's baseband pool will provide centralized processing and connect to RRUs. It will manage operations and maintenance, signaling processing, and the system clock. It will provide physical ports to connect the base station to the transport network for information exchange, CPRI ports for connectivity with RRUs, and ports for communication with environmental monitoring devices. The RRUs, on the other hand, modulate, demodulate, combine, and divide baseband and

TABLE 7.9
5G Base Station/Small Cell Specs

Attributes	5G Macro Base Station	5G Small Cell
Physical dimensions (H*W*D)	900 mm × 600 mm × 450 mm	120–200 mm × 130–220 mm 30–40 mm
Capacity	One 20 MHz LTE carrier Three 50 MHz 2 × 2 MIMO (5G carriers)	One 50 MHz 2 × 2 MIMO (5G carrier)
Operating frequencies	1800 MHz for LTE 28/37/39 GHz for 5G	28 or 37 or 39 GHz for 5G
Power consumption (Watts)	40 W, 80 W, 120 W per carrier	13–26 dBm per TX channel
Power supply	−48 V DC, +24 V DC 110 V AC, 220 V AC	100 V – 240 V AC
Operating temperature	−40°C to +55°C	−5°C to +45°C
Baseband units/pool	4–6	N/A
RRUs per BBU	4–10	N/A
MTBF	≥120,000 hours	≥100,000 hours

RF signals, and support dual/multimode operations. A hypothetical 5G base station/small cell specification is shown in Table 7.9.

7.5 CONCLUSION

A telecom network consists of a number of products and each such product could contain a number of components and ICs. This chapter briefly described certain key telecom products, namely multi-standard base stations, femto cells, and SSGN-MME nodes. These products are expected to continue in 5G networks in one form or other.

PROBLEMS

1. Define the term Product Development?
2. What are multi-standard radios as per 3GPP?
3. What are the key components of a base station?
4. What are the key functions of a base station (radio network) controller?
5. How does RRM manages channel allocations?
6. Define the key functions of controller OMU?
7. Define small cells and give two examples?
8. What are femtocells and why is wired broadband connectivity required for their operation?
9. What are the three options of HeNB connectivity?
10. Define the function of SGSN-MME and its key units?
11. Define the key characteristics of a 5G base station?

REFERENCES

1. WebFinance, Inc. 2014. Product Development. http://www.businessdictionary.com/definition/product-development.html
2. Asif, S.Z. 2011. *Next Generation Mobile Communications Ecosystem: Technology Management for Mobile Communications.* Wiley Inc., UK.
3. ABIresearch 2010. Software Defined Multi-standard Base Stations. http://www.abiresearch.com/research/1002717-Software+Defined+Multi-standard+Base+Stations
4. 3GPP TS 37.104 (V11.3.0) 2012 E-UTRA, UTRA and GSM/EDGE; Multi-Standard Radio (MSR) Base Station (BS) Radio Transmission and Reception. Technical Specification (Release 11), Technical Specification Group Radio Access Network, 3GPP, December.

5. Huawei 2009. 3900 Series Base Station.
6. Nokia Siemens Network 2009. Flexi Multiradio BTS.
7. ZTE 2011. ZXUR 9000.
8. Nokia Siemens Network 2011. Multicontroller RNC mcRNC1.0.
9. Huawei 2011. SRAN6.0 BSC6900.
10. Wikipedia. http://en.wikipedia.org/wiki/Main_Page.
11. Small Cell Forum Ltd. 2012. Small cells—What's the big idea?
12. Jean-Christophe, N. and Barry, S. 2012. Small Cells Call for Scalable Architecture. Freescale.
13. HSL 2010. HSL 2.75G Femtocell.
14. Samsung 2008. Samsung HSPA UbiCell.
15. Alcatel-Lucent 2011. Alcatel-Lucent 9361 Home Cell V2.
16. NetGear 2011. 3G Femtocell Ethernet WAN to 3G HSPA MF100H.
17. Ubee–AirWalk 2012. EdgePoint.
18. 3GPP TR 23.830 (V9.0.0) 2009. Architecture Aspects of Home NodeB and Home eNodeB. Technical Report (Release 9), Technical Specification Group Services and System Aspects, 3GPP, September.
19. Nokia Siemens Network 2011. Improving 4G Coverage and Capacity Indoors and at Hotspots with LTE Femtocells.
20. Femtoforum 2011. HeNB (LTE Femto) Network Architecture.
21. Airspan 2016. AirVelocity & AirDensity.
22. Sercom FDD-LTE Small Cell. http://www.sercomm.com/contpage.aspx?langid=1&type=prod3&L1id=2&L2id=1&L3id=1&Prodid=63
23. BTI Wireless 2015. Small Cell Solutions. http://www.btiwireless.com/products/small-cells/
24. Ericsson 2014. Ericsson SGSN-MME. http://www.ericsson.com/ourportfolio/products/sgsn-mme
25. Huawei 2010. USN9810 Unified Service Node.
26. ZTE 2011. ZXUN uMAC.
27. PICMG 2014. AdvancedTCA® Overview. https://www.picmg.org/openstandards/advancedtca/
28. National Instruments 2016. mmWave: Battle of the Bands.
29. Fierce Wireless Tech 2016. 5G: What To Expect Before 2020.
30. Zheng, M.A. et al. 2015. Key Techniques for 5G Wireless Communications: Network Architecture, Physical Layer, and MAC Layer Perspectives. *Science China Information Sciences*, 58:041301:1–041301:20.

8 Network Architecture, Mobility Management, and Deployment

The architecture of mobile networks has gone through a number of evolutionary changes. The changes have happened over the course of the last three decades and during this duration, these networks have transitioned from one generation to another. The first generation systems were analog, circuit switched, and short lived. The second-generation systems, based on digital technology, started to appear in the 1990s and are still in use. The 3G systems were introduced in the early 2000s, while 4G (LTE-Advanced) systems are currently rolling out. The 5G systems, based on ITU-R's IMT-2020 requirements, are in the research and standardization phase and commercialization is expected in the 2020s. It may be noted that all the mobile network architectures consist of four areas, namely radio access network, core network, network operations, and connectivity between these three which is supported through the transport network.

This chapter will focus on the network architectures of 2G GSM, 3G UMTS*, LTE, and 4G (LTE Advanced) technologies. An in-depth look at 5G network architecture is presented in Section 8.4, while mobility management and deployment aspects of networks are illustrated in Sections 8.5 and 8.6, respectively.

8.1 2G GSM NETWORK ARCHITECTURE

The 2G GSM network is comprised of three key interconnected systems, namely base station subsystem (BSS), NSS, and operation support system (OSS). The BSS (i.e., radio access area) consists of BTSs and BSC (base station controllers). The NSS (i.e., core network area) is comprised of MSC (mobile switching center), and databases, including HLR (home location register), VLR (visitor location register), EIR (Equipment Identity Register), and AuC (authentication center). The MSC is connected to some of these data bases directly and to the PSTN (public switched telephone network) for landline connectivity.

Later, packet switching was added to support data services with the help of GPRS. GPRS brings SGSN and GGSN (Gateway GPRS Support Node) to the network architecture. GSSN is connected to an operator's internal application servers for value added services[†] and external packet data networks (e.g., Internet). Figure 8.1 shows a high level architecture of the GSM-GPRS architecture. The details on various components can be found in [1,2].

8.2 3G UMTS NETWORK ARCHITECTURE

The key change in UMTS from the TDM based GSM was the introduction of an innovative multiple access scheme called CDMA. CDMA technique uses the same frequency, but unique codes to distinguish the users. CDMA is the basic ingredient which is used in all the major 3G standards (CDMA2000, TD-SCDMA, and UMTS), excluding WiMAX [1–4].

---

* Details on other 2G (IS-54, IS-95) and 3G (CDMA2000 (3GPP2 3G standard), TD-SCDMA (China's 3G standard) can be found in [1].
† Some Value Added Services are discussed in Chapter 13 Mobile Applications.

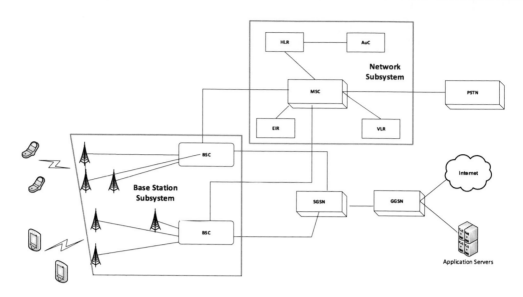

FIGURE 8.1   GSM-GPRS network architecture.

Starting from the 3GPP Release-99, some changes have been made to the architecture of 3G UMTS in almost every following release. Figure 8.2 shows the transition of architecture from Rel-99 to Rel-7.

Rel-99, which was functionally frozen* in December 1999, brought WCDMA (Wide band CDMA). WCDMA supports 2 Mbps in downlink using a 5 MHz channel. The radio portion of Rel-99 called UTRAN (Universal Terrestrial Radio Access Network) is connected to the core network which is essentially the core network of GSM-GPRS. The UTRAN consists of BTS (NodeB) and RNC (radio network controller) where RNC performs the same functions as BSC, but only for UMTS. Rel-4 came in next and added TD-SCDMA (Time Division Synchronous CDMA) and EDGE to the picture using the same core network and was functionally frozen in March 2001. The key architectural difference from R-99 is the division of MSC (Mobile Switching Center) into MSC-Server and MGW (Media Gateway).

Rel-5 brought IMS (IP Multimedia Subsystem) as part of the core network. IMS defines a standard framework for the deployment of next generation IP-based applications and services. It is access independent, communicates over IP using SIP (Session Initiation Protocol), and is applicable to both fixed and mobile networks. IMS only implements signaling procedures and application-common functions, but do not offer services itself. Thus, IMS is an open signaling system, based on standard Internet technology, which supports the migration of Internet applications (like VoIP (voice over IP), video conferencing, messaging, etc.) to the mobile environment and offers enhanced service control capabilities [3,4,6].

Rel-5, which was functionally frozen in June 2002, added HSDPA (High Speed Downlink Packet Access), supporting a peak data rate of 14.4 Mbps in the downlink. Rel-6 introduced HSUPA (High Speed Uplink Packet Access), supporting a peak data rate of 5.76 Mbps in the uplink and was functionally frozen in March 2005. Another key enhancement of Rel-6 is MBMS (Multimedia Broadcast and Multicast Service) for enabling mobile TV to users through UMTS networks. MBMS is enabled by introducing a functional node called BM-SC (Broadcast Multicast Service Center) in the core network.

---

* After "freezing," no additional functions can be added to it. However, detailed protocol specifications (stage 3) may not yet be complete. A "frozen" Technical Specification is one which can have no further category B or C (new or modified functionality) Change Requests, other than to align earlier stages with later stages [5].

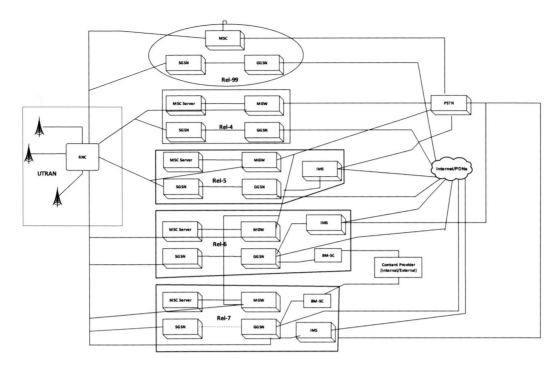

FIGURE 8.2   3GPP UMTS network architecture (Rel-99 through Rel-7).

The Rel-7 brought HSPA+ (Evolved HSPA) supporting 28 Mbps in downlink and 11.5 Mbps in the uplink. It introduced the concept of Direct Tunneling (DT) in the core network. DT enables a split between the control plane and user plane toward the packet core network. It allows the SGSN to establish a direct user plane between RAN and GGSN within the packet switched domain. The SGSN manages control plane signaling and decides whether one or two tunnels are needed. When only one tunnel is created, the user plane traffic bypasses the SGSN, making SSGN a signaling-only entity [7].

8.3 LTE AND 4G (LTE-ADVANCED) NETWORK ARCHITECTURE

3GPP Rel-8 introduced flat IP architecture under the umbrella of EPS (Evolved Packet System). EPS consists of E-UTRAN (Evolved UMTS Terrestrial Radio Access Network) which is commonly known as LTE (Long Term Evolution) and EPC (Evolved Packet Core). Rel-8 was functionally frozen in December 2008.

E-UTRAN only consists of one physical network element type, that is, eNodeB, and there is no separate entity in the form of BSC or RNC. All the typical radio functions and a MAC layer, RLC (Radio Link Control Layer), and RRC (Radio Resource Control) are part of eNodeB. EPC consists of three main elements, namely MME, SGW (Serving Gateway), and PGW (Packet Gateway). The main function of MME is to manage UE (user equipment)* mobility and UE identity and it is connected to E-UTRAN via S1-MME interface. SGW performs mobility anchoring for inter-eNodeB handovers and inter-3GPP systems. PGW is the mobility anchor for movement between 3GPP and non-3GPP access systems. PGW connects to external PDNs (packet data networks), operators' IMS and non-IMS IP services, and provides access for trusted and nontrusted non-3GPP IP networks.

---

* UE and device are interchangeably used. Device may include UE, CPE (customer premise equipment), and handsets, and includes things as described in Internet of Things.

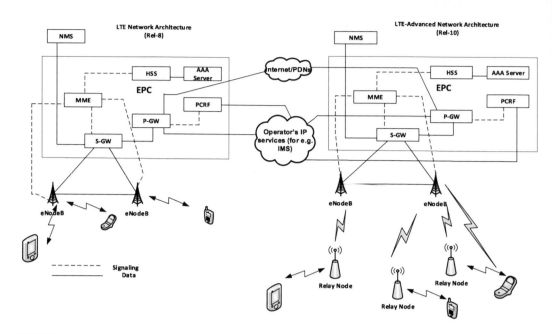

FIGURE 8.3   LTE and LTE-Advanced network architecture.

EPC also retains HSS (home subscriber server) for keeping user subscription information, AAA server (authentication, authorization, and accounting) for determining the identity and privileges of a user, tracking activities, and PCRF (policy and charging rules function) to enforce charging and QoS policies [3].

Rel-9, which was functionally frozen in December 2009, did not bring any architectural changes to EPS. LTE-Advanced (4G) was introduced in Rel-10, which was functionally frozen in June 2011. LTE-Advanced uses the same EPC as the core network while making some enhancements to air-interface and E-UTRAN. LTE-Advanced brings the concept of relay node in E-UTRAN for wider coverage and better QoS. The relay node is connected to the donor eNodeB via the radio air-interface Un, which is modified from the air interface Uu. Donor eNodeB not only serves its own UEs in its serving cell, but also shares the radio resources with the relay nodes [8].

Rel-11, Rel-12, and Rel-13, which were functionally frozen in June 2013, March 2015, and March 2016, respectively, neither brought new radio access technology nor any architectural changes. The architecture of EPS covering both LTE and LTE-Advanced is shown in Figure 8.3.

8.4   POTENTIAL 5G NETWORK ARCHITECTURE

5G represents a paradigm shift for cellular/broadband networks. It literally pushes toward virtualization and away from the implementation of monolithic network entities. It brings a number of challenges and unknowns that the telecom sector has not so far embraced. The transformation from 3G/4G to 5G will be profound and challenging as users are creating demand requiring new approaches for connectivity, bandwidth, and network architecture. This section will articulate three architectural views from a very high level to a much deeper degree for the understanding of the readers [9–13].

3GPP in collaboration with its seven organizational partners* (or SDOs) is working to produce 5G specifications which will include network architecture as well. This work is expected to be completed

---

* Association of Radio Industries and Businesses (ARIB) of Japan, Alliance for Telecommunications Industry Solutions (ATIS) of the U.S., China Communications Standards Association (CCSA), European Telecommunications Standards Institute (ETSI), Telecommunications Standards Development Society, India (TSDSI), Telecommunications Technology Association (TTA) of Korea, Telecommunication Technology Committee (TTC) of Japan.

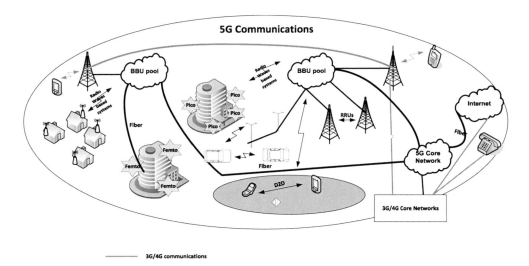

FIGURE 8.4 30,000 ft 5G network view.

in three releases, that is, Rel-14, Rel-15, and Rel-16. 3GPP is planning to submit such specifications (proposals) as per ITU-R timelines. Proposals are also expected from IEEE-SA and others.

8.4.1 30,000 Ft Bird's Eye View

The design of 5G network is still evolving and the sector will take some time to thrash out the details. Nevertheless, it is expected that the architecture will utilize the cloud functionality and small cells to the deepest extent. A 30,000 ft bird's eye view, shown in Figure 8.4, highlights the key segments of the architecture. It shows that small cells are connected with either view optical fiber cable or radio waves to the cloud radio access network (BBU pool). The pools of BBU are connected (which is called the fronthaul) to the RRUs (small cells or macro cells) and are backhauled to the core network. 5G core network in the form of cloud is integrated with existing 3G/4G core networks and provides connectivity to the Internet.

8.4.2 High Level 5G Network View

5G networks will consist of thousands of small cells making networks ultradense. This network densification is required to meet the latency and throughput requirements. Small cells are expected to carry a major portion of traffic with overall data volume expected to grow exponentially as predicted by many studies. Bringing small cells closer to the user will reduce latency and increase overall network efficiency by creating subnetworks. These subnetworks can have the functionality to route data traffic locally for the local users' calls who are communicating with each other while sending the signaling to the main network as shown in Figure 8.5 [14]. The net result from local switching is in the savings of network resources and cheaper data plans for the customers.

C-RAN, which was discussed in Chapter 5, will bring cloud functionality and likely cognitive capability to the RAN. In such flexible RAN, RRUs do not need to stick to a particular BBU, but instead they can talk to any of the BBUs of a specific pool. Cloud functionality will probably not exist in transmission networks including backhaul and will continue to rely on fiber and radios waves based systems. Additionally, DSL may continue for connectivity with indoor small cells. 5G networks are expected to integrate NFV and SDN along with separate user and control planes. Nevertheless, at least for now, four areas of cellular networks, namely radio access, transport, core networks and operations, will continue to exist in 5G architectures as shown in Figure 8.6.

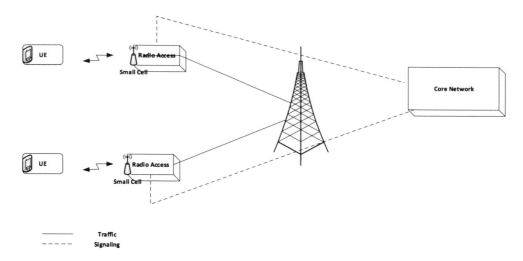

FIGURE 8.5   Local routing.

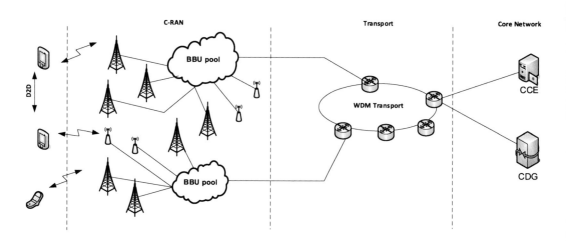

FIGURE 8.6   High level 5G network architecture.

A major challenge for 5G radio access networks is the requirement for efficient integration of an additional layer of small cells with the existing macro cell network [11]. The continued technological upgrades and growth in the number of macro cells will further augment this challenge.

In the case of 5G, small cells are deployed as light RRUs along with macro RRUs. The light RRU can be installed on streetlight poles and underground subways whereas macro RRU is suitable for traditional BTS towers. Both types of RRUs will connect to pools of BBUs through a fronthaul interface. Fronthaul is the new area in the transport network which connects the radio heads to the centralized pool of BBUs through fiber. The current choices for fronthaul interfaces are CPRI, ORI (Open Radio equipment Interface) or OBSAI (Open Base Station Architecture Initiative). However, these may encounter capacity bottlenecks and thus require the players to look for alternatives. The BBUs' pools are connected to the core network through a combination of wired and wireless technologies. These include fixed links such as dedicated fiber, together with wireless options such as microwave radios and radio over fiber. The core network will be connected to Internet and to connect different metropolitan cities, existing long haul WDM (wave division multiplexing) connectivity can be upgraded.

The core network can be split into control and user plane domains where a combined control entity (CCE) integrates the control functions of the 5G core network with the core functions of MME, SGW-C, and PGW-C. Similarly, a combined data gateway (CDG) performs data forwarding functions of 5G core network, SGW-D, and PGW-D.

8.4.3 5G Network Architecture (In-Depth View)

The 5G mobile network will include both physical and virtual functions and cloud deployments. It is also clear that 5G needs to account for existing 3G/LTE/LTE-Advanced network deployments. An in-depth view of 5G network architecture is shown in Figure 8.7.

At a high level, two options have been envisioned for RATs. One difficult possibility is having a single unified RAT that is optimized for different frequencies, various use cases (IoT, V2X [vehicle-to-everything], etc.) and a variety of services. Designing such a RAT is highly challenging and costly, requiring refarming [9] of the existing spectrum used for legacy LTE and LTE-Advanced technologies. Spectrum refarming may not be ideal in many cases, thus a better approach may be 5G carrier aggregation with LTE/LTE-Advanced carriers during the initial period of 5G rollout. Alternately, multiple RATs could complement each other where one using a high frequency band may provide capacity and high data rates in dense urban areas, whereas another RAT operating in a lower frequency band may be used for IoT and in rural areas for extended coverage.

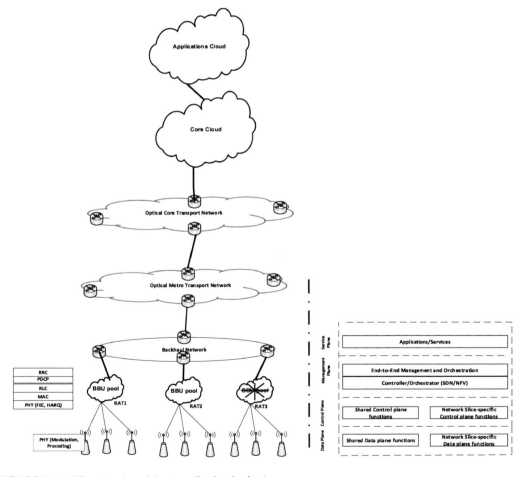

FIGURE 8.7   5G network architecture (in-depth view).

The transport network has to be flexible and dynamic to address the needs of future networks. Fronthaul, backhaul, metro, long distance, and international connectivity have to be insured by the transport network. Most of the connectivity will be provided through optical fiber cable, but some will require radio waves based systems.

3GPP envisions a logical split between RAN and core network to allow independent evolution of either one. Furthermore, a split of control and user planes is foreseen for 5G networks. In RAN, for example, it will allow macro cells to handle the control plane while small cells (particularly in the context of millimeter waves) handle the user plane. However, the exact nature of such a split is still under investigation.

A typical split could be between physical and MAC layers as shown in Figure 8.7 where the fronthaul is primarily supported by fiber. In this option, RRU supports modulation and precoding functions of the physical layer, while BBUs will support MAC, RLC, PDCP and RRC, and FEC and HARQ (Hybrid Automatic Repeat Request) functions of the physical layer. This is a centralized approach without the need of a high bandwidth support in the fronthaul.

NFV and SDN are essential in 5G networks to reduce costs and bring added value to network infrastructure. NFV is the process of moving/forwarding tasks such as load balancing, firewalls, and so on away from dedicated hardware into a virtualized environment [15]. NFV enables the execution of software-based network function on general purpose hardware by leveraging virtualization techniques. The virtualization technologies allow breakup of the software of network functions from dedicated hardware [16]. Softwarization allows implementation of network functions in software, including virtualization of such functions and programmability by setting appropriate interfaces. It is an approach to use software programming to design, implement, and maintain network equipment and services. In SDN, the control plane is decoupled from the data plane and is managed by a logically centralized controller that has a holistic view of the network [11]. Softwarization in RAN may allow some functions such as PDCP and RRC to be implemented as VNFs. Softwarization can also be used to implement certain core and transport functions. The original aim of combining NFV and SDN was to decouple services from physical resources allowing flexibility and adaptability in the network. When NFV and SDN come together, they provide the additional benefit of detaching lifecycle management from physical constraints [16].

Network slice, as discussed in Chapter 5, supports the connectivity of a particular use case through a collection of 5G network functions, and specific configurations in RAN, transport, and core networks. Network functions provide connectivity, storage, and computation. Details on network functions can be found in [11]. Finally, 5G is not all about connectivity, but also demands high end computation and storage. Computation and storage requirements also vary among the different network areas and elements. For example, a BBU pool may have less stringent needs as compared to a core network packet gateway. Similarly, the transport network encompasses several aggregation nodes that need to offer computing and storage capabilities.

8.5 MOBILITY MANAGEMENT

Mobility Management (MM) covers a lot of ground and has been exhaustively discussed by the wireless telecom industry at large. MM is one of the major functions of cellular systems that allows mobile phones to work across homogenous and heterogeneous networks. The aim of MM is to track where the subscribers are, allowing calls, messages, data, and other mobile phone services to be delivered to them anywhere, anytime.

In mobile networks, the locations of the devices are tracked so that the information can be transferred in an efficient fashion. Typically, MM procedures include location update and paging. When the device moves from one location to another, a location update procedure is used by the device to report to the network about its new position. Similarly, when there is incoming data/information for the device, the network uses the paging process to identify the location of the device [17]. MM normally includes personal mobility, session mobility, and network and terminal mobility [1,3]:

- *Personal Mobility* defines the ability of the user to access his/her personal services (for example, bookmarks, calendar, etc.) while away from the home network.
- *Session Mobility* is about maintaining seamlessness in sessions when moving from one network to another (e.g., movement from LTE to LTE-Advanced, EV-DO to EPS, LTE-Advanced to 5G, etc.) and also within the same network.
- *Network and Terminal Mobility* defines the capability of both the network and the terminal to support roaming. Terminal mobility allows mobiles to access services from different locations, while the networks shall have the capability to identify and locate such mobiles.

Designing MM for all-IP networks brings a number of challenges in the context of networks, devices, and protocol itself (IPv4, IPv6). The procedures associated with the mobility management of networks and devices are primarily prescribed by 3GPP, 3GPP2, and OMA standard development organizations. IETF, on the other hand, introduced mobility as an inbuilt feature in next generation, IPv6, to address the shortcomings of IPv4, such as a shortage of IP addresses.

8.5.1 EPS MOBILITY MANAGEMENT

The 3GPP based technologies such as GSM, UMTS, and LTE networks employ the GPRS Tunneling Protocol (GTP) for MM. All packet sessions are transported or relocated through the GPRS tunnel by the GPRS MM procedure according to the Packet Data Protocol context* [18]. The MM functionality is optimized to support both circuit-switched and packet-switched network architecture. However, due to high signaling overhead, the process is not efficient for packet-switched services [20]. To overcome this challenge, 3GPP has introduced PMIP (Proxy Mobile IP) protocol, [21] standardized by IETF, and it is highly effective in reducing the signaling cost for MM for various architectural options [20]. The S5 and S8 reference points in the EPC architecture have been defined to have both a GTP and PMIP variant. The GTP variant is documented in TS 23.401 [22] while the PMIP variant is documented in TS 23.402 [21].

The MME is responsible for MM functions in EPS. The EPS Mobility Management (EMM) and EPS Connection Management (ECM) are defined by 3GPP in [23]. The EMM describes the MM states (EMM-Deregistered and EMM-Registered) that result from procedures such as Attach and Tracking Area Update. The ECM describes the signaling connectivity between the UE and the EPC and consists of two states, namely ECM-IDLE and ECM-CONNECTED. Normally, the ECM and EMM states are independent of each other; however, some mechanisms do require interaction of these four states. For example, for transition from EMM-DREGISTERED to EMM-REGISTERED, the UE has to be in the ECM-CONNECTED state.

8.5.2 VOICE MANAGEMENT IN **LTE**

LTE is a data technology and it does not have what is required to support circuit switched voice and VoIP. To address this challenge, 3GPP identified two techniques to support voice, namely CSFB (Circuit Switched Fall Back) [24] and SRVCC (Single Radio Voice Call Continuity) [25].

---

* PDP context is a data structure present on both SGSN and GGSN that contains the subscriber's session information (IP address), IMSI (International mobile subscriber identity), etc. when the subscriber has an active session [19].

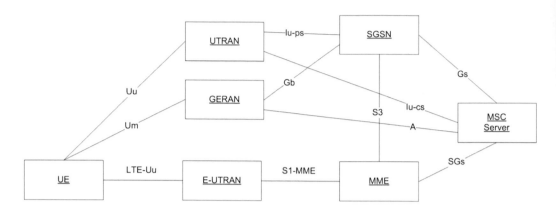

FIGURE 8.8 EPS architecture for CSFB and SMS over SGs. (From 3GPP TS 23.272 (V 12.1.0) 2013 CSFB in Evolved Packet System (EPS); Stage 2. Technical Specification (Release 12), Technical Specification Group Services and System Aspects, 3GPP, December [24].)

8.5.2.1 CSFB

Circuit Switched Fallback (CSFB) was introduced in Rel-8 by establishing a signaling interface SG between 2G/3G MSC server and LTE MME network elements (Figure 8.8). This allows devices attached to the packet-switched LTE network to change over to a circuit switched network (for example, WCDMA) for incoming and outgoing voice calls. During this fallback, the ongoing data sessions in LTE also switch over to 3G/HSPA network, and when the voice calls ends, the handset returns to LTE [26–28].

CSFB can be implemented in every Mobile Switching Center Server (MSS) or in a dedicated MSS (known as CSFB overlay). The drawback of CSFB is that it requires handsets with either dual-mode/ single-standby or dual-mode/dual-standby capabilities. The dual-mode handsets drain the battery power quickly and require complex terminal configuration [29].

CSFB has no impact on SMS or packet switched data, which is handled in parallel with voice on the 2G/3G network. CSFB does not cause interruptions in data sessions when subscribers make a voice call. The data connection returns to the LTE network seamlessly when the voice call is over [30].

8.5.2.2 SRVCC

The SRVCC or VoIP over LTE (or VoLTE) is a voice telephony solution comprised of IP Multimedia Subsystem (IMS). It allows a VoIP/IMS call in an LTE packet-switched domain to be transferred to a legacy circuit-switched domain (GSM/UMTS or CDMA2000). This technique uses a single radio solution* in the user's device, with cost, size, and battery efficiency advantages over dual radio solutions, enabling calls from LTE to non-LTE coverage areas.

Along with IMS, VoLTE is also comprised of Multimedia telephony (MMTeL†) service as defined by GSMA [31]. Based on IMS/MMTel, voice services can be further enriched with video [32] and combined with several other enhanced IP-based services such as high definition voice, presence, location, and Rich Communication Suite (RCS) additions like instant messaging, video share, and enhanced/shared phonebooks.

VoLTE, under the umbrella of SRVCC, was standardized in 3GPP Rel-8 to improve voice coverage by handing over the voice session from LTE to 2G/3G circuit-switched domain. The CS domain can

---

* Single radio mode terminal refers to the ability of the terminal to transmit or receive on only one of the given radio access networks at a given time.

† The Multimedia Telephony (MMTel) is a global standard based on the IMS and was developed jointly by the 3rd Generation Partnership Project, (3GPP), European Telecommunications Standards Institute/Telecoms and Internet Converged Services and Protocols for Advanced Networks (ETSI/TISPAN) standardization bodies. It offers converged, fixed, and mobile real-time multimedia communications that allow users to communicate using voice, video, and chat.

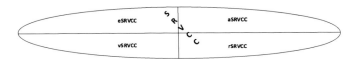

FIGURE 8.9 Forms of SRVCC.

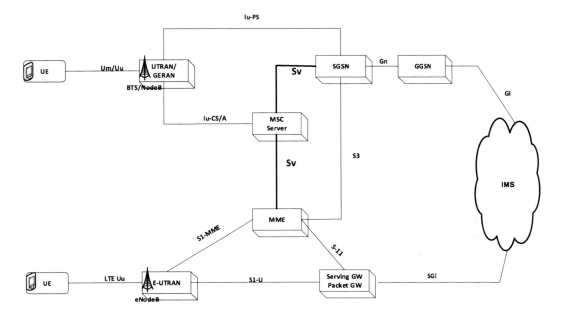

FIGURE 8.10 SRVCC 3 GPP R10 network architecture.

be part of UTRAN/GERAN or 3GPP2 1xCS (CDMA2000 1X). In Rel-9, to support e911 service over VoLTE, features like emergency services, location services, and emergency warning broadcast services were introduced.

In Rel-10, architecture enhancement for SRVCC (called eSRVCC) was introduced that supports a mid-call (inactive sessions or sessions using the conference service) feature during SRVCC handover which ultimately improves handover performance. Rel-10 also supports call transfer in the alerting phase (aSRVCC). In Rel-11, the SRVCC feature was further enhanced with the priority handover (eMPS* aspect of SRVCC), SRVCC from 2G/3G CS to LTE/HSPA (rSRVCC), and video SRVCC from LTE to UMTS (vSRVCC) [26]. The four forms of SRVCC are shown in Figure 8.9 [33–35].

The SRVCC architecture as defined in Rel-10 is shown in Figure 8.10. To support this functionality, a new interface called Sv was introduced between MME and the MSC server for LTE-UTRAN/GERAN, and between SGSN and the MSC server for HSPA-GERAN sessions. Another interface S102 (not shown) was introduced between MME and IWS (Interworking System) for SRVCC from E-UTRAN to CDMA 2000 1xRTT.

The SRVCC handover process involves IRAT (Inter Radio Access Technology) handover and session transfer steps. These steps need to meet the 3GPP voice interruption target of less than 0.3 seconds as defined in TS 22.278 [30]. The Rel-10 configuration allows signaling to follow the shortest possible path, minimizing voice interruption time caused by switching from the PS core to the CS core, whether the user's device is in its home network or roaming. Rel-10 configuration minimizes the voice interruption time by simultaneously initiating the two procedures so these can run in parallel [23].

SRVCC requires additional functionality in both the source (LTE) system and the target (legacy) system. SRVCC functionality is enabled through software upgrades to the CS core (MSC Server),

---

* Enhancements for Multimedia Priority Service (eMPS) is a feature in Rel-10 for IMS sessions and EPS bearer sessions.

TABLE 8.1
CSFB versus SRVCC

Attribute	CSFB	SRVCC
IMS anchoring	Not required	Mandatory
Device capability	Dual mode	Single mode
	More complex circuitry	Less complex circuitry
Cost (network perspective)	Less	More due to the requirement of IMS
Voice call setup time	More	Less
Switching to CS domain	For every mobile originating and terminating call that increases signalling load	Only when the terminal roams out of LTE coverage area results in less signalling load

the IMS subsystem, E-UTRAN, and EPC (MME). On the other hand, no upgrades are required to the legacy GSM/WCDMA RAN target radio access.

8.5.2.3   CSFB versus SRVCC

CSFB is a short term solution and SRVCC is a longer term solution. Some key differences between the two techniques are shown in Table 8.1.

8.5.3   MOBILITY MANAGEMENT IN HETNETS (AN EXAMPLE)

HetNet is a mobile communications network that consists of a combination of different cell types (such as macro, femto cells, etc.) and different access technologies (such as HSPA, LTE, etc.). In such HetNets, MM becomes more complicated as compared to homogenous networks.

Consider the case as shown in Figure 8.11 [36] where it is assumed that a macro is deployed in carrier f1, while small cells are deployed on a separate carrier f2 utilizing a carrier aggregation technique. The RRC connected devices are assumed to have a downlink connection from the macro layer which means that the macro cell is their primary cell (PCell). The devices which are in the vicinity of a small cell can have that particular cell configured as their secondary cell (SCell) and thereby benefit from intersite carrier aggregation to achieve a higher data rate due to the higher accessible bandwidth for the user.

Thus, while the device/UE is moving under the same microcell, it will always have a stable PCell connection while adding, removing or changing SCells on the small cell layer at f2. Macros and small cells are assumed to be interconnected, via either the X2 interface or fibers using other

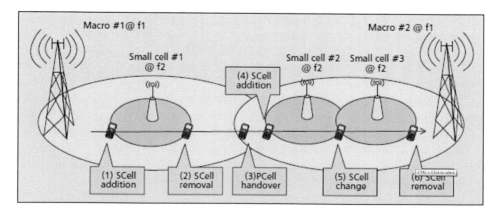

FIGURE 8.11   Basic principle of mobility for a HetNet scenario with inter-site carrier aggregation. (From Pedersen, K.I., Michaelsen, P.H. and Rosa, C. 2013 *IEEE Communications Magazine*, 51(5):64–71 [36].)

protocols. As shown in Figure 8.11, when the UE is entering the coverage area of small cell #1 at position (1), it will add this cell as an SCell, and when leaving the small cell coverage at position (2), the SCell connection is removed. At position (3), the UE will be subject to PCell handover from macro #1 to macro #2. An SCell change from small cell #2 to small cell #3 happens at position (5) while having the macro #2 as its PCell. Handovers in heterogeneous wireless networks are referred to as vertical handoffs which can be either mobile or network controlled.

As per 4G LTE-Advanced standard, the network will have to send a new RRC message to the device/UE whenever adding, removing or changing an SCell. Such network actions are triggered by measurement reports from the device/UE (sent via uplink RRC signaling). For scenarios with many small cells, the RRC signaling from SCell operations can, therefore, be significant compared to managing PCell mobility. Second, frequent device/UE assisted and network controlled SCell management constitutes an additional burden (and signaling delays) on the network as nonstationary device/UE devices are likely to be subject to relatively the frequent appearances and disappearances of small cells moving along certain trajectories.

To overcome such issues, proposals like the use of enhanced MSE (mobility state estimation), optimization of long DRX (discontinuous reception) for mobility, and improved inter-frequency small cell discovery are under investigation.

8.5.4 FUTURISTIC MOBILITY MANAGEMENT

Current mobile networks employ centralized mobility anchoring within the hierarchical network architecture. For example, in GPRS/UMTS networks, the GGSN, SGSN, and RNC constitute a hierarchy of anchors. In LTE deployments, P-GW and S-GW constitute another hierarchy of anchors [37] as shown in Figure 8.12.

8.5.4.1 Distributed Mobility Management

Centralized MM is nonoptimal with the evolving flatter network architectures. The forwarding of data through centralized anchors often results in nonoptimal routes which results in increasing

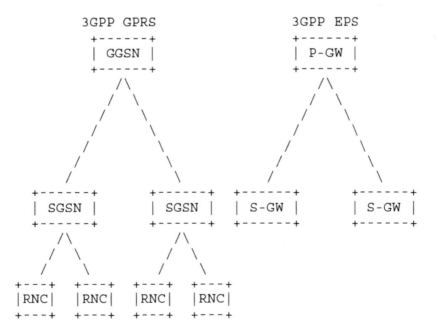

FIGURE 8.12 Centralized mobility management. (From IETF 2014 RFC 7333: Requirements for Distributed Mobility Management, August [37].)

end-to-end delay. Second, the centralized anchor (such as P-GW) also becomes a single point of failure for the huge amount of data that is traversing to/from it. Third, there are certain services that require a closer network of service-anchoring gateways such as multimedia content distribution.

To address this shortcoming, Distributed Mobility Management (DMM) has been proposed by IETF [37,38] and in literature at large [39,40]. On a similar note, although not DMM, 3GPP specified Local IP Access and Selected IP Traffic Offload (LIPA-SIPTO) techniques in Rel-10 and the LIPA Mobility and SIPTO at the Local Network (LIMONET) in Rel-12 [41] to avoid binding the IP connections to core gateways. These are primarily data offload solutions to reduce traffic load on core networks [42,43].

DMM is based on a flatter architecture where mobility anchors are placed closer to the user, while separating user and data planes. It also envisions mobility of devices between mobility anchors (such as PGWs of LTE) which is currently not practiced in centralized architectures [44]. DMM can come in two types, including partial DMM where the data plane is distributed while the control plane is centralized, and in full DMM where both data and control planes are distributed.

Considering the evolution of LTE toward DMM architecture, three core network entities, namely SGW, PGW, and MME, and LGWs (Local Gateways) and home eNodeB gateways in the RAN need to be revisited. The SGW terminates the interface toward the RAN, which means devices can only maintain connection at the IP layer or above beyond this gateway. LTE standard supports intra-SGW handover through S1 interface and also between distinct SGWs. The SGWs are primarily located in centralized locations, thus it would be difficult to support a DMM-type functionality [45].

The PGWs, on the other hand, are very few even in large scale networks, provide large scale coverage, and thus need no support from inter-PGW mobility. For example, the PGW in capital city Islamabad, Pakistan, serving northern and adjoining areas, does not need to talk to the PGW situated in the southern port city Karachi, Pakistan; they are approximately 1500 km apart. Similarly, a U.S. operator likely maintains separate PGWs for its New England and West Coast operations, but it will not deploy separate PGWs for each city of New England. For such reasons, IP mobility between PGWs is not supported and even a device cannot keep the same IP address if it changes PGWs [44].

MME manages the control plane in current LTE networks and performs various functions. It maintains information about device state, and performs idle mode procedures such as paging and device tracking. It also selects the SGW and PGW for the device and performs management of device associated bearers [3,44]. The role of MME and its evolution in support of DMM is under investigation. Unlike PGW, the LTE standard supports inter and intra MME handovers and MME usually resides with SGW.

LIPA breakout takes place at the Local Gateway (LGW) and is applied in residential and enterprise deployment. SIPTO breakout can take place either at LGW or at an HeNB Gateway. These two architectures to a certain extent reduce the load on SGWs and PGWs [46]. However, currently, inter LGW mobility is not allowed and thus the connection will get disconnected once the device moves from one LGW to another [44].

8.5.4.1.1 *Potential DMM-Based Architecture for a Mobile Network in Pakistan*

This section presents the possible DMM-based architecture of a mobile network in Pakistan. The country has four provinces and a federal capital, Islamabad. The capitals of these four provinces (Karachi, Quetta, Peshawar, and Lahore) and Islamabad are the five key cities where most of the broadband activity takes place. Karachi, the largest city, has a population of over 20 million, Lahore has over 10 million habitants, Peshawar about 2 million, and Quetta more than 1.2 million, while Islamabad has close to 2 million.

Figure 8.13 shows evolution from LTE/LTE-Advanced toward a DDM-based architecture comprised of multiple PGWs and LGWs in each city supporting inter-PGW* and inter-LGW mobility. The set of applications can be available within each city, however, connectivity to the global Internet

---

* Inter-PGW mobility is likely not required.

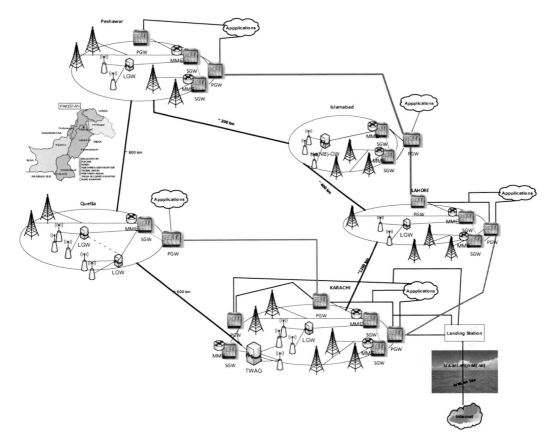

FIGURE 8.13 DMM-based mobile network in Pakistan.

will be provided through optical fiber submarine cables (SEA-ME-WE, I-ME-WE, etc.) through the landing station(s) installed in the port city of Karachi. One or more landing stations are expected to be deployed in the developing port city of Gawadar, Balochistan, for redundancy and disaster management in the coming years. The nitty gritty details are not shown since there are many time consuming and challenging technical issues that still need to be resolved in the standard bodies. These include support for inter-PGW and inter-LGW mobility, placement of LGWs either as it is or above SGW, and many others.

The technical challenges will likely be sorted out in due time, however, the cost of having a plethora of local gateways and small cells can turn out to be the prohibitive factor. The cost associated challenge becomes more daunting for the world's low ARPU (Average Revenue Per User) markets such as Pakistan.

8.6 NETWORK DEPLOYMENT

The deployment of a green field network or upgrading an existing network involves a number of steps. There are several pieces in the puzzle that need to be solved and clarified before it can be handed over to the network operations team for day to day management. The key phases of deployment include planning and engineering, implementation, optimization, and network management which are dealt with by the respective teams.

The end-to-end network includes RAN, transport network, core network, applications/value-added services, and operations sphere (excluding external networks such as Internet). A high-level network deployment process is shown in Figure 8.14, whereas the details of network deployment can be found in [1].

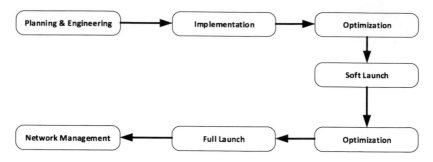

FIGURE 8.14   High level network deployment process.

8.6.1   PLANNING AND ENGINEERING

The planning and design phase primarily consists of network planning, dimensioning, and future expansion planning. Preliminary radio planning consists of cell site allocation, traffic planning, and extensive radio network planning. The process of network planning and design includes estimation of traffic intensity and traffic load. If a network of a similar nature already exists, traffic measurements of such a network can be used to calculate the exact traffic load. If there are no similar networks, then the network planner may use telecommunications forecasting methods to estimate the expected traffic intensity.

During the dimensioning exercise, the required number of network elements and their capacities are determined to meet grade of service. Dimensioning involves planning for peak hour traffic, that is, that hour during the day during which traffic intensity is at its peak. A dimensioning rule is that the planner must ensure that the traffic load should never approach a load of 100%. To calculate the correct dimensioning to comply with the above rule, the planner must take ongoing measurements of the network's traffic, and continuously maintain and upgrade resources to meet the changing requirements. The requirements of interconnects with other operators shall also be addressed. The exercise for disaster management shall be made an essential part of the exercise. The planning and dimensioning are usually conducted using specialized software tools.

The last step in network planning is the forecast for the future expansion of the network for additional subscribers, newer features, and better technologies [19,47–50].

8.6.2   IMPLEMENTATION

The implementation involves the physical rollout of equipment, installation of cell towers/small cells, enabling power, laying fiber in the ground, and so on. It also consists of hardware/software upgrades of network elements.

Cell sites are designed and selected carefully so that they can conform to the local regulatory guidelines. The radio coverage must be such that smooth handovers can take place between cell sites to cover the entire area. Radio planning software and drive testing are used in the cell site placement process.

The interoperability with other domestic operators, international operators, and connectivity with Internet is all part of this process.

8.6.3   OPTIMIZATION

The network optimization is used to deliver the best quality of service to the customers. Optimization is executed post soft launch on a cluster by cluster basis. Optimization is also performed on an as needed basis, post full launch activity. The network optimization is a process that measures the quality of the baseline network and adjusts it to perform at maximum efficiency and quality.

The RF optimization process includes field tests conducted to tune all aspects of air-interface network performance. The initial pass of network optimization is performed on a handful of clusters of sites and is intended to adjust parameters and identify and document coverage and capacity problems. And the second, more detailed tweaking phase occurs after the successful completion of all the cluster tests and once all cell sites (including small cells) in the network are activated. The system wide optimization is built upon cluster testing and is an iterative process to fix coverage and capacity related issues.

Compared to network engineering, which adds resources such as links, routers, and switches into the network, traffic engineering targets changing traffic paths on the existing network to alleviate traffic congestion or accommodate more traffic demand. This is critical when the cost of network expansion is prohibitively high and the network load is not optimally balanced. During optimization, areas of poor network performance are identified and changes are implemented to fix the problems. These adjustments can include antenna changes, antenna tilting, antenna orientation, transmission power, and other parameters. It is an iterative process which is repeated until the objectives of network planning are achieved; subsequently, the network is handed over to operations for day-to-day network management.

8.6.4 NETWORK MANAGEMENT

In the past, each network element had its own corresponding operation support system to provide management capabilities. Each of these management systems had a different user interface and different computing platform for each of the network elements. This made the task of network management for the operations team quite tedious and expensive, particularly when installed network elements are from different OEMs (original equipment manufacturers).

As service providers moved into 3G/4G and a mixed-supplier environment, they were not able to afford separate management systems for different network elements of different suppliers. To address this issue, the TM Forum (TeleManagement Forum) has provided guidelines in the form of Frameworx for efficient management of cellular networks [51].

Network management is about managing the network on a 24/7, 365 days of the year basis.

8.7 CONCLUSION

The evolution of mobile networks from 2G to 4G was briefly described in this chapter. 5G network architecture is still evolving and details are sketchy, however, considerable details have been provided, which may assist researchers to take the development to the next level.

The topic of MM, in particular for EPS (EUTRAN-EPC), was explained including two techniques (CSFB and SRVCC) for supporting voice in LTE networks. The DMM approach, which may likely be on an upgrade path for the current centralized architecture and a potential DMM-based architecture for mobile networks in Pakistan, was discussed. Finally, a snapshot of the key four stages of network deployment, namely planning and engineering, implementation, optimization, and network management, were briefly presented.

The succeeding three chapters will cover the various aspects of radio access, transport, core, and operations areas of the network architecture.

PROBLEMS

1. Describe the main components of a GSM network?
2. Describe the key architectural changes that were made in the UMTS architecture from Rel-99 to Rel-7?
3. What are the key differences in LTE and LTE-Advanced architectures?
4. What are the expected key components of 5G networks?

5. What is fronthaul?
6. Discuss whether more than one RAT is needed for 5G?
7. Define mobility management?
8. Briefly describe CSFB?
9. Describes SRVCC and its different forms?
10. What is Centralized Mobility Management?
11. Describe some disadvantages of Centralized Mobility Management?
12. What is Distributed Mobility Management?
13. Based on the concepts described in the chapter, portray a 5G Network for your country?
14. Describe the steps involved in network deployment?

REFERENCES

1. Asif, S.Z. 2007. *Wireless Communications Evolution to 3G and Beyond.* Artech House, Inc., Norwood, MA, USA.
2. 3GPP TS 23.002 (V 12.3.0) 2013. Network Architecture. Technical Specification (Release 12), Technical Specification Group Services and System Aspects, 3GPP, December.
3. Willie, W.L. 2002. *Broadband Wireless Mobile.* John Wiley & Sons, New York, NY, USA.
4. Asif, S.Z. 2011. *Next Generation Mobile Communications Ecosystem: Technology Management for Mobile Communications.* Wiley Inc., UK.
5. 3GPP Releases. http://www.3gpp.org/specifications/67-releases
6. Khartabil, H. et al. 2006. *The IMS: IP Multimedia Concepts and Services.* John Wiley & Sons, New York.
7. 3GPP TS 23.919 (V 13.0.0) 2005. Direct Tunnel Deployment Guideline. Technical Report (Release 13), Technical Specification Group Services and System Aspects, 3GPP, December.
8. Dong, Y. 2013. *LTE-Advanced: Radio Access Network Resource Management.* Master Thesis, University of Bremen.
9. NGMN Alliance 2011. 5G White Paper.
10. METIS 2015. Final Report on the METIS 5G System Concept and Technology Roadmap. Document Number: ICT-317669-METIS/D6.6. Mobile and Wireless Communications Enablers for the Twenty-Twenty Information Society (METIS), April.
11. 5GPPP 2016. View on 5G Architecture, 5G PPP Architecture Working Group, July.
12. Zhang, M.A. 2015. *Key Techniques for 5G Wireless Communications: Network Architecture, Physical Layer, and MAC Layer Perspectives.* Science China Information Sciences, Science China Press and Springer-Verlag, Berlin Heidelberg, 58(041301)1:20.
13. Huawei Technologies Co. LTD 2013. 5G: A Technology Vision.
14. Nokia Solutions and Networks 2013. Looking ahead to 5G, December.
15. SearchSDN (c/o TechTarget). What Is the Difference between SDN and NFV? http://searchsdn.techtarget.com/answer/What-is-the-difference-between-SDN-and-NFV.
16. Americas 2015. NFV and SDN Networks, November.
17. Liou, R-H. and Lin, Y-B. 2013. An Investigation on LTE Mobility Management. *IEEE Transactions on Mobile Computing*, 12(1):166–176.
18. 3GPP TS 23.060 (V 7.2.0) 2006. General Packet Radio Service (GPRS); Service Description; Stage 2. Technical Specification (Release 7), Technical Specification Group Services and System Aspects, 3GPP, October.
19. Wikipedia. http://en.wikipedia.org/wiki/Main_Page.
20. Wang, M., Michael, G. and Rahim, T. 2008. Signalling Cost Evaluation of Mobility Management Schemes for Different Core Network Architectural Arrangements in 3GPP LTE/SAE. *Proceedings of the Vehicular Technology Conference*, Marina Bay, Singapore, May 11–14, 2008, pp. 2253–2258.
21. 3GPP TS 23.402 (V 9.0.0) 2009. Architecture Enhancements for Non-3GPP Accesses. Technical Specification (Release 9), Technical Specification Group Services and System Aspects, 3GPP, March.
22. 3GPP TS 23.401 (V 9.0.0) 2009. General Packet Radio Service (GPRS) Enhancements for Evolved Universal Terrestrial Radio Access Network (E-UTRAN) Access. Technical Specification (Release 9), Technical Specification Group Services and System Aspects, 3GPP, March.
23. 3GPP TS 23.401 (V 10.10.0) 2013. General Packet Radio Service (GPRS) Enhancements for Evolved Universal Terrestrial Radio Access Network (E-UTRAN) Access. Technical Specification (Release 10), Technical Specification Group Services and System Aspects, 3GPP, March.

24. 3GPP TS 23.272 (V 12.1.0) 2013. Circuit Switched (CS) Fallback in Evolved Packet System (EPS); Stage 2. Technical Specification (Release 12), Technical Specification Group Services and System Aspects, 3GPP, December.
25. 3GPP TS 23.216 (V 12.0.0) 2013. Single Radio Voice Call Continuity (SRVCC); Stage 2. Technical Specification (Release 12), Technical Specification Group Services and System Aspects, 3GPP, December.
26. Americas 2013. 4G Mobile Broadband Evolution: 3GPP Release 11 & Release 12 and Beyond, February.
27. Nokia Solutions and Networks 2013. Evolve to Richer Voice with Voice over LTE (VoLTE), November.
28. Qualcomm 2012. VoLTE with SRVCC: The Second Phase of Voice Evolution for Mobile LTE Devices, October.
29. Shwetha, V. 2011. Single Radio Voice Call Continuity (SRVCC) with LTE. Radisys, September. http://www.slideshare.net/allabout4g/wpsrvccwithlte.pdf
30. Nokia Siemens Networks 2012. CS Fallback (CSFB).
31. GSMA 2010. IR.92 IMS Profile for Voice and SMS V3.0.
32. GSMA 2011. IR.94 IMS Profile for Conversational Video Service V1.0.
33. Ahmadi, S. 2014. *LTE-Advanced*. Elsevier Inc., Oxford, UK. http://books.google.com.pk/books?id=DTsTAAAAQBAJ&pg=PA90&lpg=PA90&dq=lte+asrvcc&source=bl&ots=zQ8N4EJfoC&sig=3N_KUB5ICBip4yBglD6jJbHSFVA&hl=en&sa=X&ei=m1QQU9GxMab9ygOx4IDwCg&ved=0CD4Q6AEwBQ#v=onepage&q=lte%20asrvcc&f=false
34. Agaur 2012. Evolution of Single Radio Voice Call Continuity (SRVCC). http://lteworld.org/blog/evolution-single-radio-voice-call-continuity-srvcc.
35. 3GPP TS 22.278 (V 12.4.0) 2013. Service Requirements for the Evolved Packet System (EPS). Technical Specification (Release 12), Technical Specification Group Services and System Aspects, 3GPP, September.
36. Pedersen, K.I., Michaelsen, P.H. and Rosa, C. 2013. Mobility Enhancements for LTE-Advanced Multilayer Networks with Inter-Site Carrier Aggregation. *IEEE Communications Magazine*, 51(5):64–71.
37. IETF 2014. RFC 7333: Requirements for Distributed Mobility Management, August.
38. IETF 2015. RFC 7429: Distributed Mobility Management: Current Practices and Gap Analysis, January.
39. Giust, F., Cominardi, L. and Bernardos, C.J. 2015. Distributed Mobility Management for Future 5G Networks: Overview and Analysis of Existing Approaches. *IEEE Communications Magazine*, 53(1):142–149.
40. Nguyen, T.-T., Bonnet, C. and Harri, J. 2016. SDN-Based Distributed Mobility Management for 5G network. *Proceedings of the IEEE Wireless Communications and Networking Conference (WCNC)*, Doha, Qatar, 3–7 April.
41. 3GPP TS 23.859 (V 12.0.1) 2013. Local IP Access (LIPA) Mobility and Selected IP Traffic Offload (SIPTO) at the Local Network. Technical Report (Release 12), Technical Specification Group Services and System Aspects, 3GPP, April.
42. America 2014. Executive Summary: Inside 3GPP Release 12: Understanding the Standards for HSPA+ and LTE-Advanced Enhancements, March.
43. Americas 2013. Mobile Broadband Explosion: The 3GPP Wireless Evolution, (Rysavy Research for 4G Americas), August.
44. Costa, R. et al. 2013. Distributed Mobility Management: A Standards Landscape. *IEEE Communications Magazine*, 51(3):80–87.
45. Panigrahi, P. 2012. LTE Handover Overview. http://www.3Glteinfo.com/lte-handover-overview/
46. Samdanis, K., Taleb, T. and Schmid, S. 2011. Traffic Offload Enhancements for eUTRAN. *IEEE Communications Surveys and Tutorials*, 14(3):884–896.
47. Ojanpera, T. and Prasad, R. 1998. *Wideband CDMA for Third Generation Mobile Communications*. Artech House, Boston, MA.
48. Lawrence, H., Levine, R. and Kikta, R. 2002. *3G Wireless Demystified*. McGraw-Hill, New York, NY.
49. Penttinen A. 1999. *Chapter 10—Network Planning and Dimensioning, Lecture Notes: S-38.145—Introduction to Teletraffic Theory*. Helsinki University of Technology, Fall.
50. Farr, R.E. 1988. *Telecommunications Traffic, Tariffs and Costs—An Introduction for Managers*. Peter Peregrinus Ltd.
51. TM Forum. Frameworx. https://www.tmforum.org/tm-forum-frameworx-2/

9 4G/5G Radio Access Network

As the title suggests, the focus of this chapter is on 4G and 5G radio access networks. 4G radio networks are designed to meet the requirements of ITU's IMT-Advanced framework. 3GPP's LTE-Advanced, which is an enhancement to LTE, is ITU's declared 4G technology. LTE-Advanced was standardized in 3GPP Release-10 in June 2011 and enhancements are continuously being made in each subsequent release. To distinguish the enhancements, 3GPP has identified Releases 8–9 as LTE, Releases 10–12 as LTE-Advanced, and Release 13 onwards as LTE-Advanced Pro. According to Global Mobile Suppliers Association (GSA), there are 194 LTE/LTE-Advanced mobile networks in 95 countries and out of these networks, 19 are LTE-Advanced Pro enabled in 15 countries. The count is increasing [1].

Today's industry buzz word is 5G, the fifth generation of mobile communications, which is under development. 3GPP is the only forum where the vast majority of stakeholders are working together to formulate and develop a complete set of 5G specifications. 3GPP has divided the development of 5G standard into two phases where 5G Phase 1 is getting developed in Release-15 while 5G Phase 2 will be specified in Release-16. The 3GPP based 5G standard is expected to meet the requirements of ITU's IMT-2020 and will be available for inspection before ITU's proposed deadline of 2019. ITU is expected to publish 5G standard(s) in 2020.

Briefly, Section 9.1 summarizes releases of 3GPP, Section 9.2 looks into LTE-Advanced, and Section 9.3 defines LTE-Advanced Pro. The technical details of 5G New Radio are summarized in Section 9.4.

9.1 3GPP RELEASES

The 3GPP was established in 1998 to develop standards for the evolution of 2G GSM systems. It currently unites seven telecommunications standard development organizations (ARIB, ATIS, CCSA, ETSI, TSDSI, TTA, TTC*) known as "Organizational Partners" and provides members with a stable environment to produce reports and specifications that define 3GPP technologies [2].

The 3GPP standards are structured as Releases. Once a release is frozen, no technical change can be made and it is then considered ready for implementation. 3GPP works on a number of releases in parallel where it starts future work well ahead of completion of the current release [2].

3GPP completed its first release, that is, Rel-99, in December 1999, which provided specifications for 3G. Currently, 3GPP is defining 5G in Rel-15/Rel-16 which are expected to be completed in the next two years. This section will only provide a snapshot (Figure 9.1) of the key features of these releases [3–5], and for details, readers can refer to 3GPP's provided relevant documents [6–17].

Key Aspects of Rel-99: 3GPP with its Rel-99, which was completed in December 1999, brought a W-CDMA based 3G system with an evolved radio network called UTRAN. The new access technique, that is, CDMA, is applicable for both FDD and TDD modes of duplex. An advanced BSC called RNC (Radio Network Controller) was introduced. The radio channels in W-CDMA are 5 MHz wide as compared to the 200 kHz channels of GSM/GPRS. It supports up to 2 Mbps in both directions in a 5 MHz channel.

---

* Association of Radio Industries and Businesses (ARIB) of Japan, Alliance for Telecommunications Industry Solutions (ATIS) of USA, China Communications Standards Association (CCSA), European Telecommunications Standards Institute (ETSI), Telecommunications Standards Development Society, India (TSDSI), Telecommunications Technology Association (TTA) of Korea, Telecommunication Technology Committee (TTC) of Japan.

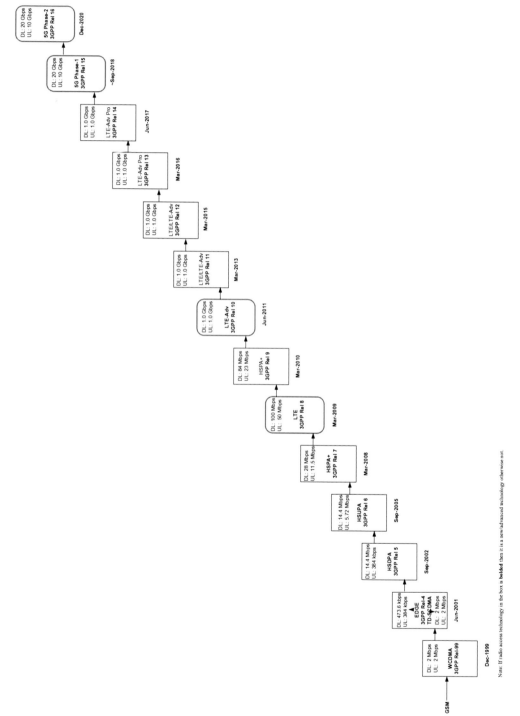

FIGURE 9.1 Migration from 3GPP Rel 99 to Rel 16.

The W-CDMA systems have the capability to handover to GSM/GPRS networks where UMTS coverage is not available. The Rel-99 inherits a lot from the GSM model on the core network side. The core network of Rel-99 is an evolution of GSM core network which is based on MAP (Mobile Application Part). Rel-99 also supports the legacy GSM BSS radio access network. This backward compatibility feature assists the legacy mobiles in operating in 3GPP Rel-99 networks in a seamless fashion.

Key Aspects of Rel-4: Rel-4 is associated with the inception of TD-SCDMA and EDGE radio technologies. EDGE is a TDM-based technology that provides the evolution for 2G IS-136 and GSM systems with the capability to support up to 473.6 kbps. TD-SCDMA or UTRA/UMTS-TDD 1.28 Mcps Low Chip Rate (LCR) is China's home grown 3G technology. TD-SCDMA supports 2 Mbps in a 1.6 MHz channel bandwidth and uses the UMTS core network.

Rel-4 features enhanced speech support, for example, with transcoder free operation and tandem free operation*. These enhancements provide transmission efficiency and cost reduction in the core network and are applicable to both GSM and UMTS systems. Rel-4 was completed in June 2001.

Key Aspects of Rel-5: The two key features of Rel-5 are IMS (IP Multimedia Subsystem) and HSDPA (High Speed Downlink Packet Access). HSDPA supports peak data rates of 14.4 Mbps and 384 kbps in downlink and uplink, respectively. IMS is a special core network subsystem, an open signaling system, based on standard Internet technology, uses Session Initiation Protocol (SIP), supports the migration of Internet applications (like VoIP, video conferencing, messaging, etc.) to the mobile environment, and offers enhanced service control capabilities.

An Iu-PS interface was enabled that allows GERAN (GSM EDGE Radio Access Network) to connect directly to the UMTS packet core network. An IP-based transport was enabled instead of just Asynchronous Transfer Mode (ATM) in the core network. A framework for end-to-end QoS was also introduced for the packet switched domain. Rel-5 was completed in September 2002.

Key Aspects of Rel-6: Rel-6 enabled the counterpart of HSDPA, that is, HSUPA, to support higher data rates in the uplink. HSUPA, using Enhanced Dedicated Channel (E-DCH) and other techniques, supports a peak data rate of 5.76 Mbps in uplink. It supports multiple receiving antennas for HSDPA terminals which improves the capacity of the UMTS networks.

Multimedia Broadcast and Multicast Service (MBMS) was introduced for supporting multimedia. MBMS is a broadcasting service and a means of delivering mobile TV to masses via UMTS networks. IMS was enhanced with several features including messaging, interworking between IMS and circuit switched and non-IMS IP networks, and IMS-based charging. Rel-6 was completed in September 2005.

Key Aspects of Release 7: Rel-7 introduced evolved HSPA (HSPA+), evolved EDGE, and enhancements to IMS and MBMS and was completed in March 2008. HSPA (High Speed Packet Access) includes both HSDPA and HSUPA systems. HSPA+ supports peak data rates of 28 Mbps in the downlink and 11.5 Mbps in the uplink. Radio enhancements to HSPA+ include 64 Quadrature Amplitude Modulation (64-QAM) in the downlink and 16-QAM in the uplink, MIMO to improve coverage/capacity and spectral efficiency, and MMSE equalizer that minimizes noise/interference to augment receive diversity. And it also introduced UMTS in the 900 MHz band and provided support for additional bands in both FDD and TDD modes.

Rel-7 also upgrades the existing EDGE technology to support peak data rates of 1 Mbps. It enhanced the uplink performance by means of 16-ary and 32-ary modulation schemes and turbo coding and provided faster signaling with Signaling Transport (SIGTRAN). Furthermore, to overcome the 200 kHz channel bandwidth limitation of GSM, a dual carrier in the downlink was recommended.

---

* Tandem Free Operation is a configuration of a speech or multimedia call for which transcoders are not utilized but are present. Transcoder Free Operation is a configuration of a speech or multimedia call for which transcoders are not present in the communications path.

Rel-7 provides several improvements for running circuit switched voice service over HSPA to boost capacity. MBMS was optimized with single frequency network (MBSFN) function, while IMS was augmented by allowing multiple UEs to register with the same Public User Identity and by supporting multimedia telephony service.

Key Aspects of Release 8: Rel-8 defines EPS which consists of E-UTRAN (LTE) and EPC. EPS is one hundred percent packet switched technology and thus is fundamentally different from 2G GSM and 3G UMTS systems. Another significant difference is that 3G systems use CDMA while LTE is based on OFDMA (Orthogonal FDMA) access method. Simultaneous use of MIMO and 64-QAM was permitted in HSPA along with dual carrier HSDPA (DC-HSDPA) wherein two downlink carriers can be combined for doubling data throughput. Rel-8 was functionally frozen in March 2009.

Key Aspects of Release 9: No new RAN or core network was defined in Rel-9, however, enhancements were made in both. Concepts of femtocells (Home eNodeB), Self-Organizing Networks (SON), and evolved multimedia broadcast and multicast service (eMBMS) were defined for LTE. Furthermore, new spectrum bands (e.g., 800 MHz and 1500 MHz) were identified for LTE operation. DC-HSDPA in combination with MIMO along with dual-carrier HSUPA was identified [18]. Rel-9 also supports regulatory features such as emergency user equipment positioning, Commercial Mobile Alert System (CMAS), and evolution of IMS architecture. Rel-9 was functionally frozen in March 2010.

Key Aspects of Release 10: LTE-Advanced (4G), which fulfills ITU's IMT-Advanced requirements, was defined in Rel-10. Key features of LTE-Advanced include carrier aggregation up to 100 MHz enhanced MIMO (8×8 MIMO in downlink and 4×4 MIMO in uplink), enhanced ICIC (Inter-Cell Interference Coordination), HetNets, and enhancements in SON (Self-Organizing Network) and MBMS. For HSPA, it supports quad-carrier operation and additional MIMO options. Rel-10 also initiated work on Minimization of Drive Tests (MDT) and was functionally frozen in June 2011.

Key Aspects of Release 11: For LTE, Rel-11 supports a Coordinated Multi-Point (CoMP) feature for coordinated scheduling, Enhanced Physical Control Channel (EPDCCH) for increasing control channel capacity, and further enhanced ICIC (FeICIC) for devices with interference cancellation. For HSPA, 8 carrier aggregation with up to 40 MHz aggregated bandwidth in the downlink, noncontiguous HSDPA carrier aggregation, 4×4 MIMO in the downlink, and 2×2 MIMO in the uplink are supported [19]. Several enhancements to MTC, IMS, WiFi related integration, Home NodeB, HeNB, and so on have been identified. Rel-11 was functionally frozen in March 2013 and it did not add any new radio access and core network technology.

Key Aspects of Release 12: For LTE, FDD-TDD carrier aggregation, device to device communication (also referred to as Proximity Services), and dual connectivity were defined. Enhancements were made for small cells/HetNets, MTC, MIMO, and SON. Furthermore, recovery mechanisms were added for MBMS and dynamic adaptation of uplink-downlink ratios was introduced for better utilization of TDD resources. For HSPA, enhancements were introduced in HetNets, Enhanced Uplink, Home NodeB, and dedicated channels. Additionally, features were introduced for improvements in MTC, public safety, WiFi integration, WebRTC (Web Real-Time Communication), and so on. [20]. Rel-12 was functionally frozen in March 2015 and it did not add any new radio access and core network technology.

Key Aspects of Release 13: LTE-Advanced Pro is the 3GPP approved LTE marker to describe appropriate specifications from Release-13 on and to show that the LTE platform has been significantly enhanced to address new markets with much better efficiency. Rel-13 completed specifications for mission critical services (such as mission critical Push-To-Talk), carrier aggregation supporting 32 component carriers, licensed assisted access (LAA) at 5 GHz, full dimension MIMO, enhanced

MTC and CoMP, and a few others. HSPA+ features include support for dual-band uplink CA. Rel-13 was functionally frozen in March 2016.

Key Aspects of Release 14: Rel-14 is primarily associated with the study items for 5G and was functionally frozen in June 2017. Rel-14 includes downlink multi user superposition transmission (MUST), support for V2X/C-V2X (Cellular V2X) services based on LTE, enhancements in CoMP, FD-MIMO LAA (License Assisted Access), NB-IoT (Narrow band Internet of Things), support for latency reduction, and many others.

Key Aspects of Release 15: Rel-15 is defining the first phase of 5G with expected completion in September 2018. It will describe the non-standalone 5G New Radio, that is, 5G NR that uses the LTE core network. It will build specifications based on ITU's IMT-2020 requirements and usage scenarios. Enhancements are expected in 4G-5G interworking, MIMO/beamforming, LAA, NB-IoT, and many others.

Key Aspects of Release 16: Rel-16 will define the second phase of 5G that will include NGC or Next Generation Core. This is called standalone 5G and does not need the LTE core network. Enhancements are expected in HSPA+ and LTE as well. Rel-16 will be completed in December 2020.

9.2 LTE-ADVANCED

3GPP Release-10 describes the LTE-Advanced (True 4G) technology that is designed to meet the diverse requirements of businesses and consumers. LTE-Advanced is both backward and forward compatible with LTE, that is, LTE devices are operational in LTE-Advanced networks and LTE-Advanced devices are operational in pre-Release-10 LTE networks. LTE-Advanced meets and/or exceeds all ITU's IMT-Advanced (4G) requirements. Details on LTE can be found in [2].

The LTE-Advanced/LTE-Advanced Pro air-interface is very similar to LTE and like LTE, consists of E-UTRAN and EPC units as shown in Figure 9.2. The E-UTRAN consists of eNodeBs (eNBs), providing the E-UTRA user plane (PDCP/RLC/MAC/PHY*) and control plane (RRC-Radio Resource Control) protocol terminations toward the UE (user terminal). The eNBs are interconnected with each other by means of the X2 interface. The eNBs are also connected by means of the S1 interface to the EPC, more specifically to the MME by means of the S1-MME interface and to the S-GW through the S1-U interface. The S1 interface supports a many-to-many relation between MMEs/S-GWs and eNBs [4].

The key new functionalities introduced in LTE-Advanced are carrier aggregation, advance multi-antenna transmissions, CoMP, and support for relay nodes. Additionally, some enhancements have been made to improve and support HetNets, MBMS, and SON features [4]. This section will briefly provide some details on certain key features of LTE.

9.2.1 Carrier Aggregation

Carrier aggregation is a feature of 4G or LTE-Advanced and was conceived in 3GPP Rel-10 to meet the IMT-Advanced peak data rate requirements. It is a costly technique to address the requirement of providing higher capacity while ensuring backward compatibility with LTE (Rel-8/Rel-9). Carrier Aggregation (CA) can be used in both FDD and TDD modes and it can also combine FDD and TDD modes (i.e., an LTE/FDD carrier with an LTE/TDD carrier) [4,5,22–25].

3GPP has referred to each aggregated carrier as a component carrier (CC) where the spacing between center frequencies of contiguously aggregated CCs is a multiple of 300 kHz. The spacing

---

* PDCP—Packet Data Convergence Protocol, RLC—Radio Link Control, MAC—Medium Access Control, PHY—Physical.

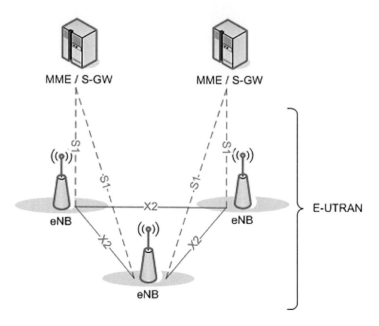

FIGURE 9.2 LTE-Advanced overall architecture. (From 3GPP TS 36.300 (V11.3.0) 2012 Evolved Universal Terrestrial Radio Access (E-UTRA) and Evolved Universal Terrestrial Radio Access Network (E-UTRAN); Overall Description; Stage 2. Technical Specification (Release 11), Technical Specification Group Radio Access Network, 3GPP, September [4].)

is required to be compatible with the 100 kHz frequency raster* of Rel-8 and at the same time to preserve orthogonality of the subcarriers with 15 kHz spacing. If the need arises for an aggregation scenario, the n*300 kHz spacing can be facilitated by insertion of a low number of unused subcarriers between contiguous CCs [26].

Contiguous CCs, noncontiguous CCs, and both symmetric as well as asymmetric CA are supported. The CA scenarios can be broadly classified into three categories as shown in Figure 9.3, namely [26]:

- Intra-band contiguous: contiguous CCs are combined together.
- Intra-band noncontiguous: CCs which are separated by a frequency gap within the same operating band are lumped together.
- Inter-band: in this case, CCs which are separated by a frequency gap in different operating bands are lumped together.
- Inter-technology: aggregation of one or more carriers of two technologies (such as one LTE carrier with one HSPA+ carrier). This type is still under investigation.

9.2.1.1 CA Technicalities

Each CC uses the 3GPP Rel-8 numerology and occupies a maximum of 110 physical resource blocks. Further, each CC corresponds to an independent data stream, one independent transport block, and one HARQ (Hybrid Automatic Repeat Request) entity (Figure 9.4). This permits maximum reuse of Rel-8 functions (such as RLC, PDCP, etc.) and better HARQ performance due to carrier component-based link adaptation. An MAC layer acts as the multiplexing entity for the aggregated CCs as they are activated or deactivated by MAC control elements. On the other hand, each CC has its own physical layer entity providing channel coding, data modulation, and resource mapping. Furthermore, the control plane architecture of LTE Rel-8 also applies to CA [26–28].

---

* The 100 kHz frequency raster means that the carrier center frequency must be an integer multiple of 100 kHz.

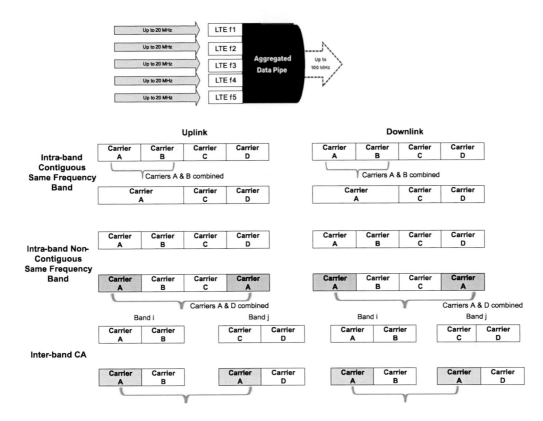

FIGURE 9.3   Carrier aggregation scenarios.

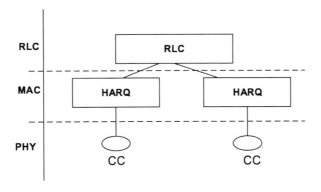

FIGURE 9.4   Carrier aggregation (at different protocol layers).

The major design challenge in CA lies in the device side because of its size and limited power. CA increases the complexity of RF circuits, duplexers, power amplifiers, diplexers, and so on. CA is sensitive to the desensing that occurs when the harmonic of a transmission signal falls in the receiver band of a paired CA band. Harmonics causes degradation in receiver sensitivity preventing the desired signal from being detected. CA also exerts pressure on duplexers to maintain transmit-receive isolation. In addition, for intra-band CA, power amplifiers need to maintain high linearity and for inter-band, switches have to provide the same [29].

Extending beyond five CA (beyond 100 MHz) is not a straight forward task as the current solution puts all uplink transmissions, including channel feedback (uplink control information), on

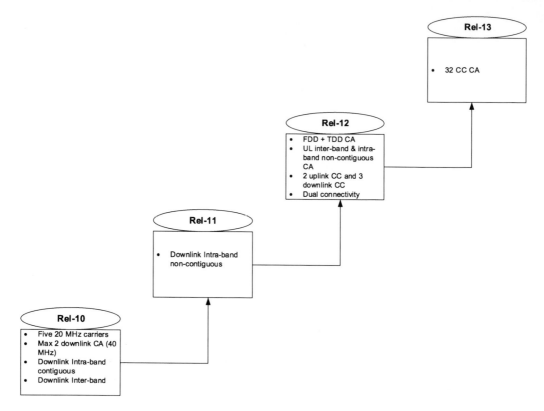

FIGURE 9.5   Carrier aggregation evolution—key features.

the primary cell (i.e., the primary carrier). This uplink control information (UCI) is carried by the Physical Uplink Control Channel (PUCCH) and increases to a large extent when there are more than five CCs. This model may not be sustainable and thus uplink traffic and channel feedback may need to be distributed over multiple carriers [30]. [31] summarizes the enhancements agreed/proposed in Rel-13 to support CA beyond five carriers.

9.2.1.2   CA Evolution

3GPP is continuously upgrading the CA technology with each release following Rel-10. At the same time, the operators are also adding CA in their networks to increase the ROI. The evolution of CA as depicted in Figure 9.5 shows the start of aggregation from Rel-10 with five CCs all the way up to 32 carriers in Rel-13. Rel-10 supports aggregation of five carriers, but most of the deployments were limited to two contiguous intra-band CCs with a maximum bandwidth of 40 MHz in downlink. Rel-11 supports more CA configurations and noncontiguous intra-band while Rel-12 specified aggregation of carriers between TDD and FDD frequency bands and supports dual connectivity* [31,32].

Rel-13 identifies support for 32 CCs with the initial focus on downlink tailoring to address the usage scenarios of IMT-2020. Further enhancements in CA are expected in following releases. In many commercial networks around the globe, two to three CAs have been reported. For example, AT&T (American Telephone & Telegraph Company) has aggregated 700 MHz with Personal Communications Service (PCS) (1900 MHz) that is, bands 2 and 17, Nippon Telegraph and Telephone (do communications over the mobile network) has aggregated band 1 (2100 MHz) with band 21 (1500 MHz), and so on. In many countries, CA has not been kicked off; for example, LTE was first launched in 2014 in Pakistan; however, no form of CA was deployed until 2017.

---

* With dual connectivity, UE can receive/transmit data from/to multiple eNBs simultaneously.

9.2.2 Antenna Diversity

During the last decade, operators started testing and implementing antenna diversity both at the base stations and devices. One of the first proof of concept tests for having more than one receiving antenna in mobile phone showed a 3 dB gain in the downlink capacity of a CDMA2000 network [33]. The research continues and antenna diversity has become an essential feature of EVDO, HSPA, LTE, LTE-Advanced, and LTE-Advanced Pro for improving the performance both in uplink and downlink.

Smart antennas involve employing multiple antennas either at the base station, the device or both. The prevailing form is MIMO involving more than one antenna at both ends. A 2 × 2 MIMO, which refers to two transmission antennas at the base station and two receiving antennas at the device, is commonly used in LTE networks. Smart antennas may also include beamforming that allows multiple antennas to shape a beam to increase the gain for a specific receiver and/or improve cell-edge performance by suppressing specific interfering signals.

The following are the key terms used to define multi antenna transmission in LTE systems [34].

- Codeword: a codeword represents user data before it is formatted for transmission. Depending on the radio conditions and transmission mode, one or two codewords can be used. For example, in downlink SU-MIMO, two codewords are sent to a single device whereas with downlink MU-MIMO, each codeword is intended for one device only*.
- Layer (Stream): this is the data stream sent in either direction. Eight layers are supported in downlink whereas there are four in uplink for spatial multiplexing in 3GPP Rel-10. The number of layers is always less than or equal to the number of antennas.
- Precoding: This modifies the data streams before transmission for reasons such as diversity, beamforming, and spatial multiplexing.
- Rank: It refers to how many independent signal paths can be recognized by the receiver. For example, a 4 × 2 MIMO system has a rank of 2 as the receiver signal can only accept two data streams.

9.2.2.1 MIMO in LTE

MIMO was introduced in LTE (3GPP Rel-8) to increase data rates through the transmission of two (or more) different data streams on two (or more) different antennas using the same resources in both frequency and time, separated only through the use of different reference signals to be received by two or more antennas. Rel-8 only supports single user MIMO in the downlink whereas Rel-10 includes multi-user MIMO. LTE-Advanced (Rel-10) supports 8 × 8 MIMO in the downlink and 4 × 4 in the uplink [22].

LTE enables multiple types of antenna Transmission Modes (TM) that are used according to radio conditions. These transmission modes differ in terms of number of layers (paths/streams), antenna ports, type of reference signal, and precoding type. Antenna ports do not correspond to physical antennas, but rather are logical entities distinguished by their reference signals. A base station can assign/map the antenna ports to the physical transmit antennas as per need.

In the downlink, there are nine different TMs, where TMs 1–7 were introduced in Rel-8, and TMs 8–10 were established in Rel-9, Rel-10, and Rel-11, respectively. In the uplink, two transmission modes are defined, that is, TM-1 and TM-2. The 8 × 8 MIMO in downlink is supported through TM-9 whereas the 4 × 4 MIMO in uplink is assisted by TM-2 that can reach peak spectral efficiencies of 30 bps/Hz and 15 bps/Hz, respectively. Thus, in summary, Rel-8 defined seven transmission modes and Rel-9 added TM-8 (dual layer beamforming). Rel-10 extended the dual layer mode of TM8 to TM9 with up to eight layers whereas Rel-11 added TM10 8-layer transmission with support for CoMP.

The details for LTE smart antennas can be found at [5,35–38].

---

* In SU-MIMO (Single User MIMO), full bandwidth of the base station is assigned to a single UE whereas MU-MIMO (Multi-user MIMO) enables multiple users to simultaneously access the same channel by providing spatial degrees of freedom.

9.2.3 Relaying

The idea of relaying or cooperative communications can be traced back to the 1970s [39]. Recent research on relaying demonstrates its benefits for wireless communications. The concept of relaying was introduced in 3GPP Rel-10 to improve coverage, cell edge throughput, in-building penetration, and to reduce coverage holes. It can also be said that relays brought the possibility of heterogeneity to LTE and 4G networks. Relays use already assigned spectrum, giving operators a cost-effective means, particularly in initial deployments when usage is relatively low and without the backhaul expense to improve coverage and throughput [22–26,40].

9.2.3.1 Architecture

The architecture involving relaying is rather simple. As per 3GPP TR 36.912 [26], relaying takes place via a relay node (RN) which is likely wirelessly connected to a donor cell (sector) of a donor eNB through the Un interface. RN is also wirelessly connected with UEs through the Uu interface as shown in Figure 9.6. On the Uu interface, all access stratum control plane (RRC) and user plane (PDCP, RLC, and MAC) protocols are terminated in RN. On the Un interface, the user plane is based on standardized protocols (PDCP, RLC, MAC) while the control plane on Un uses RRC (for the RN in its role as UE).

RNs can be considered as low power nodes on top of the conventional macro base stations in a cellular system. UEs can communicate directly with the eNBs or with the assistance of RNs. The UEs at cell edge are better served due to this cooperation, thus capacity can be enhanced [41]. Cooperation takes place when relays assist the transmission from source to destination by transmitting different copies of the same signal from different locations. This allows the destination to get independently faded versions of the signal that can be combined to obtain an errorless signal.

9.2.3.2 Types of Relays

3GPP has defined two types of relays in LTE-Advanced, namely Type 1 and Type 1a RNs. A Type 1 RN can assist a UE which is located far away from an eNB to access the eNB by having a separate Physical Cell ID (PCI), transmitting their own common reference signal and the control information. The key goal of a Type 1 relay is to extend coverage. Type 1 RN is an in-band RN meaning that the eNB-RN link shares the same carrier frequency with RN-UE links. This may result in self-interference in the RN as Uu and Un are transmitting at the same time on the same frequencies. Type 1a essentially has the same features as Type 1 except it operates outband. In RNs that operate outband, the eNB-RN link does not use the same carrier frequency as the RN-UE links [26].

During the course of standardization, the following types of relays were identified within the protocol architecture as shown in Figure 9.7 [24,42].

Layer 1 Relay: Similar to a repeater, it receives the signal from the source (BS or MS), amplifies it, and transmits it to the destination (UE or eNB). Layer 1 (L1) relay is simple and easy to implement through RF amplification with relatively low latency. However, it amplifies inter-cell interference and noise along with the desired signal components. This harmful amplification reduces SINR (Signal to Interference plus Noise Power Ratio) and network performance. L1 relay can be called amplify-and-forward relay.

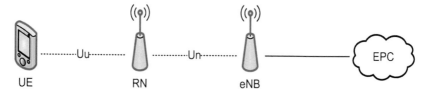

FIGURE 9.6 Relays. (From 3GPP TR 36.912 (V14.0.0) 2014 Feasibility Study for Further Advancements for E-UTRA (LTE-Advanced). Technical Report (Release 11), Technical Specification Group Radio Access Network, 3GPP, March [26].)

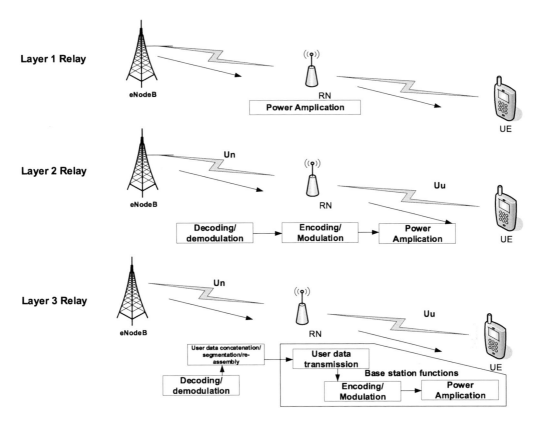

FIGURE 9.7 Types of relays. (From Iwamura, M., Takahashi, H. and Nagata, S. 2010 *NTT Docomo Technical Journal*, 12(2), 29–36 [42].)

Layer 2 Relay: Layer 2 (L2) relay is a decode and forward type of relay where RF signals received from the base station/eNB on the downlink are demodulated and decoded and then encoded and modulated again before being sent to the UE. This radio signal processing helps to overcome the inter-cell interference and noise as witnessed in L1 relays. However, L2 relays cause delay to the associated radio signal processing.

Layer 3 Relay: Layer 3 (L3) relay or self-backhauling has less impact on the eNB design but introduces more overhead as compared to L2 relay. Besides L2 functions, L3 relay also performs ciphering and user data concatenation/segmentation/re-assembly. Similar to L2 relay, it improves throughput by incorporating the same functions as base station/eNB. L3 relays also feature a unique PCI on the physical layer different from that of the base station. The PCI assists the UE to identify that a cell provided by a relay differs from a cell provided by the base station/eNB.

9.2.3.3 Relaying Functionality

The resources have been partitioned for the operation of RNs. For inband relaying, a few time-frequency resources are set aside for the backhaul link (Un) and cannot be used for the access link (Uu).

3GPP [26,43] defines the following general principles for resource partitioning at the relay node:

Resource partitioning at the RN

    In downlink: eNB → RN and RN → UE links are time division multiplexed in a single carrier frequency and only one is active at any time

    In uplink: UE → RN and RN → eNB links are time division multiplexed in a single carrier frequency and only one is active at any time

Multiplexing of backhaul links[*]

eNB $\rightarrow$ RN transmissions are carried out in the downlink frequency band, whereas RN $\rightarrow$ eNB transmissions take place in the uplink frequency band for FDD systems.

eNB $\rightarrow$ RN and transmissions are carried out in the downlink subframes of the eNB and RN, where RN $\rightarrow$ eNB, while eNB transmissions are carried out in uplink subframes of the eNB and RN for TDD systems.

For simultaneous operation of RN $\rightarrow$ eNB (Un) and RN $\rightarrow$ UE (Uu) links, antenna separation is needed to avoid interference. An RN can select the most appropriate air-interface beam and/or frequency with one of the following methods:

- The RN may be comprised of a signal processing device for determining one or more attributes of a wireless signal received from an eNB and a memory unit for storing a reference list that includes a list of one or more preferred frequency bandwidths to be received, amplified, and transmitted. The RN is further comprised of a controller coupled with the signal processing device for comparing the one or more attributes of the wireless signal from the eNB with the reference list, where the controller selects a first active set of one or more frequency bandwidths from the preferred frequency bandwidths based on the comparison. The RN may still further include a first reception filter, where the first reception filter, in response to control signal(s) from the controller, filters out signals and communicates to the first reception filter that has frequencies outside the first active set of frequency bandwidths [44].
- The RN is comprised of a signal processing device for determining one or more attributes of a wireless signal received from a UE and a memory unit for storing a reference list that includes one or more preferred frequency bandwidths to be repeated. The RN also comprises a controller coupled with the signal processing device for comparing the one or more attributes of the wireless signal with the reference list, wherein the controller selects a first active set of one or more frequency bandwidths from the preferred frequency bandwidths based on the comparison. The RN still further comprises a first reception filter coupled with the controller, where the first reception filter, responsive to control signal(s) from the controller, filters out signals and communicates to the first reception filter, having frequencies outside the first active set of frequency bandwidths [45].
- An RN may include a donor antenna, a coverage antenna, a mobile station modem, a processor, and data storage. The donor antenna will receive a plurality of air-interface beams on the downlink from an eNB. The coverage antenna will pass each downlink beam received to a UE being served by the RN. Further, the donor antenna will pass each downlink air-interface beam to the MSM (mobile station modem) where the MSM will apply a rake receiver to identify a signal characteristic of the downlink of each beam received, and the processor will record in the data storage the signal characteristic corresponding to each beam. Given this data, the RN will select an air-interface beam with the most preferable signal characteristic and will keep only this beam active for uplink communications [46].

Another method to avoid interference is to operate the relay such that the relay is not transmitting to terminals when it is supposed to receive data from the donor eNodeB by creating gaps in the RN $\rightarrow$ UE transmission. These gaps, during which UEs (including Rel-8 UEs) will not receive any relay transmission, can be created by configuring MBSFN subframes as shown in Figure 9.8. RN $\rightarrow$ eNB transmissions can be carried out by prohibiting any UE $\rightarrow$ RN transmissions in some subframes [26,43].

A relay may be part of the donor cell or control cells of its own. In the case where the relay is part of the donor cell, the relay does not have a cell identity of its own. To some extent, some part

---

[*] FDD: LTE frame length 10 ms, divided into 20 individual slots. Each subframe consists of 2 slots. TDD: LTE frame length 10 ms, which comprises of two half frames, each 5 ms long. Subframes are 1 ms long.

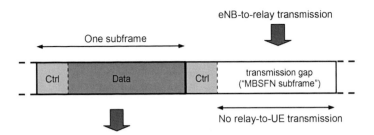

FIGURE 9.8 Example of relay-to-UE communication using normal subframes (left) and eNodeB-to-relay communication using MBSFN subframes (right). (From 3GPP TR 36.912 (V14.0.0) 2014 Feasibility Study for Further Advancements for E-UTRA (LTE-Advanced). Technical Report (Release 11), Technical Specification Group Radio Access Network, 3GPP, March [26].)

of the RRM (Radio Resource Management) is controlled by the eNodeB to which the donor cell belongs, while parts of the RRM may be located in the relay. Smart repeaters, decode-and-forward relays, and different types of Layer 2 relays are examples of this type of relaying. In the case where the relay is in control of cells of its own, the relay controls one or several cells and a unique PCI is provided to each of the cells controlled by the relay. The same RRM mechanisms are available in both relay and cells controlled by eNodeBs and from a UE perspective there is no difference in accessing either one. Layer 3 relay, Type 1 relay nodes, Type 1a relay nodes, and so on, use this type of relaying [42,43,47].

9.2.3.4 Challenges/Solutions

One of the key challenges is to select and pair nearby RNs and UEs to achieve the relay/cooperative gain. The selection of cooperative partners is a key element for the success of the relaying strategy [47]. Thus, for efficient working of relaying architecture, suitable relay transmission protocol and relay selection methods are required [48–51].

The relaying protocols are initiated by source or destination where relays can be selected prior to data transmission. Such protocols require additional control/handshake messages which may result in additional overhead. Relay selection that intends to identify the most suitable relay(s) for assisting transmissions is also a difficult task. Relay selection can be performed either by source or destination and both approaches incur some overhead.

9.2.4 SON Enhancements

SON is an automation concept designed to make the planning, configuration, management, optimization, and healing of mobile radio access networks simpler and faster. SON improves network quality and reduces OPEX.

SON was introduced in Rel-8 to extend LTE coverage in the existing 2G/3G networks. As LTE's coverage is becoming ubiquitous like 2G/3G, requiring operators to spend more resources to optimize capacity and coverage in a heterogeneous environment with multiple carriers per radio access technology, several enhancements have been made to SON. Some of these SON enhancements are as follows:

SON solutions can be divided into three categories, namely self-configuration (Rel-8), self-optimization (Rel-9), and self-healing (Rel-10). The self-configuration functionality assists the eNB to configure the PCI, transmission frequency, and power that leads to quicker cell planning and rollout. The self-optimization focuses on improvement of coverage, capacity, handover, and interference through functions such as mobility load balancing, mobility robustness optimization, and RACH (Random Access Channel) optimization. Self-healing includes features such as automatic detection and removal of failures and automatic adjustment of parameters. SON can assist with the following [4,23,26,52,53]:

Coverage and Capacity Optimization: Capacity and coverage tradeoff is a complex matter with no easy answer. Network performance (coverage/capacity/quality) can be improved through measurements derived via drive testing. Based on these measurements, the network can optimize the performance by trading off capacity and coverage.

Mobility Robustness Optimization: This is a solution for automatic detection and correction of errors, particularly during handovers. The focus is on the optimization of mobility related parameters to reduce the number of handover related radio link failures. Too late or early handovers or handover to an incorrect cell can cause can cause radio link failures.

Mobility Load Balancing: This feature allows transfer of load from cells suffering from congestion to cells that have spare resources. It authorizes load report sharing between eNBs for smooth transfer of traffic and minimizing the number of handovers.

Energy Savings: Some cells can be switched off when capacity is not needed to save energy. The cells can be re-activated on an as needed basis by the cells that remain on for providing coverage. This can be performed by giving wake-up calls to the sleeping ones.

Minimization of Drive Tests (MDT): UEs can be used to provide the same information which is collected during drive testing. This reduces costs and resources associated with drive testing. UE can also provide data from indoor environments which is a great advantage. Rel-10 initiated work on defining automated solutions, including involving UEs in the field to reduce the operator costs for network deployment and operation. Rel-10 specified that the MDT data reported from UEs and the radio network may be used to monitor and detect coverage problems in the network including coverage hole, weak coverage, pilot pollution, capacity shortage, and so on. [54].

9.2.5 eMBMS

MBMS is a broadcasting service, is a means of delivering mobile TV to multiple viewers via 3GPP networks, and was introduced in 3GPP Rel-6. In Rel-7, multicast/broadcast, MBSFN function was included for optimization of MBMS capabilities. In MBSFN, a time-synchronized common waveform is transmitted from multiple cells for a given duration through a dedicated carrier. An MBSFN transmission from multiple cells within the MBSFN area is seen as a single transmission by a UE. When transmissions are delivered through the LTE network, the MBMS service is referred to as eMBMS (Enhanced MBMS).

Rel-10 added an MBMS counting function to allow calculating the number of connected mode[*] users that are either receiving a particular MBMS service or are interested in receiving a particular MBMS service. This function is controlled by the Multi-cell/multicast Coordination Entity (MCE) and it allows the MCE to enable or disable MBSFN transmission [21,23].

In Rel-10, the prioritization of different MBMS services through MCE is enabled as MCE is responsible for deciding the allocation of radio resources for MBSFN transmissions. In this way, the MCE can pre-empt radio resources used by an ongoing MBMS service(s) in the MBSFN area according to Allocation and Retention Priority (ARP) of different MBMS radio bearers [23].

9.2.6 LATENCY IMPROVEMENT

The user plane latency is defined as the one-way transit time for a packet from source (UE/eNB) to destination (eNB/UE). The control plane latency refers to the transition time from the idle to active state. The ITU's user and control planes' latency requirements for IMT-Advanced systems are 10 ms and 100 ms, respectively [55,56].

---

[*] In connected mode, the mobility of radio connections has to be supported.

At this moment in time, LTE has the lowest latency of any mobile communications technology and it could be as low as 7 ms (between UE and eNB), exceeding ITU recommended requirements [5]. LTE not only fulfills the latency requirements of ITU, but has added several mechanisms that could be used to further reduce the c-plane latency such as [26]:

- Combining RRC connection request and NAS (non-access stratum) service request messages can reduce overall latency from idle mode to connected mode by approximately 20 ms.
- By decreasing the RACH scheduling period from 10 ms to 5 ms, the average waiting time for the UE to initiate the procedure to transit from idle mode to connected mode can be reduced by 2.5 ms.
- Shorter PUCCH cycles may reduce the transition from dormant to active state in the connected mode. It specifically may reduce the average waiting time for a synchronized UE to request resources in connected mode.

LTE provides u-plane latency below 10 ms for synchronized UEs. When a UE does not have a valid scheduling assignment or when it needs to synchronize and obtain a scheduling assignment, a reduced RACH scheduling period, shorter PUCCH cycle, and so on can be used to improve the latency as compared to LTE Rel-8 [26].

9.2.7 ENHANCED INTER-CELL INTERFERENCE COORDINATION (eICIC)

LTE supports various forms of interference cancellation and interference avoidance techniques. One such technique, ICIC, was defined in Rel-8, and its enhanced version, that is, eICIC in Rel-10, while further enhancements (FeICIC) were made in Rel-11. ICIC includes a frequency as well as a time domain component.

9.2.7.1 ICIC versus eICIC/FeICIC

ICIC is a multi-cell radio resource management function that manages the radio resources such that this interference remains under control. For the time domain ICIC, beginning with Rel-10, subframe utilization across different cells are coordinated in time through Almost Blank Subframes (ABS) [57]. Rel-10 and Rel-11 provide support to manage the interference in the HetNet scenario in the time domain through eICIC and FeICIC and in the frequency domain with carrier-aggregation based ICIC [58].

ICIC decreases the interference between the neighboring macro base stations by reducing the power of some subchannels in the frequency domain. Since their power is reduced, these subchannels can only be received in the close proximity of the base station. The same subchannels can also be used in the neighboring cells. eICIC, on the other hand, is designed to provide interference management in HetNets since by design, the coverage of small cells overlaps with the macro cell. Thus, ICIC is more like a macro cell interference mitigation scheme while eICIC/FeICIC are designed to handle interference in HetNets [5,23,59].

9.2.7.2 eICIC in HetNets

In HetNets, there is large disparity (for example $\geq 10 \times$ dB) between the transmit power levels of macro and small cells. This makes the downlink coverage of a small cell much smaller than that of a macro base station. The interference from macrocell signals to the transmission of small cells may diminish the capability to offload traffic from macrocells. Thus, load balancing between macro and small cells is needed and can be provided by extending the coverage of small cells and subsequently increasing cell splitting gains. This concept is called range expansion [60].

Rel-8 and Rel-9 ICIC support range expansion up to about 3 dB bias, while larger biases toward small cells can be supported through ABS (almost blank subframe) which was introduced in Rel-10. ABS allows macro base stations to reserve some subframes for the associated small cells. Capacity may be increased as each small cell can use the empty subframes without interference from the other small cells. Rel-11 enabled devices support Cell specific Reference Symbol (CRS) interference cancellation [58,61]. Such devices can cancel interference on common control channels of ABS caused by interfering cells such as CRS signals of high power macro cells. Without such capability, heterogeneous networks' eICIC can only work for noncolliding CRS cases. Qualcomm [62] claims that the CRS interference cancellation is necessary with eICIC to obtain worthwhile improvements in cell data throughputs.

9.2.8 COORDINATED MULTI-POINT TRANSMISSION AND RECEPTION (COMP)

CoMP is a technique to improve coverage, throughput, and spectral efficiency of LTE networks. CoMP was studied in Rel-10 and standardized in Rel-11 to combat inter-cell interference that mainly occurs with cell edge devices [63].

In principle, CoMP optimizes the transmission and reception from multiple distribution points in both homogenous and heterogeneous network deployments. The devices at the cell edge can receive signals from multiple transmitters and if these transmitters can coordinate their transmissions, performance of the downlink can be improved. For the uplink, signals from devices received at multiple reception points can significantly improve the link performance. Both links benefit from techniques such as joint transmission, joint reception, coordinated beamforming, and so on [5,57,64].

CoMP transmissions can be either intra-site, inter-site or within distributed radio architecture where RRUs are housed at different locations than the baseband units areas shown in Figure 9.9. The primary benefit of intra-site CoMP is that a significant portion of information exchange can take place between the cooperating cells since backhaul connections between eNBs are not involved. The inter-site CoMP, on the other hand, involves the coordination of multiple sites for CoMP transmission requiring backhaul transport. The inter-site CoMP brings additional burden and requirements to the backhaul design. The distributed RAN architecture is a form of inter-site CoMP where RRUs of a single eNB are installed on the tower mast and primarily connected via fiber cable [5,64].

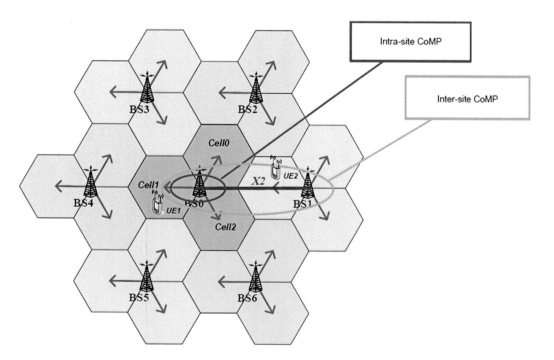

FIGURE 9.9 Coordination types of CoMP. (From 4G Americas 2012 4G Mobile Broadband Evolution: Release 10, Release 11 and Beyond – HSPA+, SAE/LTE and LTE-Advanced [23].)

9.2.8.1 Downlink CoMP

Downlink CoMP refers to different types of coordination in the downlink transmission from multiple geographically separated transmission points [26]. It can be assisted with techniques such as Coordinated Scheduling/Beamforming (CS/CB) or Joint Transmission (JT). The CS/CB is dynamically coordinated among cells in order to reduce interference between different transmissions and to increase the served device's signal strength. The JT enables simultaneous data transmission from multiple points to a single device. This transmission will be coordinated as a single transmitter with multiple antennas at separate locations requiring stringent backhaul capabilities but allowing higher performance [64].

9.2.8.2 Uplink CoMP

The uplink CoMP refers to the reception of the transmitted signal at multiple geographically separated points. The interference can be controlled by coordinating scheduling decisions among cells [26,64]. Techniques such as joint reception (JR) can be applied to implement UL CoMP where PUSCH transmitted by the device is received jointly at multiple points (part of or the entire CoMP cooperating set*) at a time, for example, to improve the received signal quality [63].

9.2.8.3 Performance Expectations

3GPP TR 36.819 [63] provides some performance gains associated with CoMP. When resource utilization is less than 35% in FDD mode, CoMP with joint processing (joint transmission) may provide a 5.8% performance gain for the mean user and a 17% gain for cell edge users relative to HetNets without eICIC in the downlink. For resource utilization of more than 35% in FDD mode, CoMP with JT may provide a 17% mean gain and a 40% cell-edge gain relative to HetNets without eICIC in the downlink. In the uplink, the gains are 15.2% (macro cell area average) and 45% (cell edge users) relative to HetNets without eICIC.

9.3 LTE-ADVANCED PRO

LTE-Advanced Pro describes a milestone (evolutionary stage) in the development of LTE and from that point on the LTE platform is considered to be considerably enhanced to address new markets as well incorporate added functionality to improve efficiency. Rel-13 and beyond are considered as LTE-Advanced Pro by 3GPP [65].

The key features/enhancements that were completed in Rel-13 while maintaining backward compatibility include support for higher data rates beyond 3 Gbps, support for both licensed (450 MHz to 3.8 GHz) and unlicensed (5 GHz) spectrum, multi-user superposition transmission in downlink, RAN sharing enhancements, small cell dual-connectivity and architecture, full dimension MIMO, and many others [65–67]. A subset of such enhancements is described in this section.

9.3.1 CARRIER AGGREGATION ENHANCEMENTS

The concept of carrier aggregation was proposed in LTE-Advanced (Rel-10) with five CCs, each having a size of 20 MHz. The total aggregated bandwidth remained 100 MHz until Rel-12. Rel-13 provides a significant upgrade and supports aggregation of up to 32 component carriers, each of 20 MHz, resulting in a total bandwidth of 640 MHz. With this much bandwidth, 3 Gbps or higher can be supported [68,69].

9.3.2 LICENSED ASSISTED ACCESS USING LTE

LTE networks primarily support deployment in paired and unpaired setups and in FDD and TDD modes in the range of 400 MHz to 3.8 GHz. 3GPP TR 36.889 [70] specifies the support of the

---

* A set of (geographically separated) points which are intended for data reception from a device.

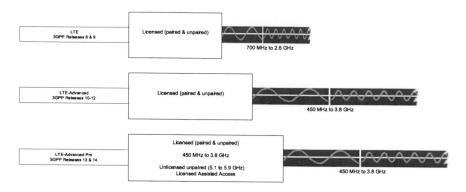

FIGURE 9.10 LTE spectrum utilization.

unlicensed 5 GHz band for LTE downlink to address the ever-increasing data demands. This enhancement of LTE in the unlicensed spectrum is called Licensed Assisted Access (LAA) and was specified in Rel-13. In the unlicensed 5 GHz band, a sizeable amount of spectrum is available which is suitable for small cells (downlink only). The evolution of LTE spectrum usage from Rel-8 to Rel-14 is shown in Figure 9.10. Rel-14 supports LAA in the uplink to increase uplink data rates and capacity [68,69].

Earlier attempts such as LTE-U (LTE in unlicensed spectrum) to standardize LTE in the unlicensed spectrum had limited success as these did not consider the right amount of regional and country-specific regulatory restrictions. Rel-13 LAA specifications produced a global framework for LAA to the unlicensed spectrum (5 GHz) while considering the regional regulatory power limits.

Under Rel-13 LAA, licensed carriers are aggregated with unlicensed carriers to opportunistically enhance the downlink data performance. It is also a highly productive method for traffic offloading as the data traffic can be split, with millisecond resolution, between licensed and unlicensed frequencies. The split allows the device to simultaneously receive transmissions on both paths. LAA enables LTE to use low power small cells in the unlicensed spectrum using CA. It uses CA in the downlink to combine LTE in the unlicensed spectrum with LTE in the licensed band. The licensed spectrum carries all of the control and signaling information while data uses the unlicensed band [71]. It mainly uses a contention protocol known as listen-before-talk (LBT) to coexist with other WiFi devices on the same band and dynamic frequency selection (DFS) for radar avoidance in certain bands [67,71].

9.3.3 MIMO ENHANCEMENTS

MIMO was introduced in LTE (3GPP Rel-8) to provide higher capacity and spectral efficiency. LTE-Advanced (Rel-10) supports 8×8 MIMO in the downlink and 4×4 in the uplink. LTE-Advanced Pro (Rel-13) introduces the next step with elevation beamforming (EB) and Full Dimensional MIMO (FD-MIMO).

FD-MIMO supports both elevation and azimuth beamforming as shown in Figure 9.11. It envisions increasing the number of antenna ports at the transmitter (base station) up to 64 to improve spectral efficiency with Rel-13. FD-MIMO incorporates the 3D model to capture elevation angles as a 2D model cannot describe elevation differences because it assumes zero elevation angles [72].

Until Rel-11/12, 3GPP mainly considered one-dimensional antenna arrays that exploit the azimuth dimension and MIMO with eight or fewer antenna ports. With Rel-13, an extensive study [73] was carried out by 3GPP to understand the performance benefits of enhancements targeting antenna array operations with 8 or more transceiver units (TXRUs) per transmission point. A large number of transceivers can be implemented with active antenna arrays. Active antenna arrays beside passive radiating elements may also include RF amplifiers, filters or digital processing functionality [68]. The key conclusions of the study are as follows:

FIGURE 9.11 FD-MIMO.

- Nonprecoded, beamformed, and hybrid CSI-RS (Channel State Information Reference Signals) based schemes demonstrated significant throughput gains
- Noncodebook based CSI reporting, DMRS (Demodulation Reference Signal) enhancements as well as SRS (Sounding Reference Signal) enhancement are beneficial for EB/FD-MIMO

9.3.4 LATENCY IMPROVEMENT

Work is ongoing in 3GPP to find ways for reducing latency in LTE systems. LTE/LTE-Advanced uses a 1 ms TTI (Transmission Time Interval) corresponding to 14 OFDM symbols and one subframe. 3GPP is planning to change this approach for LTE-Advanced Pro systems by reducing the frame length. The TTI can be reduced by decreasing the number of symbols from 14 to 2 as shown in Figure 9.12. This approach is viable in maintaining backward compatibility and usability in existing LTE bands. LTE/LTE-Advanced 1 ms TTI produces approximately 15–20 ms latency (round trip time), while an LTE-Advanced Pro may provide less than 2 ms round trip time.

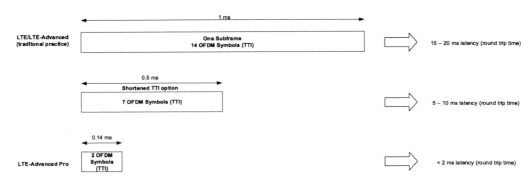

FIGURE 9.12 TTI and latency.

9.4 5G NR

The 5G NR is currently under the defining stage. According to ITU-R, IMT-2020 (including 5G) are mobile systems that include new radio interface(s) that support new capabilities of systems beyond IMT-2000 and IMT-Advanced [56,74]. The ITU-R has envisioned the following three usage scenarios for IMT-2020 [5,75]

- eMBB: providing higher speeds for applications such as web browsing, streaming, and video conferencing.
- URLLC: enables mission-critical applications, industrial automation, new medical applications, and autonomous driving that require very short network traversal times.
- mMTC: extends LTE IoT capabilities to support a huge number of devices with enhanced coverage and long battery life.

9.4.1 Requirements

The ITU-R has published a schedule [76] for the standardization of IMT-2020 and set the following timelines as shown Figure 9.13:

- 2016–2017: Determine technical performance requirements, evaluation criteria, and identification of spectrum bands.
- 2018–2019: Submission and evaluation of proposal.
- 2019–2020: Release of IMT-2020 based radio specifications.

In line with its schedule, the ITU-R published/approved minimum technical performance criteria for IMT-2020 candidate radio interface technologies in 2017 [56,77]. It defines eight key "Capabilities for IMT-2020," which forms a basis for the 13 technical performance requirements presented in Table 9.1. The same recommendation also acknowledges that these key capabilities will have different relevance and applicability for the various usage scenarios of IMT-2020.

In comparison to IMT-Advanced (4G), the requirements are tougher and more diversified. For instance, downlink peak data is 1 Gbps in 4G while it is 20 Gbps in IMT-2020, recommended control plane latency is 100 ms in 4G while it is 20 ms in IMT-2020, and many other changes.

3GPP is expected to be the key SDO to answer ITU-R's call for proposals on IMT-2020. 3GPP is a well-known consortium of seven SDOs, namely, ARIB, ATIS, CCSA, ETSI, TSDSI, TTA, and

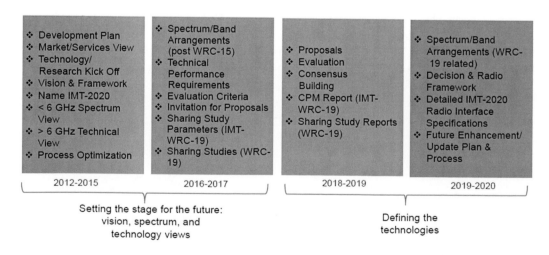

FIGURE 9.13 IMT-2020 standardization process. (From ITU 2016 Emerging Trends in 5G/IMT2020. Geneva Mission Briefing Series, September [76].)

TABLE 9.1
IMT-2020 (5G NR) Requirements

	Key Performance Indicator	Minimum Target	Usage/Applicability
1	Downlink peak data rate	20 Gbps	eMBB
	Uplink peak data rate	10 Gbps	
2	Downlink peak spectral efficiency	30 bits/s/Hz	eMBB
	Uplink peak spectral efficiency	15 bits/s/Hz	
3	User experience data rate	No specific target	
4	Downlink 5% percentile user spectral efficiency	0.3/0.225/0.12 bit/s/Hz	eMBB
		0.21/0.15/0.045 bit/s/Hz	
	Uplink 5% percentile user spectral efficiency	(indoor/dense urban/rural)	
5	Downlink average spectral efficiency	9/7.8/3.3 bit/s/Hz/TRxP	eMBB
	Uplink average spectral efficiency	6.75/5.4/1.6 bit/s/Hz/TRxP	
		(indoor/dense urban/rural)	
		TRxP: Transmission point	
6	Area traffic capacity (downlink)	10 Mbps/m^2 (indoor hotspot)	eMBB
7	Latency	4 ms	eMBB
	User plane latency (one way)	1 ms	URLLC
	Control plane latency	20 ms	eMBB and URLLC
8	Connection density (i.e., total number of devices fulfilling a specific QoS per unit area)	1,000,000 devices per km^2	mMTC
9	Energy efficiency	No specific target (radio technology may support high sleep ratio and long sleep duration)	
10	Reliability (i.e., relates to the capability of transmitting a given amount of traffic within a predetermined time duration with the high success probability)	1–10^5	URLLC
11	Mobility	Up to 500 km/h (for high speed trains)	eMBB
		1.5 bit/s/Hz – 10 km/h	
		1.12 bit/s/Hz – 30 km/h	
		0.8 bit/s/Hz – 120 km/h	
		0.45 bit/s/Hz – 500 km/h	
12	Mobility interruption time (i.e., the shortest time duration supported by the system during which a device cannot exchange user plane packets with any base station during transitions)	0 ms	eMBB and URLLC
13	Bandwidth (i.e., maximum aggregated system bandwidth)	100 MHz (up to 1 GHz bandwidth for above 6 GHz)	

Source: ITU-R 2017 Draft New Report ITU-R M. [IMT-2020. Tech Perf Req] Minimum Requirements Related to Technical Performance for IMT-2020 Radio Interface(s) [56].

TTC, known as Organizational Partners and a few industry forums. 3GPP has over 500 members (commercial organizations as well as academia) associated with one of these organizational partners. So, in a nutshell, all the key stakeholders of ICT as well representatives from the transportation, energy and other sectors are present in 3GPP, working to improve/innovate the telecom landscape.

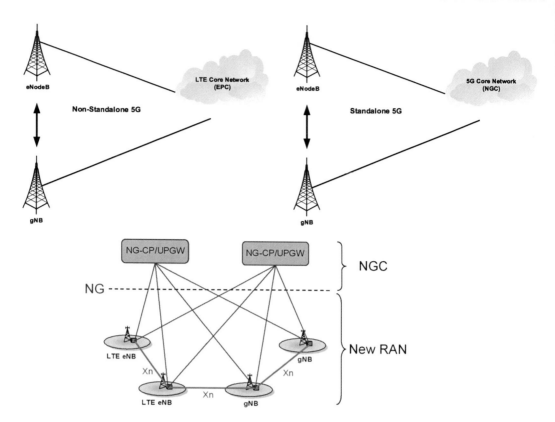

FIGURE 9.14 Overall layer 2 structure for NR. (From 3GPP TR 38.912 (V14.0.0) 2017 Study on New Radio (NR) Access Technology. Technical Specification (Release 14), Technical Specification Group Radio Access Network, 3GPP, March [78].)

9.4.2 3GPP PHASES FOR IMT-2020

3GPP is working to introduce NR access technology to address the requirements set by ITU-R for IMT-2020. 3GPP has divided this work into two phases, where phase 1 may introduce the NR while retaining the LTE core network, that is, the EPC and is called non-standalone. In this type of architecture, UP (user plane) traffic is either routed from the NR base station, called gNB, through eNB to the EPC or routed directly from gNB to the EPC. In the second phase, the 5G Core Network architecture termed as NGC is established to which NR can connect directly. Also, LTE eNB can also connect to NGC and may not require EPC [67,78]. The high-level architecture of these phases is shown in Figure 9.14.

This two-phased program started in Rel-14 followed by Rel-15 (phase 1) and Rel-16 (phase 2) with the goal to have everything in line with IMT-2020 requirements and to be completed by the ITU-R deadline. Rel-14, which was completed in June 2017, focuses on both RAN technologies and new System Architecture (SA) aspects. It either started or completed studies on certain aspects of NR including overall RAN requirements, technologies, and potential solutions. Rel-15 will provide specifications for the phase-1 of 5G, that is, non-standalone NR, which is likely to emphasize eMBB and sub-40 GHz operation and is scheduled for completion in September 2018. Rel-16 will design specifications for phase-2, that is standalone NR, which is expected to be finalized by 2020. The phase-2 specifications are expected to fully comply with the ITU-R's IMT-2020 requirements enabling standard-based 5G ready for deployment in a year or so.

9.4.3 NR Air-Interface

The critical element of 5G is 5G NR air-interface which is required to enable the three usage scenarios specified by ITU-R. Thus, it will not only enhance mobile broadband, but may need to deliver massive IoT and mission-critical services. The features such as waveforms, multiple access techniques, physical layer, layer 2, and RRC are expected to get formalized in Rel-15. Rel-16 and beyond may introduce support for frequency bands above 40 GHz and NGC.

NR is under investigation and development thus a number of issues are still not fully resolved. The focus of this section is to present certain aspects including some challenges of NR as highlighted by 3GPP up through 2017.

9.4.3.1 NR Numerology and Frame Structure

The physical layer of NR supports multiple numerologies where numerology is defined by subcarrier spacing and cyclic prefix (CP) overhead. This is in contrast to LTE that supports a fixed numerology of 15 kHz. 5G NR introduces scalable OFDM numerology supporting subcarrier spacing from 15 kHz to 480 kHz. Similarly, the TTI, which refers to the duration of a transmission on the radio link, is still applicable in NR. As stated in [78], the TTI duration corresponds to a number of consecutive symbols in the time domain in one transmission direction. LTE supports a fixed TTI duration of 1 ms and there is ongoing research to have TTIs of various durations to support lower latency requirements of 5G NR. Research is also underway to introduce service ware TTI. For example, a high QoS eMBB service may utilize a 500 μs TTI while some other latency sensitive service further shortens it to 140 μs.

Additional key attributes of NR numerology are as follows:

- Maximum channel bandwidth per carrier is 400 MHz in Rel-15.
- Numerology selection is independent of frequency band; however, very low subcarrier spacing may not be suitable for high carrier frequencies.
- Maximum number of subcarriers per NR carrier is 3300 or 6600 in Rel-15.
- Subframe duration is fixed to 1 ms and frame length is 10 ms.
- Subcarrier spacing is varied from 15 kHz to 480 kHz defined by 15 kHz * 2^n (n is a non-negative integer).
- Number of subcarriers per PRB (physical resource block) is 12.
- Physical layer supports extended CP (duration of 16.67 microseconds) and there will be only one in the given subcarrier spacing in Rel-15. In Rel-15, the LTE scaled extended CP (cyclic prefix) is supported, at least for 60 kHz subcarrier spacing.
- No explicit DC (direct current) subcarrier is reserved both for downlink and uplink.
- A slot consists of 7 or 14 OFDM symbols for the same carrier spacing of up to 60 kHz and 14 OFDM symbols for the same carrier spacing higher than 60 kHz with normal CP.
- Slot aggregation is supported, that is, data transmission can be scheduled to span one or multiple slots.
- Mini slots with a length of 1 symbol above 6 GHz are supported.

9.4.3.2 NR Frequency Analysis

WRC-15 identified the 24.25 GHz to 86 GHz frequency range for IMT-2020/5G systems. The investigations are ongoing to find the right set of frequencies, suitable spectrum bandwidth, subcarrier spacing, and other RF parameters. The initial analysis at the 3GPP provides the following key findings:

- Frequency range was divided into three ranges, that is, 30 GHz, 45 GHz, and 70 GHz.
- A significant difference between LTE and NR is that several subcarrier spacings are defined for NR as compared to LTE that only supports 15 kHz. As stated above, the subcarrier

spacing varies from 15 kHz to 480 kHz for NR. The potential candidates for subcarrier spacing are as follows:

- <6 GHz: 15 kHz, 30 kHz and 60 kHz
- >6 GHz: 60 kHz, 120 kHz, and 240 kHz.

• The maximum channel bandwidth from the physical layer perspective is 400 MHz while from the RF perspective it is as follows:

- <6 GHz: 100 MHz to 200 MHz under investigation
- >6 GHz: 100 MHz to 1 GHz for under investigation.

• The maximum spectrum utilization of LTE is 90%. For example, the transmission bandwidth of a 20 MHz channel in LTE in 18 MHz. For NR, initial studies indicate that transmission bandwidth utilization can be above 90%. Hence NR is expected to provide better spectrum utilization than LTE.

9.4.3.3   PHY Layer: OFDM-Based Transmission

OFDM-based waveform is supported for at least up to 40 GHz for eMBB and URLLC services in downlink and uplink. The scalable OFDM numerology enables support for diverse frequency bands and deployment options as shown in Figure 9.15. The sample shown in this figure represents a potential scenario and by no means depicts a standardized solution.

In addition to CP-OFDM, DFT-S-OFDM (Discrete Fourier Transform Spread OFDM) is supported in uplink only for eMBB up to 40 GHz. CP-OFDM waveform supports single stream and multi-stream (i.e., MIMO) transmissions whereas DFT-S-OFDM supports only single stream transmissions. The decision on waveform selection rests with the network. It is mandatory for devices to support both types of waveforms. Additionally, synchronous scheduling-based orthogonal multiple access is supported for both downlink and uplink transmissions for eMBB. Uplink is required to support nonorthogonal multiple access for mMTC. Some additional aspects of the physical layer are provided in 3GPP TR 38.912 where all technical outcomes of study item NG NR Access technology are collected; this is illustrated in Table 9.2.

9.4.3.4   Layer 2 and RRC

The layer 2, as in LTE, consists of MAC, RLC, and PDCP sublayers with similar functions. A new sublayer, called the AS sublayer, is added to layer 2 of NR. The key functions of the AS sublayer are mapping of QoS flow to a data radio bearer (DRB) and marking of QoS flow ID for both downlink and uplink packets. On the receiving side, it performs packet delivery to the corresponding PDU

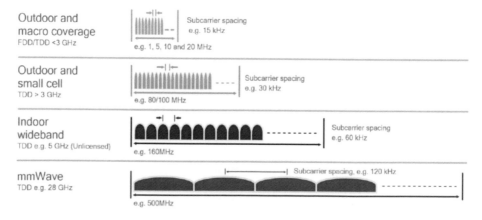

FIGURE 9.15   Example usage models, channel bandwidths, and subcarrier spacing. (From Qualcomm Technologies, Inc. 2016 Making 5G NR a reality [79].)

TABLE 9.2
NR Physical Layer Aspects

Attributes	Downlink	Uplink
Modulation	QPSK, 16-QAM, 64-QAM and 256-QAM modulation with the same constellation as in LTE	QPSK, 16-QAM, 64-QAM and 256-QAM modulation with the same constellation as in LTE 0.5 pi-BPSK[a] is also supported for DFT-S-OFDM in uplink
Physical channels	Physical downlink shared channel, PDSCH Physical broadcast channel, PBCH Physical downlink control channel, PDCCH	Physical uplink shared channel, PUSCH Physical uplink control channel, PUCCH Physical random access channel, PRACH
Physical signals	Demodulation reference signals, DM-RS Channel-state information reference signal, CSI-RS Primary synchronization signal, PSS Secondary synchronization signal, SSS	Demodulation reference signals, DM-RS Sounding reference signal, SRS
Multiplexing	Multiplexing different numerologies within the same NR carrier bandwidth supported Multiplexing of transmissions with different latency and/or reliability requirements for eMBB/URLLC in DL is supported by using the same subcarrier spacing with the same CP overhead or using different subcarrier spacing Dynamic resource sharing between different latency and/or reliability requirements for eMBB/URLLC in DL supported	Multiplexing different numerologies within the same NR carrier bandwidth supported

Source: From 3GPP TR 38.912 (V14.0.0) 2017 Study on New Radio (NR) Access Technology. Technical Specification (Release 14), Technical Specification Group Radio Access Network, 3GPP, March [78].

[a] Binary phase-shift keying (BPSK).

tunnel/session according to QoS flow ID. The overall L2 structure along with key functions of each sublayer is shown in Figure 9.16 [78].

The functions of the RRC layer under NR are also similar to RRC. Some key functions include establishment, configuration, maintenance, release of signaling radio bearers and data radio bearers, and release of RRC connection between the UE and NR RAN. RRC also supports certain mobility and QoS management functions. Details can be found in [78].

9.4.4 5G RAN Architecture

5G RAN will consist of logical nodes, namely, gNBs and eNBs, that are connected to each other with an Xn interface and toward the NGC by means of an NG interface as shown in Figure 9.14.

The Xn interface connects two gNBs or two eNBs or one gNB with an eNB. The Xn control plane interface (Xn-C) and Xn user plane interface (Xn-U) are defined between two NEW RAN nodes for respective purposes. The NG interface supports signalling between gNB/eNB and NGC, control and user planes' separation, and many other functions. This interface is divided into control and user planes' interfaces, namely, NG control plane interface (NG-C) and NG user plane interface (NG-U). The former is defined between NR gNB/eNB and NG-CPGW (NG control plane gateway) while the latter is identified between gNB/eNB and NG-UPGW (NG user plane gateway). The NG interface supports one-to-many relationships between NGC and new RAN nodes.

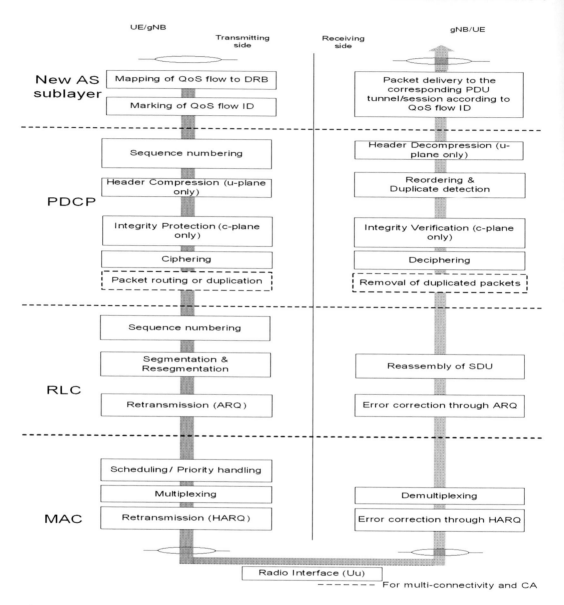

FIGURE 9.16 Overall layer 2 structure. (From 3GPP TR 38.912 (V14.0.0) 2017 Study on New Radio (NR) access technology. Technical Specification (Release 14), Technical Specification Group Radio Access Network, 3GPP, March [78].)

The protocol stack of these interfaces is shown in Figure 9.17. The SCTP (Stream Control Transmission Protocol) layer sits on top of the IP/Transport layer, providing guaranteed delivery of application layer messages in the control plane. In the user plane, GTP-U (GPRS tunneling protocol user plane) provides nonguaranteed delivery of PDUs between the respective network elements. Lastly, the application layer signaling protocol is defined as Xn-AP (Xn Application Protocol).

9.5   CONCLUSION

This chapter discussed LTE-Advanced, LTE-Advanced Pro, and 5G NR radio access technologies. Details on various features of LTE-Advanced and LTE-Advanced Pro were provided. It also looked into the first phase of 5G (non-standalone version) as well as the second phase (standalone version)

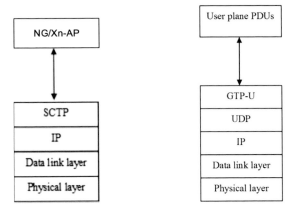

FIGURE 9.17 Interfaces' protocol stack. (From 3GPP TR 38.912 (V14.0.0) 2017 Study on New Radio (NR) access technology. Technical Specification (Release 14), Technical Specification Group Radio Access Network, 3GPP, March [78].)

which are expected to be standardized by 3GPP in 2018 and in 2020, respectively. ITU is expected to endorse the 5G standard in 2020. IMT-2020 requirements, 5G NR air-interface, and 5G RAN architecture were also discussed.

PROBLEMS

1. What are 3GPP Releases?
2. What is LTE-Advanced?
3. What are the possible scenarios of Carrier Aggregation?
4. What are the key challenges associated with Carrier Aggregation?
5. What is beamforming?
6. What is the key difference between LTE and LTE-Advanced MIMO?
7. How many transmission modes of MIMO are defined?
8. What are Relay Nodes?
9. What are Type 1 and Type 2 relays?
10. What are L1, L2, and L3 relays?
11. How can interference be avoided in relaying?
12. What is SON?
13. Which key features of SON are defined by 3GPP?
14. Define eMBMS?
15. Define user plane and control plane latency?
16. What is eICIC?
17. Define CoMP and associated gains?
18. What is LTE-Advanced Pro?
19. How is LTE utilizing the unlicensed spectrum?
20. What is Full Dimensional MIMO?
21. What is IMT-2020?
22. What are the three usage scenarios of IMT-2020?
23. Describe the key differences between IMT-Advanced and IMT-2020?
24. What is non-standalone and standalone 5G NR (New Radio)?
25. Define NR numerology and frame structure?
26. Describe OFDM based transmission?
27. Describe 5G RAN Architecture?

REFERENCES

1. Global Mobile Suppliers Association 2017. LTE-Advanced—LTE-Advanced & LTE-Advanced Pro Networks. https://gsacom.com/paper/lte-advanced-lte-advanced-pro-networks/
2. 3GPP 2017. About 3GPP Home. http://www.3gpp.org/about-3gpp/about-3gpp
3. Asif, S. 2007. *Wireless Communications Evolution to 3G and Beyond*. Artech House, USA.
4. 3GPP TS 36.300 (V11.3.0) 2012. Evolved Universal Terrestrial Radio Access (E-UTRA) and Evolved Universal Terrestrial Radio Access Network (E-UTRAN); Overall Description; Stage 2. Technical Specification (Release 11), Technical Specification Group Radio Access Network, 3GPP, September.
5. 5G Americas 2017. LTE to 5G: Cellular and Broadband Innovation. Rysavy Research, LLC.
6. ETSI Mobile Competence Center 2004. Overview of 3GPP Release 99—Summary of all Release 99 Features.
7. ETSI Mobile Competence Center 2004. Overview of 3GPP Release 4—Summary of all Release 4 Features.
8. ETSI Mobile Competence Center 2003. Overview of 3GPP Release 5—Summary of all Release 5 Feature.
9. ETSI Mobile Competence Center 2006. Overview of 3GPP Release 6—Summary of all Release 6 Features.
10. ETSI Mobile Competence Center 2007. Overview of 3GPP Release 7—Summary of all Release 7 Features Incomplete Draft for TSG#37.
11. ETSI Mobile Competence Center 2014. Overview of 3GPP Release 8, V0.3.3, September.
12. ETSI Mobile Competence Center 2014. Overview of 3GPP Release 9, V0.3.4, September.
13. ETSI Mobile Competence Center 2014. Overview of 3GPP Release 10, V0.2.1, June.
14. ETSI Mobile Competence Center 2014. Overview of 3GPP Release 11, V0.2.0, September.
15. ETSI Mobile Competence Center 2015. Overview of 3GPP Release 12, V0.2.0, September.
16. 3GPP 2015. Release 13 Analytical View Version, RP-151569, September.
17. ETSI Mobile Competence Center 2014. Overview of 3GPP Release 14, V0.0.1, September.
18. 3GPP 2017. Release 9. http://www.3gpp.org/specifications/releases/71-release-9
19. Nakamura, T. 2013. LTE Release 12 and Beyond. LTE Africa 2013, 3GPP.
20. 5G Americas 2015. Mobile Broadband Evolution towards 5G: Rel-12 & Rel-13 and Beyond.
21. Asif, S. 2011. *Next Generation Mobile Communications Ecosystem: Technology Management for Mobile Communications*. Wiley Inc., UK.
22. Jeanette, W. 2013. LTE-Advanced. Available via 3GPP. http://www.3gpp.org/technologies/keywords-acronyms/97-lte-advanced
23. 4G Americas 2012. 4G Mobile Broadband Evolution: Release 10, Release 11 and Beyond—HSPA+, SAE/LTE and LTE-Advanced.
24. 4G Americas 2014. Beyond LTE: Enabling the Mobile Broadband Explosion. Rysavy Research, LLC.
25. 4G Americas 2012. Mobile Broadband Evolution: The 3GPP Wireless Evolution. Rysavy Research, LLC.
26. 3GPP TR 36.912 (V14.0.0) 2014. *Feasibility Study for Further Advancements for E-UTRA (LTE-Advanced)*. Technical Report (Release 11), Technical Specification Group Radio Access Network, 3GPP, March.
27. Zhang, H. et al. 2012. Analysis and Application of Carrier Aggregation Technology in Wireless Communications. *2012 National Conference on Information and Computer Science (CITCS)*, Lanzhou, China, November 16–18, 2012, pp. 303–307.
28. Rohde & Schwarz. LTE-Advanced Carrier Aggregation. https://www.rohde-schwarz.com/us/solutions/wireless-communications/lte/in-focus/lte_advanced_carrier_aggregation_73018.html?rusprivacypolicy=1
29. Miller, L. 2016. *Carrier Aggregation Fundamentals for Dummies*, Qorvo Special Edition. John Wiley & Sons, Inc., USA.
30. Mobile Society Type Pad 2015. Moving Beyond the Aggregation of 5 LTE Carriers. http://mobilesociety.typepad.com/mobile_life/2015/04/moving-beyond-the-aggregation-of-5-lte-carriers.html
31. Bharmi, A., Hooli, K., and Lunttila, T. 2016. Massive Carrier Aggregation in LTE-Advanced Pro: Impact on Uplink Control Information and Corresponding Enhancements. *IEEE Communications Magazine*, 54(5):92–97, May.
32. Nokia Corporation, NTT DoCoMo Inc., Nokia Networks 2014. RP-142286 LTE Carrier Aggregation Enhancement Beyond 5 Carriers. *3GPP TSG RAN Meeting #66*, Maui, Hawaii (US), December 8–11, 2014.
33. Asif, S.Z. 2004. Mobile Receive Diversity Technology Improves 3G Systems Capacity. *IEEE Radio and Wireless Conference*, Atlanta, GA, USA, September 19–22, 2004.
34. Rumney, M. 2013. *LTE and the Evolution to 4G Wireless: Design and Measurement Challenges*, 2nd Edition. Wiley Inc., UK.

35. Keysight Technologies, Inc. Antenna Ports and Transmit-Receive Paths (LTE). http://rfmw.em.keysight.com/wireless/helpfiles/89600b/webhelp/subsystems/lte/content/lte_antenna_paths_ports_explanation.htm
36. The 3G4G Blog 2012. LTE 'Antenna Ports' and their Physical Mapping. http://blog.3g4g.co.uk/2012/05/lte-antenna-ports-and-their-physical.html
37. Schulz, B. 2015. *LTE Transmission Modes and Beamforming*. Rohde & Schwarz.
38. 4G Americas 2013. MIMO and Smart Antennas for Mobile Systems.
39. Cover, T.M. and Gamal, A.E. 1979. Capacity Theorems for the Relay Channel. *IEEE Transactions on Information Theory*, 25(5):572–584.
40. Qualcomm 2014. LTE Advanced—Evolving and Expanding in to New Frontiers, August.
41. Qian, Y. et al. 2012. Cooperative Communications for Wireless Network: Techniques and Applications in LTE-Advanced Systems. *IEEE Wireless Communications*, 19(2).
42. Iwamura, M., Takahashi, H. and Nagata, S. 2010. Relay Technology in LTE-Advanced. *NTT Docomo Technical Journal*, 12(2):29–36.
43. 3GPP TR 36.814 (V9.0.0) 2010. Evolved Universal Terrestrial Radio Access (E-UTRA); Further Advancements for E-UTRA Physical Layer Aspects. Technical Report (Release 9), Technical Specification Group Radio Access Network, 3GPP, March.
44. Asif, S.Z. et al. 2009. Radio Frequency Repeater with Automated Block/Channel Selection. United States Patent 7,480,485.
45. Asif, S.Z. et al. 2007. Radio Frequency Repeater with Automated Block/Channel Selection. United States Patent 7,299,005.
46. Asif, S.Z. et al. 2009. Wireless Repeater and Method for Managing Air Interface Communications. United States Patent 7,480,486.
47. Yang, Y. et al. 2009. Relay Technologies for WiMAX and LTE-Advanced Mobile Systems. *IEEE Communications Magazine*, 47(10):100–105.
48. Jamal, T., Mendes, P. and Zuquete, A. 2012. Wireless Cooperative Relaying Based on Opportunistic Relay Selection. *International Journal on Advances in Networks and Services*, 5(1&2):116–128.
49. Nam, S., Vu, M. and Tarokh, V. 2008. Relay Selection Methods for Wireless Cooperative Communications. *42nd Annual Conference on Information Sciences and Systems (CISS)*, March 19–21, 2008, Princeton University, NJ, USA.
50. Marchenko, N., Yanmaz, E., Adam, H. and Bettstetter, C. 2009. Selecting a Spatially Efficient Cooperative Relay. *IEEE Global Telecommunications Conference*, Hawaii, USA, November 30–December 4, 2009.
51. Hu, C. et al. 2008. Network Coding in Cooperative Relay Networks. *IEEE 19th International Symposium on Personal, Indoor and Mobile Radio Communications (PIMRC)*, Cannes, France, September 15–18, 2008.
52. Nohrborg, M. SON Self—Organizing Networks. 3GPP. http://www.3gpp.org/technologies/keywords-acronyms/105-son
53. Rhode & Schwarz, SON—Self-Organizing Networks. https://www.rohde-schwarz.com/us/solutions/wireless-communications/lte/in-focus/self-organizing-networks_229096.html
54. 3GPP TR 37.320 (V11.1.0) 2012. Universal Terrestrial Radio Access (UTRA) and Evolved Universal Terrestrial Radio Access (E-UTRA); Radio Measurement Collection for Minimization of Drive Tests (MDT); Overall description; Stage 2. Technical Report (Release 11), Technical Specification Group Radio Access Network, 3GPP, September.
55. ITU-R 2008. Report ITU-R M.2134 Requirements Related to Technical Performance for IMT-Advanced Radio Interface(s).
56. ITU-R 2017. Draft New Report ITU-R M. [IMT-2020.Tech Perf Req] Minimum Requirements Related to Technical Performance for IMT-2020 Radio Interface(s).
57. Roessler, A. and Kottkamp, M. 2013. LTE-Advanced (3GPP Rel.11) Technology Introduction (1MA232_1E), Rohde & Schwarz.
58. 4G Americas 2013. Mobile Broadband Evolution, The 3GPP Wireless Evolution. Rysavy Research, LLC.
59. Mobile Society Type Pad 2012. What's the Difference between LTE ICIC and LTE-Advanced eICIC? http://mobilesociety.typepad.com/mobile_life/2012/03/whats-the-difference-between-lte-icic-and-lte-advanced-eicic.html
60. Qualcomm Incorporated 2011. LTE Advanced: Heterogeneous Networks.
61. Anpalagan, A., Bennis, M. and Vannithamby, R. 2016. *Design and Deployment of Small Cell Networks*. Cambridge University Press, Cambridge, UK.
62. 3GPP TSG-RAN WG1 #66bis, R1-113566 2011. ICIC Evaluations for Different Handover Biases, Qualcomm Inc., October 10–14, 2011, Zhuhai, China.
63. 3GPP TR 36.819 V11.1.0 2012. Coordinated Multi-Point Operation for LTE Physical Layer Aspects. Technical Report (Release 11), Technical Specification Group Radio Access Network, 3GPP.

64. 4G Americas 2012. 4G Mobile Broadband Evolution: 3GPP Release 10 and Beyond. HSPA+, SAE/LTE, and LTE-Advanced.
65. 3GPP 2015. LTE-Advanced Pro Ready to Go. http://www.3gpp.org/news-events/3gpp-news/1745-lte-adv anced_pro
66. RF Wireless World. LTE Advanced Vs LTE Advanced Pro-Difference between LTE Advanced and LTE Advanced Pro. http://www.rfwireless-world.com/Terminology/LTE-Advanced-vs-LTE-Advanced-Pro.html
67. 5G Americas 2017. Wireless Technology Evolution Towards 5G: 3GPP Release 13 to Release 15 and Beyond.
68. Nokia Networks 2015. LTE-Advanced Pro—Pushing LTE Capabilities Towards 5G.
69. Qualcomm Technologies, Inc. 2017. The Essential Role of Gigabit LTE & LTE Advanced Pro in a 5G World.
70. 3GPP TR 36.889 (V13.0.0) 2015. Study on Licensed-Assisted Access to Unlicensed Spectrum. Technical Report (Release 13), Technical Specification Group Radio Access Network, 3GPP, June.
71. Qualcomm Technologies, Inc. 2016. Progress on LAA and its relationship to LTE-U and MulteFire.
72. Zhang, C. et al. 2014. *Full-Dimension MIMO: Status and Challenges in Design and Implementation.* Samsung.
73. 3GPP TR 36.897 (V13.0.0) 2015. Study on Elevation Beamforming / Full-Dimension (FD) Multiple Input Multiple Output (MIMO) for LTE. Technical Report (Release 13), Technical Specification Group Radio Access Network, 3GPP, June.
74. ITU-R 2015. RESOLUTION ITU-R 56-2—Naming for International Mobile Telecommunications.
75. ITU-R 2015. Recommendation ITU-R M.2083-0—IMT Vision—Framework and Overall Objectives of the Future Development of IMT for 2020 and Beyond.
76. ITU 2016. Emerging Trends in 5G/IMT2020. Geneva Mission Briefing Series, September.
77. ITU 2017. Press Release: ITU Agrees on Key 5G Performance Requirements for IMT-2020. http://www. itu.int/en/mediacentre/Pages/2017-PR04.aspx
78. 3GPP TR 38.912 (V14.0.0) 2017. Study on New Radio (NR) Access Technology. Technical Specification (Release 14), Technical Specification Group Radio Access Network, 3GPP, March.
79. Qualcomm Technologies, Inc. 2016. Making 5G NR a Reality.

10 Transport Network

5G will bring a number of connectivity options for the end users and devices. The connectivity between users, between things or between users and things is provided through a transmission network. A transport or transmission network, as the name suggests, is responsible for the delivery of all types of traffic from source to destination in mobile networks.

The transport networks can be broken down into three major areas, namely backhaul, metro, and core transport. This bifurcation is present in 2G, 3G, LTE, and 4G networks and is expected for 5G as well. There is no standardized definition of these domains; however, these are widely used and their high level conceptualization view is shown in Figure 10.1. The following definitions can be considered for these domains [1].

Backhaul: the link from the base stations to the first traffic aggregation location (or transmission hub) is called mobile backhaul.

Metro Transport: the connection from the first aggregation hub to the core node of cellular networks is considered as metro transport (or intra city connectivity). The core node can house equipment such as MSC, MGW, SGSN, MME, and Gateways which are described in Chapter 11.

Core Transport: the core nodes which are not in the same city can be connected through core transport systems. The intercity connections, connectivity to neighboring countries via terrestrial means, and long-distance locations through the sea fall into the domain of core transport.

10.1 5G TRANSPORT REQUIREMENTS

The identification of 5G traffic capacity requirements is a prerequisite for the development of transmission networks. This capacity identification will lead to the development of transmission technologies and standards needed to fulfill such requirements. It is well established that 5G will bring tremendous traffic demands and various types of traffic (eMBB, URLLC, and mMTC),* thus a one size fits all policy will not work. The traffic demands will be different for different networks in a country, within networks, and across nations. This section will provide high level estimations of the 5G traffic demands that transmission systems will need to carry. The estimations are calculated using LTE as a baseline.

The connectivity from the access node (base station) to the core node has been a big challenge since the inception of mobile data networks. The data capacity of such links will be further aggravated in 5G as compared to predecessor mobile generations due to the abundance of small cells, telemetry operations, and other connectivity options which have not been conceived so far. Figure 10.2 shows a typical mobile network deployment scenario where the traffic is summed up at each aggregation level before reaching the core node. The aggregation levels are defined on the basis of certain geographical parameters (boundaries), for example, aggregation node 1 (level 1) may combine the traffic of certain base stations and small cells within a local vicinity. Level 2 may add up the traffic of some level 1 locations, and perhaps the suburban level and level 3 at the city level will accumulate the traffic of level 2 hubs before sending the traffic to the mobile/core node. From the core node, the traffic can be routed to other cities within the country and also globally.

Today, the transmission of traffic in LTE networks is configured very similarly if not exactly as the one described above. This configuration is expected to continue in 5G networks as well. It is

---

* eMBB: enhanced Mobile Broadband; URLLC: Ultra-Reliable and Low-Latency Communications; mMTC: massive Machine Type Communications

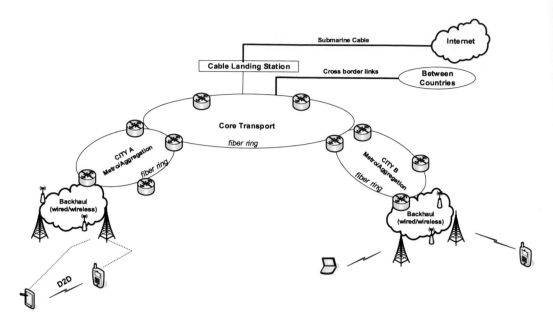

FIGURE 10.1   Transport network.

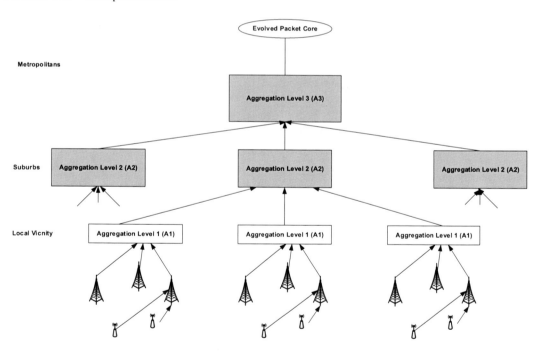

FIGURE 10.2   Transmission dimensioning topology.

common that each macro LTE base station (i.e., eNode B) or cell site consists of three sectors where the total base station throughput is the sum of all individual sector throughputs that are controlled by that specific eNode B. The macro cell sites may also support small cells, so in that case, throughput will be even greater. In hindsight, the traffic output of such individual cell sites accumulates at an aggregation point. This connection from the cell site to the first aggregation point as stated earlier is called as backhaul. Furthermore, aggregation levels are identified with A1, A2, and A3 labels and where A2 and A3 aggregate the traffic of their predecessor hubs.

The challenge has been and will be to correctly predict the traffic throughput levels for today and for at the least next few years. After that, the challenge is to find the right mix of wired and wireless transmission technologies and solutions to handle the traffic. Hundreds of studies if not thousands have been made to determine the LTE cell site throughput and transmission capacities, and results have been successfully implemented. Table 10.1 represents the traffic requirements for the LTE network shown in Figure 10.2 for comparison with 5G. Still using this model, it is expected that operators need to plan for at least a hundred times more capacity than today in the early years of 5G.

This configuration requires identification of peak data rates and particularly the data throughput of the cell sites and a number of others parameters that are described in this section.

The cell site throughput can be found by multiplying the total frequency spectrum with spectral efficiency (bps per Hertz). The throughput of any radio base station heavily depends on three factors, namely spectrum bandwidth, modulation scheme, and antenna configuration, and it also varies during the quiet and busy times in the network. During busy times, many users/devices will be served by each sector and their individual spectral efficiencies will vary depending on the quality of their radio links. Since some users will have better connections than others and at the same time not all users will have bad radio links, the cell site throughput will average out. However, in quiet times, perhaps only one user/device will be served by the entire cell site and thus the spectral efficiency and throughput will entirely depend on that user/device. If the user/device has a good link, it can use the entire frequency spectrum. This is the condition which represents the media headline figures for peak data rates. Thus, the two key factors setting the required backhaul link capacity are peak data rate and average throughput. It may be kept in mind that dimensioning at peak rates could lead to the backhaul network being largely empty, but it would ensure the advertised speed. On the other hand, planning backhaul at average throughput will be more cost effective; however, with this, the peak speed cannot be guaranteed [2–6].

Furthermore, the backhaul traffic is comprised of additional components beside peak speed and average throughput which are shown in Figure 10.3. So far, S1 (eNode B to Core) user plane traffic is primarily discussed in the above sections. The other factors include user plane traffic on X2

TABLE 10.1
LTE versus 5G Transport Capacity Requirements

Network Attributes	LTE with Small Cells (year 2015–16)	5G with Small Cells (year 2020–21)
Configuration (MHz)	20 MHz	100 MHz
Modulation	Modulation schemes provided by standard	Modulation schemes provided by standard
Antenna configuration	2 × 2 MIMO	4 × 4 MIMO
Average throughput per tri-sector eNodeB (Mbps)	100	–
Average throughput per tri-sector 5G Base Station (Mbps)	–	1000
Average throughput per tri-sector eNodeB with 5 small cells (Mbps)	350	–
Average throughput per tri-sector 5G base station with 5 small cells (Mbps)	–	4000
Spectral efficiency (bps/Hz/per Base Station)	2.19	5
X2 Overhead (4%)	1.04	1.04
Transport overhead with IPsec (25%)	1.25	1.25
Overbooking factor	30%	30%
Estimated base station average throughput (Mbps)	455	5200
Estimated average throughput @ first aggregation level (Mbps)	4550	52,000
Estimated average throughput @ second aggregation level (Mbps)	18,200	208,000
Estimated average throughput @ core node (Mbps)	36,400	416,000

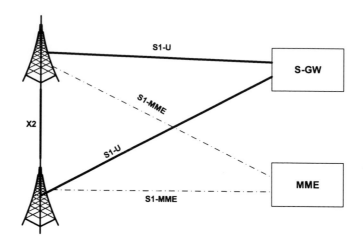

FIGURE 10.3   LTE with key interfaces.

(eNode B to eNode B) interface, control plane traffic on both S1 and X2, transport overhead, and overbooking methodology.

- The X2 interface supports the interconnection between eNode Bs by allowing signal information exchange between these entities along with forwarding user traffic. This interface predominantly carries user traffic forwarded during handovers between eNode Bs. Though the volume of X2 traffic is not significant and varies by single percentage points, it is an important interface that needs to be considered in the backhaul design.
- The control plane signaling on both S1 and X2 can be considered negligible in comparison to associated user plane traffic and can be ignored. The same is true for OAM (Operations, Administration, and Maintenance) and synchronization signalling.
- The transport overhead is associated with data encapsulation as the traffic is carried through EPC. The volume of this encapsulation also varies and could be set around 25% with and 15% without IPsec. IPSec is specified by 3GPP for data protection and safety [7,8].
- The LTE network deployments have been based on a parameter called the overbooking factor. The overbooking factor relates to the average number of subscribers that can share a given unit of channel. As the LTE devices are statistically distributed within the radio sectors and not expected to download at maximum peak rates all the time, backhaul capacity can be split and overbooked among individual sectors. The busier a macro cell is, the lower this overbooking factor needs to be. For low usage sites, the factor can be safely increased.

This chapter will primarily look into a handful of key emerging technologies for transport networks. The mobile backhaul is the key identified challenge for 5G networks. This chapter will focus on high capacity microwave radios or e-band radios and popular layer 2 MPLS-TP (Multi Protocol Label Switching-Transport Profile) technology. A case study of e-band radios for Pakistan is also presented. Finally, options for the convergence of IP and optical along with challenges and enablers will be presented.

10.2   MOBILE BACKHAUL

Backhaul is a challenging area for any current mobile network and has been also identified as a major area of development for 5G networks. To handle 1000s of Mbps of capacity per 5G cell site as shown in studies and indicated in Table 10.1, it is clear that the telecom sector needs to look beyond today's traditional microwave radios.

These traditional microwave radios, which operate from 6 to 42 GHz, cannot handle gigabits of capacity as required by 5G networks. These lower fixed service bands have been divided into fixed channel rasters that are normally 56 or less MHz wide and capable of carrying only a few 100 Mbps per modem, making them unsuitable for 5G networks. It should also be noted that these traditional microwave frequency bands are becoming heavily congested and obtaining 56 MHz or 112 MHz bandwidth channels is becoming increasingly difficult.

Additionally, the penetration of fiber into the space of backhaul in current mobile networks is highly limited at the global level. This makes the case more complicated and challenging for future IMT or 5G networks. However, one possible solution is perhaps the use of millimeter wave bands operating in the range of 60–100 GHz. In such bands, the radio channel bandwidth can be much larger (e.g., 250 MHz) and can deliver significantly more capacity than traditional microwave bands. One such possibility, that is, e-band radios that operate in the 71–76 and 81–86 GHz spectrum, will be discussed in the next section.

10.3   E-BAND MICROWAVE RADIOS

Fiber cannot be laid down for every physical structure and for every cell site which would almost guarantee the best broadband experience. This means that a wireless medium would still be needed for backhauling 5G traffic which will amount to gigabits on a per cell site basis.

The microwave radios that operate in the above mentioned frequency band can be used to address the backhaul needs of 5G. This frequency spectrum is part of the millimeter band that starts at 30 GHz and ends at 300 GHz (ITU EHF [Extremely High Frequency]) and where the corresponding free space wavelengths range from 10 to 1 mm. The research on millimeter wave technology started in the 1890's and a lot of work was done through World War II [9]. The very famous curves showing average atmospheric absorption of millimeter waves were published by Rosenblum in 1961 [10]. These curves, as shown in Figure 10.4, have doubtlessly been used in scores of articles, reports, and books since their inception.

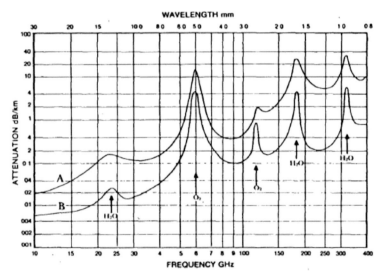

Fig. 2.   Average atmospheric absorption of millimeter waves (from [40]).

A: Sea level   B: 4 km
$T = 20°C$ $T = 0°C$
$P = 760$ mm $\rho_{H_2O} = 1$ g/m³
$\rho_{H_2O} = 7.5$ g/m³

FIGURE 10.4   Average atmospheric absorption of millimeter waves. (From Rosenblum, E.S. 1961. *Microwave Journal*, 4:91–96, Mar. 1961 [10].)

However, it was not until the beginning of the current century that the world started to see standardization activities on millimeter waves. On the regulatory front, the U.S. FCC allocated 71–76, 81–86 and 92–95 GHz frequencies for commercial use in 2003. During 2007, the United Kingdom's Ofcom also established rules to promote private sector development for these frequencies. The development of 70/80 GHz radios for mobile communications started in the 1990s and become a reality in the 2000s.

The rest of this section will look into technology, radio, and product aspects, link engineering, testing, application scenarios, and present a case study.

10.3.1 E-Band Radio Technology

The ECC (Electronic Communications Committee) within CEPT (the European Conference of Postal and Telecommunications Administrations) and ETSI took the lead in defining the frequency channel arrangements, equipment attributes, and other technical characteristics*. Particularly, the ECC Recommendation (05)07 [11] published in 2009 defines channel arrangements for fixed point-to-point systems operating in the stated frequency band. The 5 GHz band in each direction is divided into nineteen (19) channels. Each such channel is 250 MHz wide and two or more channels can be combined to form a single channel of larger size. This channel size is quite large as compared to the widely deployed microwave radio units which operate with 7–56 MHz wide channels. The specified channels can be used to form either TDD or FDD systems within the single band or in combination with the two bands, respectively.

The principle of using the channels from within bands 71–76 and 81–86 GHz in a single duplex FDD arrangement is described in Figure 10.5.

The alternative approach is channel aggregation where multiples of 250 MHz channels are aggregated into FDD channels with duplex separation equal to or more than 10 GHz as shown in Figure 10.6.

10.3.1.1 Radio Interface Capacity

The second technology element is radio interface capacity (RIC) which largely depends on the modulation scheme. The ETSI EN 302 217-3 [12] specifications come in handy for explaining the characteristics and requirements for point-to-point digital fixed radio systems operating in higher frequency bands (including the frequencies under discussion). The standard defines six different modulation states as system spectral efficiency classes of operation as shown in Table 10.2, however, there is no definite limit on capacity and modulation format as shown in [13].

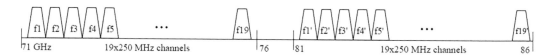

FIGURE 10.5 Combining the channels from 71–76/81–86 GHz bands into a single FDD arrangement with a duplex separation of 10 GHz. (From ECC Recommendation (05)07 2009. Radio Frequency Channel Arrangements for Fixed Service Systems Operating in the Bands 71–76 and 81–86 GHz, ECC with CEPT [11].)

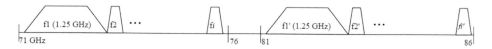

FIGURE 10.6 Example of aggregating multiple 250 MHz channels, possibly alongside original 250 MHz wide channels. (From ECC Recommendation (05)07 2009. Radio Frequency Channel Arrangements for Fixed Service Systems Operating in the Bands 71–76 and 81–86 GHz, ECC with CEPT [11].)

---

* ECC develops regulations for the effective use radio frequency spectrum in Europe, whereas ETSI develops standards for radio communication systems and equipment.

TABLE 10.2
System Spectral Efficiency Classes of Operation Used in This Clause

Spectral Efficiency Class	128 States Modulation	64 States Modulation	32 States Modulation	16 States Modulation	4 States Modulation	2 States Modulation
	5 (128)	5 (64)	4H	4L	2	1

Source: ETSI EN 302 217-3 (V1.3.1) 2009. Fixed Radio Systems; Characteristics and Requirements for Point-to-Point Equipment and Antennas; Part 3: Equipment Operating in Frequency Bands where Both Frequency Coordinated or Uncoordinated Deployment Might be Applied; Harmonized EN Covering the Essential Requirements of Article 3.2 of the R&TTE Directive, ETSI [12].

Note: Modulation format is only for reference (other modulation schemes could be used).

Table 10.3 shows the RIC values in one direction for such radios in the FDD mode. The TDD systems would match that capacity as the sum of the capacities in both directions. The channel raster specified 250 MHz slots to form channel widths from 250 MHz to 4.75 GHz. A maximum of 19 Gbps is possible in a 4.75 GHz channel with 128-QAM and with equipment compliant to the ETSI standard. Thus, if an operator has access to a 750 MHz bandwidth, it can carry a maximum of 3 Gbps of traffic in one direction of a point-to-point radio link.

TABLE 10.3
Typical RIC Values for the Spectral Efficiency Classes

Aggregate Channel (MHz)	Typical RIC Values (Mbit/s)					
	Class 5(128)	Class 5(64)	Class 4H	Class 4L	Class 2	Class 1
250	1000	900	750	600	300	150
500	2000	1800	1500	1200	600	300
750	3000	2700	2250	1800	900	450
1000	4000	3600	3000	2400	1200	600
1250	5000	4500	3750	3000	1500	750
1500	6000	5400	4500	3600	1800	900
1750	7000	6300	5250	4200	2100	1050
2000	8000	7200	6000	4800	2400	1200
2250	9000	8100	6750	5400	2700	1350
2500	10,000	9000	7500	6000	3000	1500
2750	11,000	9900	8250	6600	3300	1650
3000	12,000	10,800	9000	7200	3600	1800
3250	13,000	11,700	9750	7800	3900	1950
3500	14,000	12,600	10,500	8400	4200	2100
3750	15,000	13,500	11,250	9000	4500	2250
4000	16,000	14,400	12,000	9600	4800	2400
4250	17,000	15,300	12,750	10,200	5100	2550
4500	18,000	16,200	13,500	10,800	5400	2700
4750	19,000	17,100	14,250	11,400	5700	2850

Source: ETSI EN 302 217-3 (V1.3.1) 2009. Fixed Radio Systems; Characteristics and Requirements for Point-to-Point Equipment and Antennas; Part 3: Equipment Operating in Frequency Bands Where Both Frequency Coordinated or Uncoordinated Deployment Might be Applied; Harmonized EN Covering the Essential Requirements of Article 3.2 of the R&TTE Directive, ETSI [12].

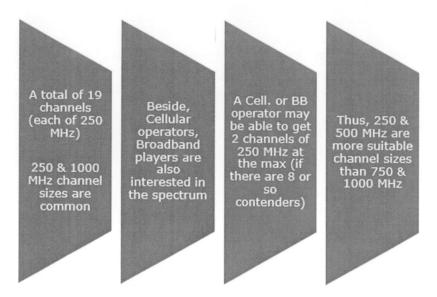

FIGURE 10.7   E-band technology channel bandwidths.

10.3.1.2   Radio Channel Spacing

A very key element that determines the performance of e-band radios is the channel size which is very large compared to traditional microwave systems. In the lower bands (6–42 GHz), the typical channel arrangements are in multiples of 3.5 MHz, that is, 14, 28, 56 and 112 MHz, as stated earlier. Thus, even with high modulation schemes, the radio link capacity per modem is less than 300 Mbps. The advantage, however, with smaller channel sizes is that many operators can be accommodated.

Contrarily, the minimum channel bandwidth of e-band radios is 250 MHz which could provide more than 150 Mbps of capacity. Until recently, only 250 and 1000 MHz channel sizes were available in the products, thus only a few operators (probably two or three cellular/broadband players per market) could get their hands on the spectrum (as shown in Figure 10.7). For example, a frequency assignment of 1000 MHz (i.e., an amalgamation of four 250 MHz channels) can be given to only four cellular/broadband players while one service provider can allow for three 250 MHz channels in a market, thus reducing the suitability of e-band radios. A cellular or broadband operator may be able to get 2 channels of 250 MHz at the maximum if there are 8 or so contenders. However, the capacity of a 250 MHz channel is less considering its size and in relative terms to traditional microwave radios.

Thus to address this key challenge, 250 and 500 MHz channel sizes started to appear in the market with better and higher capacities as claimed by some manufacturers. For example, manufacturers are claiming to support 2.5 Gbps in a 500 MHz channel with 64-QAM modulation. Thus, if it is proven, the industry will have a reasonable number of channels and capacity to address the demands of the market.

10.3.1.3   Other Key Parameters

The ETSI EN 302 217-3 standard has also defined receiver power density levels (RSL) for BER (bit error rate) $\leq 10^6$ and $\leq 10^8$ and cochannel and adjacent channel interference sensitivity. Some other key parameters include [12]:

- Maximum EIRP (Equivalent Isotropic Radiated Power)*: $\leq +85$ dBm (Decibel-milliwatt) for $G_{ant} \geq 55$ dBi (Decibel-isotropic)
- Minimum Antenna gain G_{ant}: 38 dBi
- Maximum transmitter output power: $\leq +35$ dBm (this is the maximum possible transmit power, including tolerances, delivered to the antenna connector)

---

* EIRP refers to the radiated power relative to an isotropic antenna for which transmit antenna gain is 1 [14].

TABLE 10.4
E-Band Link Licensing Sample

Country	E-Band License Structure	Typical E-Band License Fee
U.S.	On-light license	$75 for 10-year license
UK	Light license	£50 per year (~$80)
Australia	Light license	AU$187 per year (~$145)
Russia	Light license	Minimal registration fee
UAE	Traditional PTP	4500 Dirhams per year (~$1225)
Jordan	Traditional PTP	Dinar 200 per year (~$280)

Source: Wells, J. 2010. *Light Licensing.* E-Band Communications Corporation, California, USA [15].

10.3.1.4 Radio Licensing

E-band microwave radios can be licensed using light licensing techniques. These techniques reflect the ease of coordinating, registering, and licensing, and set license fees that cover administrative costs, but do not penalize the high data rates and bandwidths that are required for 5G services [15].

The assignments in e-band can be made on a link by link basis on the basis of first come first serve. The regulator can manage the database of the approved links on the public domain. New applicants can apply for assignments taking into account the existing assignments and protecting the earlier assignments. The regulator can arbitrate where cases of interference are identified. The traditional licensing fee based on the amount of data transmission or bandwidth usage will result in tariffs that can be extraordinarily high for such systems. Thus, light licensing presents a simple approach to tackle this challenge.

The U.S. was one of the first countries to open the e-band frequencies for commercial use. After three years of public consultation, the FCC in 2005 initiated a 'light licensing' scheme that enables links to be registered online in a few minutes. Link registration fees were set as low as $75 for a period of 10 years. Table 10.4 shows some examples of e-band license structure.

10.3.2 E-Band Product Aspects

A radio system is designed with methods and technologies that can generate the desired frequency with sufficient power and modulation methods to carry the desired information. A typical microwave radio consists of an IDU (indoor unit), ODU (outdoor unit), and antenna (dish). The ODU is connected to the IDU via IF (Intermediate Frequency) cable and also attached to the dish for connectivity with other radios. The ODU consists of baseband circuitry and modulator/demodulator while the IDU communicates with BTS (e.g., eNodeB). The typical distance between IDU and ODU can be in the range of 70–150 ft [16,17].

Contrary to traditional microwave units, an e-band radio does not have an IDU. The entire circuitry is in ODU which is attached to a 30 or 60 cm antenna (dish). The antennas have large gain as compared to lower frequency band antennas with similar sizes which helps in overcoming losses associated with rain. A power over Ethernet cable (one cable connection) runs from the unit via power injector* to the BTS (for example, eNodeB) as shown in Figure 10.8. The cable carries three sets of information, namely operations and monitoring information, Ethernet traffic, and power. Another possible setup is to run an optical cable from ODU to BTS for traffic and monitoring information and a separate DC power cable (−48V) from ODU to the power cabinet.

The ODU with all the interfaces, electronics, and required circuitry is enclosed in a weather-sealed chassis. The overall radio system consists of two radios that transmit to each other in a full duplex channel pair (in the case of FDD), providing point-to-point Ethernet connectivity between two locations.

---

* Inject power without affecting data.

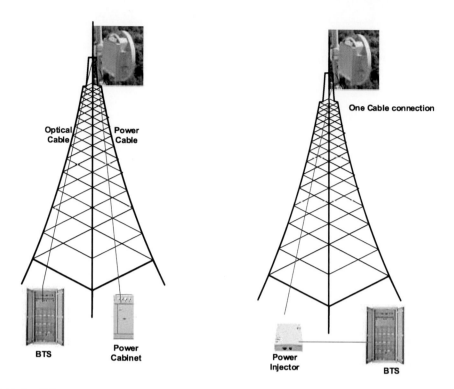

FIGURE 10.8   E-band radio setups. (From Asif, S.Z. 2015. *International Journal Wireless and Microwave Technologies*, 2015(4):37–46 [17].)

The narrow beams of millimeter wave links allow for deployment of multiple independent links in close proximity. For example, [18] shows that using an equivalent antenna, the beam width of a 70 GHz link is four times as narrow as that of an 18 GHz link, thus providing as much as 16 times the density of e-band millimeter wave links in a given area.

10.3.2.1   Product Specifications

The very first products came at the beginning of the current century. The key pioneers were three U.S. based suppliers, namely Bridgewave Communications, Gigabeam Corporation, and E-Band Communications. These radios were primarily supporting 1 Gbps capacity in a 1.0 GHz channel. These radios were expensive, tailored toward enterprise applications, and had few carrier class characteristics [19].

Later on, in the late 2000s, the tier-1 vendors like Alcatel-Lucent, Huawei, Nokia Siemens Networks, and NEC started to offer these radios under their own brands as resellers. The resale agreements with the above mentioned suppliers made these first generation products more expensive with the same features, but with more powerful branding.

Now, in the 2010s, Huawei, NEC, SIAE Microelettronica, and a handful of others have also started to produce their in-house radios. These second-generation products have been typically designed to handle the needs of mobile operators. They support higher capacity in excess of 1.0 Gbps and smaller channel bandwidths. The key performance attributes of these radios are shown in Table 10.5.

10.3.3   E-Band Radio Link Engineering

The e-band link engineering is similar to the designing of other microwave radio links except the engineers have to allow for additional losses. The key factors that contribute to the fading of a millimeter wave radio signal propagating through the atmosphere are free space path loss,

TABLE 10.5
E-Band Product Specifications

Attributes	First Generation	Second Generation Additional Attributes
Manufacturers	U.S. based Bridgewave Communications, Gigabeam Corporation and E-Band Communications. Tier-1 vendors as resellers	Huawei, NEC, and SiAE
Timeline	2000s	2010s
Channel spacing	250 and 1000 MHz	500 and 750 MHz
Modulation	BPSK, QPSK (4-QAM)	16 /64-QAM
Adaptive code and modulation	Limited	Good strength
Capacity	120, 240, 600, 1200 Mbps	2.5 Gbps (with 64-QAM and 500 MHz)
Spectral efficiency	1 bps/Hz	5 bps/Hz
Ethernet ring protection	Not available	ITU-T G.8032 v2 ethernet ring protection switching (<50 msec)

Following attributes more or less remain the same

Quality of service	Shaping/Scheduling/Congestion-Avoidance Techniques	
Synchronization	SyncE, IEEE 1588 v2	
OAM	802.1ag/802.3ah/Y.1731	
Protection	1 + 0, 1 + 1 HSB (hot standby)	
Antenna	Parabolic, 30 or 60 cm diameter, 42–52 dBi gain	

Source: Asif, S.Z. 2015. *International Journal Wireless and Microwave Technologies*, 2015(4):37–46; Bridgewave. 80GHz Wireless. http://www.bridgewave.com/products/80ghz.cfm; E-Band Communications, LCC. 4G Evolution Series. http://www.e-band.com/ProductOverview; Huawei. E-band Microwave. http://www.huawei.com/en/solutions/broader-smarter/hw-196711.htm; NEC. iPasolink EX. http://www.nec.com/en/global/prod/nw/pasolink/products/ipasolinkEX.html; SiAE. ALFOplus80 series. https://www.siaemic.com/index.php/products-services/telecomsystems/microwave-product-portfolio [17, 20–24].

atmospheric gaseous losses, and rain attenuation losses. For millimeter waves, there are standardized prediction methods for measuring attenuation due to hydrometeor precipitation (basically rain), clouds and fog, and atmospheric gases. The main recommendation for point to point links is ITU-R P.530 [25], which provides prediction methods required for the design of terrestrial line-of-sight systems. Another recommendation, ITU-R P.838-3 [26], provides a specific attenuation model for rain that can be used in prediction methods. Some additional details on link calculations can be found in [27,28].

10.3.3.1   Key Signal Loss Factors

10.3.3.1.1   *Free Space Loss*

The free space loss (FSL) is a frequency and distance dependent attenuation loss between two isotropic antennas. The isotropic antenna is a theoretical point source that radiates energy equally in all directions. The loss is given by the following Equation 10.1 [14]:

$$L_{FSL} = 92.4 + 20 \log f + 20 \log R \tag{10.1}$$

where f is the frequency in GHz and R is the line of sight range between transmit and receive antennas in km.

It can be seen from Equation 10.1 that radio systems operating at 80 GHz will have a much higher FSL as compared to 38 GHz microwave systems. For a distance of 2 km, the FSL at 38 GHz is 130 dB whereas at 80 GHz it is 136 dB, so there is an additional 6 dB loss. However, it has been shown by studies that the link distances at 38 and 80 GHz are pretty similar due to the use of large gain antennas in e-band systems.

10.3.3.1.2 Atmospheric Gaseous Losses

Signal attenuation occurs when millimeter waves traveling through the atmosphere are absorbed by molecules of oxygen, water vapor, and other gaseous atmospheric constituents. These losses are greater at certain frequencies, coinciding with the mechanical resonant frequencies of the gas molecules. Figure 10.9 shows qualitative data on gaseous losses. It shows several peaks that occur

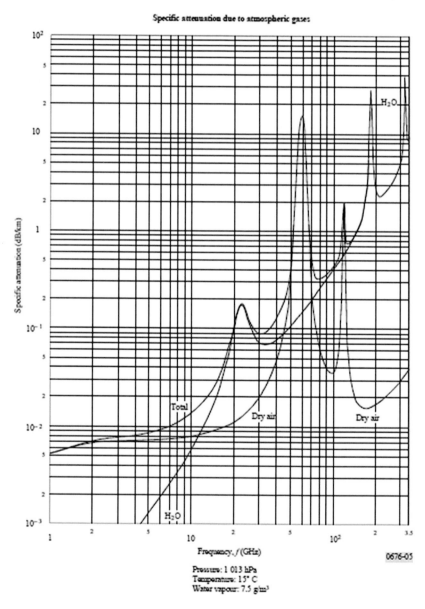

FIGURE 10.9 Specific attenuation due to atmospheric gases. (From ITU-R 2001. Recommendation ITU-R P.676-5 Attenuation by atmospheric gases [29].)

due to absorption of radio signals by water vapor (H_2O) and dry air. At these frequencies, absorption results in high signal attenuation, therefore, a short propagation distance. For example, it can also be witnessed that dry air absorption is at its peak at 60 GHz with an attenuation of 15 dB/km compared to less than 1 dB/km at 76 GHz.

10.3.3.1.3   Rain Attenuation Losses

Rain also impacts the performance of e-band radios. Raindrops are roughly the same size as the radio wavelengths, and therefore, cause scattering of the radio signal. Figure 10.10 shows that heavy rainfall at the rate of 25 mm/hour (1″ per hour) causes just over 10 dB/km attenuation at e-band frequencies. This attenuation increases to 30 dB/km with tropical rain which is about 100 mm/hour (4″ per hour) for the same set of frequencies.

Rain rates in all regions of the world have been measured and documented extensively. The rain rate maps, normally called ITU rain zones, refer to actual rain rates by using an alphabetical nomenclature. The world regions having the lowest rain rates are classified in rain zone A whereas the regions with highest rain rates are classified in rain zone Q in shown in Table 10.6. For example, a microwave radio system installed in rain zone K and/or E needs to overcome a rain rate of 42 mm/h

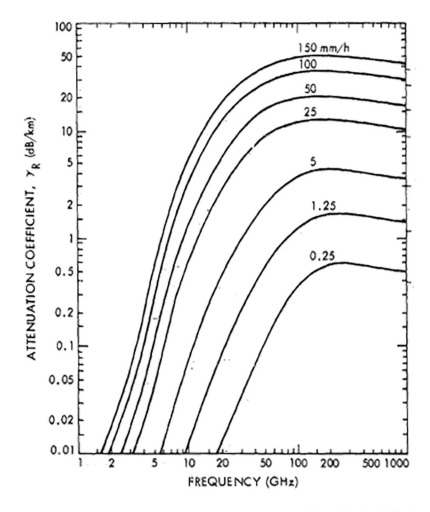

FIGURE 10.10   Attenuation due to rain at various frequencies. (From CCIR (ITU-R) 1978. Attenuation and Scattering by Precipitation and Other Atmospheric Particles, Report 721, in Vol. V, Propagation in Non-ionized Media, Recommendations and Reports of the CCIR, 1978, pp. 107–115 [30].)

TABLE 10.6

ITU Rain Zone Regions Along with Associated Rain Rates (mm/hr) and Duration

Percentage of Time (%)	A	B	C	D	E	F	G	H	J	K	L	M	N	P	Q
1.0	<0.1	0.5	0.7	2.1	0.6	1.7	3	2	8	1.5	2	4	5	12	34
0.3		0.8	2	2.8	4.5	2.4	4.5	7	4	13	4.2	7	11	15	34	49
0.1		2	3	5	8	6	8	12	10	20	12	15	22	35	65	72
0.03		5	6	9	13	12	15	20	18	28	23	33	40	65	105	96
0.01		8	12	15	19	22	28	30	32	35	42	60	63	95	145	115
0.003		14	21	26	29	41	54	45	55	45	70	105	95	140	200	142
0.001		22	32	42	42	70	78	65	83	55	100	150	120	180	250	170

Source: Recommendation ITU-R PN.837-1 1994. Characteristics of Precipitation for Propagation Modeling (Question ITU-R 201/3) [31].

and/or 22 mm/h, respectively, to maintain the radio link availability figure of 99.99%. These rain rates and corresponding duration ultimately determine the suitability of an e-band radio link in a specific rain zone.

The rain fade can be calculated using the following equation.

$$L_{\text{rain}} = \gamma_R D_{\text{rain}} \tag{10.2}$$

where

L_{rain} = loss in signal strength due to rain (dB)
D_{rain} = path length through the troposphere (km)
γ_R = specific attenuation (dB/km)
$\gamma_R = kR^\alpha$ (where k and α are various frequency coefficients for horizontal and vertical polarization). The values of these coefficients are defined in ITU-R P.838-2 [26]

10.3.3.1.4 Link Budget

The link budget is a balance sheet of gains and losses. It takes into account all of the gains and losses, from the transmitter, through the medium, to the receiver in a radio communications system. The link budget for a radio system can be written as shown in Equation 10.2.

$$P_R = P_T + G_T - (L_T \mp L_{\text{FSL}} \mp L_M + L_R) + G_R \tag{10.3}$$

where

P_R = received power (dBm); P_T = transmitter output power (dBm)
G_T = transmitter antenna gain (dBi); L_T = transmitter losses (coax, connectors) (dB)
L_{FSL} = free space path loss (dB)
L_M = miscellaneous (atmospheric, fading, rain) losses (dB)
G_R = receiver antenna gain (dBi); L_R = receiver losses (coax, connectors) (dB)

The product $P_T G_T$ is called EIRP or effective radiated power with respect to an isotropic antenna. EIRP refers to the radiated power relative to an isotropic antenna for which the transmit antenna gain is 1. After calculating the received power, it needs to be determined whether this power is sufficient for radio communications. Thus, a system operating margin or fade margin is added into the calculation. SOM or (system operating margin) or link/fade margin is the difference between the signal a radio is actually receiving versus what it needs for good data recovery (i.e., receiver sensitivity). This margin is calculated by subtracting the receiver sensitivity value (dB) from the received power calculated via Equation 10.4.

$$SOM = P_R - S_R \qquad (10.4)$$

where
   SOM = System Operating Margin (dBm)
   P_R = received power (dBm)
   S_R = Receiver Sensitivity (dBm)

The second factor after determining the received power is signal to noise ratio or SNR. SNR is the ratio (usually measured in dB) between the signal level received and the noise floor level for that particular signal. A link fade margin of 20 dBm or more is considered to be quite adequate, while values less than 10 dBm are unacceptable. The receiver sensitivity or bit error rate of e-band radios hovers around −50 to −75 dBm as compared to traditional microwave units which range from −60 to −85 dBm, depending on the modulation scheme. Table 10.7 shows the relationship between fade margin and link availability where 20 dBm accounts for a downtime of around 70 hours per year.

The combination of Equations (1) to (4) can be used to determine the range (distance) between transmitting and receiving ends. The key drawback of using higher frequencies is that it limits the distance or the range that can be covered by the radio. Similarly, e-band radios achieve high throughputs, but can typically cover less than 2 km. However, for LTE, 4G, and 5G networks this range is quite satisfactory.

10.3.4 EXPERIMENTAL TESTING

The e-band radios have been under research and investigation for quite some time in many parts of the world. This section will cover two such assessments that were conducted by Telenor Group for the course of a few years [17,33,34]. The analysis was started in Norway with the pioneers of these radios in the mid-2000s. Later in the 2010s, its subsidiary in Pakistan also trialed first and second generation radios. In this section, the focus will be on the various testing activities that took place in Oslo and Islamabad.

TABLE 10.7
Fade Margin for Rayleigh Fading

Fade Margin (dBm)	Availability (%)	Downtime (per year)
8	90	876 hours
18	99	88 hours
28	99.9	8.8 hours
38	99.99	53 minutes
48	99.999	5.3 minutes
58	99.9999	32 seconds

Source: 4Gon Solutions 2013 Path Loss, Link Budget & System Operating. http://www.4gon.co.uk/solutions/technical_path_loss_link_budget_som.php; Freeman, R.L. 2006 *Radio System Design for Telecommunications*, Third Edition. Wiley Inc., UK [27,32].

10.3.4.1 Testing in Oslo (2008–2011)

A 3.4 km* long experimental radio link was set up in the inner part of the Oslo fjord centered at 10.6 °E and 59.9 °N between Hovik and Fornebu. The transmitter and receiver antenna heights were 70 and 32 m above sea level at the Høvik and Fornebu ends, respectively.

The aim of the measurement program was two-fold. First, to understand technology and evaluate first generation products of multiple vendors, and second, to analyze prediction methods accounting for propagation effects such as attenuation due to hydrometeor precipitation. The ITU-R has recommended propagation prediction methods, however, the extent of available long-term data from radio link deployment or test beds is limited.

The radio signal strength and meteorological data were measured at the receiving end of the 83.5 GHz link which was located at Fornebu. The signal strength measurements were taken at about 1 s sampling time where the actual sampling time varied between about 1 and 1.5 s. A meteorological station at Fornebu gave air temperature, pressure, relative humidity, wind speed and direction, and hydrometeor precipitation with 1 minute sample values. Another set of minute precipitation data from a nearby station at Lilleaker was provided by the Norwegian meteorological institute.

10.3.4.1.1 Results and Observations

The results showed that propagation impairment can be dealt with, but the current ITU-R recommended precipitation attenuation method apparently under predicts [33,34]. The key observations are as follows:

- Hydrometeor Precipitation Attenuation: It was noted that the predicted attenuation distribution is significantly more optimistic than the observed attenuation.
- Effects of Atmospheric Gases: Gaseous attenuation varies with a maximum of about 1 dB/km during warm summer months, that is, in the period with the heaviest rainfall, and less than 0.3 dB/km during the winter period.
- Effects of Wet Snow and Ice: it was observed that when the antenna was covered with wet snow and ice, the signal degraded severely. Wet snow in the air causes significantly more attenuation than the same amount of liquid water and ITU-R P.530 included a method to account for this.
- Antenna Wetting: The antennas had a hydrophobic coating to minimize snow and ice accumulation, but the protection was reduced with time. When antenna covers were sprayed with water, a 2 dB additional attenuation was recorded, which means that if wetting happens under a period with rainfall, the antenna wetting could account for 3–4 dB loss in total.
- The effects of air turbulence, wind, fog, and clouds were not quantifiable due to the lack of proper instruments.

10.3.4.2 Testing in Islamabad (2012–2014)

Figure 10.11 shows a typical mobile backhaul segment of the cellular networks of Pakistan. One to three microwave radio hops are deployed to reach a core location in urban cities. The core location holds BSC/RNC, may also include core elements like MGW and SGSN (serving GPRS support node), and connects to MSC and GGSN through optical fiber. In the case of LTE, the EPC location consists of MME and S-PGs.

Simulations and studies have shown e-band radios tend to be effective as long as the hop length is less than 2 km. Once the distance crosses this threshold, it gets difficult to meet the criteria of 99.999% (annual downtime 5 minutes) radio link availability [35]. The limitation factors are the high frequency band and high rainfall rate in the country. Pakistan faces heavy monsoon rains that last for about 3 months (July–September) every year. During this time, the total rain accumulation is around 800 mm, while daily it could exceed 80 mm [36].

---

* Although the path length is somewhat longer than a typical application, it is useful for the purpose of testing the possible maximum length.

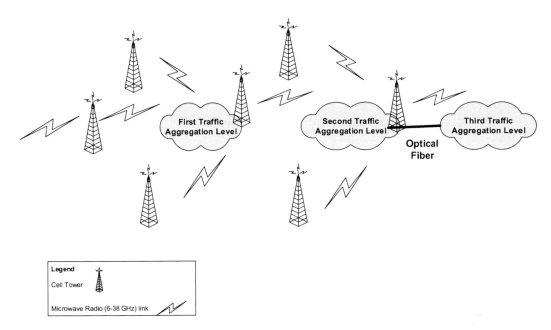

FIGURE 10.11 Mobile backhaul (Pakistan cellular market). (From Asif, S.Z. 2015 *International Journal Wireless and Microwave Technologies*, 2015(4):37–46 [17].)

In major metropolitan cities, the link distance between sites and between aggregation sites ranges from close to a kilometer to 3–4 kilometers. There are only a handful of core locations in urban cities making those hard to reach via e-band radios from the BTS sites. However, sites that require high transmission capacity and are less than 2 km from each other or from traffic aggregation hubs or from core locations can use such microwave radios.

Keeping the above mentioned considerations in mind, 1.0 km or less hop lengths were selected in an urban city. The service provider evaluated both first and second generation products at the same and different times of the year including the monsoon season for about 2 years. Table 10.8 shows a comparison between the two products.

10.3.4.2.1 First Generation Product Testing

The first generation e-band radio (a reseller product) was analyzed against certain standard microwave radio parameters and some specific to such radios. A 1.0 km point-to-point microwave radio link was set up in 1 + 0 configuration with a 0.3 m dish for assessment. The product supports 120, 240,

TABLE 10.8

ePasolink and iPasolink EX Attributes

Attributes	First Generation Product	Second Generation Product
Transmit power (dBm)	+18	+9 to +18
Channel bandwidth (MHz)	250 and 1000	250
Modulation	BPSK, QPSK	QPSK, 16/32/64/128/256-QAM
Capacity (Mbps)	120–120	390–1560
Antenna size (cm)	30	30
Antenna gain (dBi)	43–44.6	43–44.6

Source: NEC. iPasolink EX. http://www.nec.com/en/global/prod/nw/pasolink/products/ipasolinkEX. html [13].

YYYY-MM-DD : HH:MM:SS	Status	Level	Source	Message
Critical : 0 Major : 41 Minor : 22			Cleared : 38 All : 101	Clear Alarm History
201x-xx-29 : 14:49:07	Cleared	Normal	Link	Link up Ethernet Copper #9
201x-xx-27 : 05:49:34	Cleared	Normal	RSL	Normal RSL = -57.0 dBm, FEH Temp=33.0C
201x-xx-27 : 06:48:53	Cleared	Minor	RSL	Minor RSL = -62.0 dBm, FEH Temp=33.2C
201x-xx-27 : 06:48:48	Cleared	Major	RSL	Major RSL = -63.0 dBm, FEH Temp=33.2C
201x-xx-27 : 06:48:43	Cleared	Minor	RSL	Minor RSL = -62.0 dBm, FEH Temp=33.2C
201x-xx-27 : 06:47:46	Cleared	Major	RSL	Major RSL = -63.0 dBm, FEH Temp=33.4C
201x-xx-27 : 06:47:36	Cleared	Minor	RSL	Minor RSL = -62.0 dBm, FEH Temp=33.6C
201x-xx-27 : 06:45:27	Cleared	Major	RSL	Major RSL = -63.0 dBm, FEH Temp=33.8C
201x-xx-27 : 06:44:21	Cleared	Minor	RSL	Minor RSL = -62.0 dBm, FEH Temp=33.8C
201x-xx-27 : 06:41:57	Cleared	Major	RSL	Major RSL = -63.0 dBm, FEH Temp=34.1C
201x-xx-27 : 06:40:55	Cleared	Minor	RSL	Minor RSL = -58.0 dBm, FEH Temp=33.8C
201x-xx-27 : 06:40:50	Cleared	Normal	RSL	Normal RSL = -57.0 dBm, FEH Temp=33.8C
201x-xx-27 : 06:38:11	Cleared	Minor	RSL	Minor RSL = -58.0 dBm, FEH Temp=33.6C
201x-xx-27 : 06:33:13	Cleared	Normal	RSL	Normal RSL = -57.0 dBm, FEH Temp=32.8C

FIGURE 10.12 Radio link failures (at major RSL). (From Asif, S.Z. 2015. *International Journal Wireless and Microwave Technologies*, 2015(4):37–46 [17].)

600, and 1200 Mbps of capacity with 250 and 1000 MHz channel sizes using BPSK (Binary Phase Shift Keying) and QPSK (Quadrature Phase Shift Keying) modulation schemes.

The RFC 2544 (Benchmarking Methodology for Network Interconnect Devices) [37] was used to check the viability of carrying Ethernet traffic. This test was performed by generating traffic through an analyzer. The results of three performance indicators, namely throughput, latency, and frame loss, were found to be satisfactory. QoS, SyncE (Synchronous Ethernet), scheduling mechanism, and adaptive modulation were tested and found to be satisfactory.

During the monsoon season, however, a number of deficiencies were observed in the first generation product. First, radio link failures were observed at 73 mm and above rainfall that lasted for a few seconds and occurred several times, resulting in dropped calls (voice) and discontinuation in data sessions. The link was operational at 1200 Mbps with QPSK having a threshold level of up to -61 dBm at 10^{-6} BER. The product only provided RSL (received signal strength) which is similar to BER as shown in Figure 10.12. Figure 10.12 is a sample alarm snapshot where a major RSL value indicated radio link failures. Higher latency and high frame error rates during some Ethernet performance tests were also witnessed. Overall, the under trial first generation radio turned out not to be a carrier grade microwave radio product.

10.3.4.2.2 Second Generation Product Testing

The second-generation product, though in beta form, gave a comparatively better performance. The radio supports capacity ranging from 390 to 1560 Mbps in a 250 MHz channel using QPSK and 16/32/64/128/256-QAM. It may be noted that the product supports higher capacities than prescribed by the standard (Table 10.3) using the same and better modulation schemes, which is obviously beneficial for the telecom industry. The testing was executed for six months including some part of the monsoon season. The overall test results were positive and the link did not fail even at 80 mm rainfall.

The interesting element of this evaluation was the test setup comprised of 3 hops connected in a ring topology. The distance between any two hops was less than 1.0 km. This configuration was used to demonstrate the feasibility of such radios as an alternative to ring topologies that are normally made with optical fiber cables. The OFC (optical fiber cable) rings are supported with 50 ms restoration time in case of a cut. Similarly, the second generation e-band radios support ERPS (Ethernet Ring Protection Switching) which also operates at the same protection level (sub-50 ms). Various tests were executed using this ring configuration and all passed successfully. A test tool capable of testing and maintaining Ethernet, gigabit Ethernet, IP, and fiber channel services was placed at one cell site.

TABLE 10.9

Gigabit Ethernet Performance Results with 256-QAM Modulation

Frame Length (bytes)	Throughput (Mbps)	Latency Roundtrip Threshold (1000 μs)	Frame Loss (# of lost frames)
64	2000	240	0
128	1820	298	0
256	1704	310	0
512	1652	335	0
1024	1624	399	0

Source: Asif, S.Z. 2015. *International Journal Wireless and Microwave Technologies*, 2015(4):37–46 [17].

The Ethernet (Layer 2) performance was measured with three key parameters, namely throughput, latency, and frame loss. This evaluation was performed using the criteria of RFC 2544. RFC 2544 specifies certain test criteria that allow the service provider and customer to reach an agreement. RFC 2544 requires the standard frame sizes (64, 128, 256, 512, 1024, 1280, 1518, and 1522 bytes) to be tested for a certain length of time and a certain number of times. The product supports multiple GbE/FE (Gigabit Ethernet/Fast Ethernet) ports up to a maximum of 3 ports where a maximum of 1 Gbps can be provided via a single port. The traffic was generated through an analyzer and the four performance indicators were measured over the ring. Various frames sizes of RFC 2544 were tested for the various supported rates, bandwidths, and modulation types. The product also supported jumbo frames such as 9000 byte frames, which was also tested on different modulation schemes. One such result with 256-QAM and 250 MHz bandwidth is shown in Table 10.9.

The above Table 10.9 shows that 2 Gbps throughout is achievable with a 250 MHz channel, which is double what was envisioned in the ETSI EN 302 217-3 standard but using a higher modulation scheme (256-QAM). With higher bandwidths, higher data throughputs than 2 Gbps can be achieved. This further validates that e-band radio is a suitable solution to address the backhaul needs of 5G networks.

10.3.5 APPLICATION SCENARIOS

The most relevant use of these radios is in the backhaul, particularly with the advancement to 5G. Some key applications of high capacity e-band radio links are as follows [17]:

Alternative to Optical Fiber: The deployment of optical fiber takes months since it requires acquisition of land and digging ground. E-band radios can be deployed until fiber is laid in the ground. It will expedite the delivery of services to customers and will speed up time to market. For example, DAS (distributed antenna system) technology standards require digitizing of the antenna signal before transmitting it to a remote antenna. This large amount of digitized data throughput is usually transported using fiber optic cables. However, millimeter wave is an ideal technological alternative when DAS signals need to be transported wirelessly [18].

Redundancy to Fiber: E-band radios can provide a redundant path for carrying traffic along with optical fiber. These are suitable for second or third level backup for existing and new fiber routes. E-band radios will come in handy in case there is a dual fiber cut.

Extending Fiber Backbone: Fiber may not be required to cover sites which are at boundaries of the city. E-band radio links cannot cover longer distances due to use higher losses associated with such higher frequencies, however, this limitation can be mitigated by the use of hybrid fiber and wireless links.

Frequency Relief: The traditional microwave radios have smaller channel bandwidths (typically, the largest is 56 MHz) as compared to e-band radios (250 MHz is the minimum) and thus have lower capacity to carry traffic. Many operators often face a shortage of such frequencies, as many channels are used at transmission hubs to support traffic backhauling of various individual (dependent) cell sites. E-band radios can thus be used to free up some traditional microwave frequencies at high capacity sites.

Tower Loading Relief: An e-band radio can accommodate a capacity equal to that of at least 2–3 traditional microwave radio units. Thus, it reduces the loading on cell towers making room for future use.

Small Cells Backhaul: Small cells are expected to be used heavily in 5G networks. These can be placed on street light poles and in congested locations where traditional BTSs cannot be accommodated. The e-band radios can be used to support backhaul for small cells.

Security Situation/Emergency Relief: In the event of natural disasters or manmade upheavals, cells-on-wheels equipped with e-band radio can be used to address relief efforts as an alternative to fiber.

10.3.6   E-Band Radios: Case Study (Pakistan)

Pakistan's mobile and fixed wireless broadband subscription as of June 2016 stood at about 32 million. The mobile wireless broadband market in the country is in the infancy stage. The 3G-UMTS service was launched by four cellular players in mid-2014, while 4G-LTE was launched by two of them in late 2014. There are five cellular operators in the country; one player is offering both 3G and 4G service, while the smallest player (in terms of subscribers) is only offering 4G service. The fixed wireless broadband market, which runs on CDMA2000 EV-DO and WiMAX, has been around for more than eight years, but its penetration rate is less than 1% [38].

The total 3G/4G and broadband (fixed and fixed wireless) subscription is expected to reach 60 million in 2020 as predicted by GSMA [39]. The expected breakdown of this subscriber's penetration by 2020 is as follows:

- 3G/4G: 55 million
- Fixed Broadband (Wired): 4 million
- Fixed Wireless Broadband*: 1 million, respectively (approx[†])

10.3.6.1   Challenge

In simple terms, the traditional microwave radio based mobile backhaul of 3G/LTE operators is not sufficient to keep up with the ongoing enormous data growth. Moreover, low ARPU (Average Revenue Per User) and high OPEX are additional key inhibiting factors to deploying fiber in the backhaul.

The traffic can be backhauled in a variety of way, but for the most part, in the cellular networks it is carried either through microwave radios or optical fiber. In the case of Pakistan, even after the launch of 3G and 4G services, the traffic is still carried through low to mid capacity microwave radios that operate in the 6–38 GHz frequency spectrum. Microwave radios are heavily deployed to take care of the traffic not only in the backhaul, but in the subsequent aggregation points (hubs) and all the way to the core node. There were around 40,000 cell sites and 30,000 microwave radio links (bidirectional) deployed in the country.

Optical fiber cable has been laid for intra and intercity communications by a few operators and capacity for the most part is available. However, extending fiber to the individual cell sites and even in backhaul has been proven to be uneconomical. The key reasons for not having a positive business case for OFC deployment in backhaul are having a very low ARPU and high OPEX due to the absence of a right-of-way acquisition policy at the national level, power (electricity) crisis, and security challenges which are external factors (over which the telecom sector has no control).

---

* Fixed Wireless is comprised of CDMA2000 EV-DO (Evolution Data Optimized) and WiMAX technologies. Wired is mainly comprised of DSL (Digital Subscriber Line) and very few fiber-to-the-home subscribers.

† This approximation was not given in the GSMA study [39] and it is based on current and preceding trends.

There is no meaningful change in terms of revenue even with a considerable increase in the number of 3G/4G subscribers and data usage. According to some industry estimates, the increase in data usage is 1000 percentage points, and in data users' is 100 percentage points, while the increase in data revenue is only in double digits after two years of 3G/4G launch. In a nutshell, the status of being one of the lowest ARPU (<$2) markets of the world has not changed.

10.3.6.2 Possible Solution: E-Band Radios

To address the challenge, Telenor Pakistan, as an innovation leader in the market, took the lead to find a solution. The company's concerned unit recommended e-band radios as a potential solution. This section will look into the efforts that were made to start the use of e-band radios in the country.

Telenor Pakistan was the first operator to get the required spectrum for research and testing purposes for Islamabad in November 2011. However, it took the company two and a half years to get the required approval from concerned government organization. There were three key reasons for this delay. First, this spectrum was not even on the radar of concerned government organization as there was no long-term strategy. Second, concerned government organization does not have the tools to plan and later monitor the same. Third, there was no concrete policy at the national level on the allocation of frequencies for research and test purposes.

Telenor Group spent considerable time and effort in educating concerned government organization about the technology, propagation effects, and products. As a result, the company became the first operator to secure the spectrum for trial. Later, after acquiring the spectrum, Telenor Pakistan was the first to test such radios in the country. The service provider tested both first and second generation radios during the course of three years (2012–2014). The test results of the second generation product were satisfactory (as discussed above) and this radio was confirmed as a carrier-grade product. The operator is now deploying e-band radios in the network.

Furthermore, the incumbent fixed line and wireless local loop operator PTCL (Pakistan Telecommunications Company Limited) which is also the SMP (Significant Market Power) is also evaluating this technology for its EV-DO and LTE networks. Later, another cellular operator, Mobilink, which has the highest number of subscribers, also joined the bandwagon.

10.3.6.3 Why E-Band Radios?

This section describes why e-band radio is a good suitable alternative to optical fiber in the areas of backhaul and at traffic aggregation sites to address the enormous growth of broadband in Pakistan.

E-band radio links will be primarily deployed in key urban cities, primarily in point-to-point, and could be in ring configurations. Broadband services today and in the coming years will be concentrated in the urban and semi-urban cities since most of the educated masses live and work in these urban centers. Close to 70% of Pakistan's population lives in villages and for the most part is uneducated.

The key reasons why e-band radios are needed by the telecom sector are:

Right of Way (RoW): One of the most daunting challenges that the sector faces every day is obtaining RoW for telecom services. It is a difficult problem to understand and solve due to a number of both legal and clandestine reasons. OFC deployments, which run for hundreds and thousands of miles, always face such timely and costly hurdles. There is no concrete policy from the federal government that can resolve the RoW challenges across the country. On the other hand, the e-band radio deployments will be quicker and with comparatively fewer RoW disputes.

Frequency Congestion: The urban centers already have thousands of fixed radio links creating frequency congestion. In addition, these current links are only capable of carrying a few 100 Mbps of data and are not tailored to support LTE and future 5G data hungry services. Thus, e-band radio technology provides a suitable alternative for carrying a high amount of data.

Cost: The scenario depicted in Figure 10.13 and Table 10.10 compares the cost of optical fiber ring with e-band radios which are also interconnected in a ring format. The ring topology

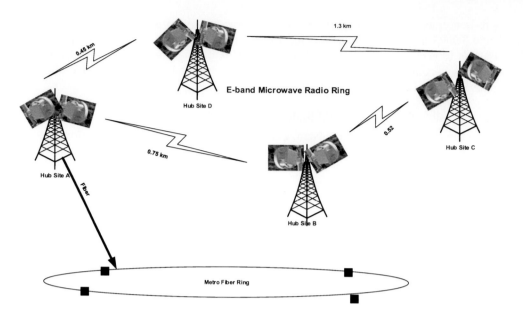

FIGURE 10.13 Potential 70/80 GHz ring topology sample.

is comprised of four hub sites—Site A, Site B, Site C, and Site D. As there are four hops in a ring setup, 8 ODUs will be required. The distance between these sites is less than 1.5 km which is achievable with these radios. Each hub site is carrying about 1.0 Gbps or more which makes it difficult to carry it with traditional microwave radios and the required capacity with ring protection is 2.0 Gbps. Following ITU-T G.8032v2 (Ethernet Ring Protection Switching) standard, traffic is sent in two directions around a ring, effectively doubling the capacity. It also offers reduced capital expense by eliminating the need for fully protected sites [40]. The setup requires eight 500 MHz channels, one for each ODU.

Though it is not an apple to apple comparison and prices are indicative, Table 10.10 still shows that these radios can provide more than 40% savings in CAPEX and 67% in OPEX, provided e-band radios can effectively operate in ring topologies.

10.3.6.4 How E-Band Radios Can Be Used

This section will look at what is needed to make these radios and frequency spectrum available to all the wireless and wireless local loop (WLL) providers. In a nutshell, to make e-band radio technology an attractive choice, the following may be considered.

Policy and Licensing: A framework is required that entails how and why the e-band radio links should be assigned. The licensing cost of fixed radio links is based on either transmission data rate or channel bandwidth size or on a combination of these two parameters. However, this formula will not work as these will result in excessive and disproportionate costs for

TABLE 10.10
Optical Fiber versus E-Band Radios

Item	Optical Fiber	Item	E-Band Radio
CAPEX for fiber path (includes fiber, equipment, power)	$270,000	CAPEX for 4 microwave radio hops, 8 ODUs (1 + 1)	$160,000
OPEX per year	$21,000	OPEX per year (including licensing)	$7000

e-band links. The light licensing techniques as described earlier will be more appropriate in this case, however, at this stage, the concerned government organization has not provided any guidelines.

Channels: As stated earlier, the technology was not appealing in the early years, as there were only two channel bandwidths available, that is, 250 and 1000 MHz. However, the availability of 500 and 750 MHz channels along with higher capacities make it a more promising alternative. Thus, the availability of the 500 MHz channel will be able to accommodate all the five cellular and three key fixed wireless broadband players.

Configurations: The e-band radios can be considered for both point-to-point and in ring configurations. The e-band equipped ring topology will be ideal for the country's 3G/4G players as fiber rollout is timely and costly and for future 5G to connect a multitude of small cells.

10.3.6.5   Future Outlook

E-band radios are expected to be deployed in the coming years both in point-to-point and to some extent in ring configurations. The data usage through 3G-UMTS and LTE has considerably increased, however, the service providers are still finding it difficult to extend the optical fiber in the backhaul, thus these radios seems to be the only alternative they can rely on, at least for the next few years.

10.3.7   E-Band Radios Evolution

The development of these microwave radios for mobile communications started in the 1990s and became a reality in the 2000s. A good amount of testing was conducted by operators in the 2010s and deployments of second generation radios are taking place in the current decade. In the coming years, Cloud-RAN topologies are expected in 5G, and e-band radios are expected to fronthaul CPRI traffic from RRUs to the BBUs. These radios may carry large capacities to support metro-ring topologies as well. Figure 10.14 shows the evolution of these radios, capturing the information stated in the earlier sections.

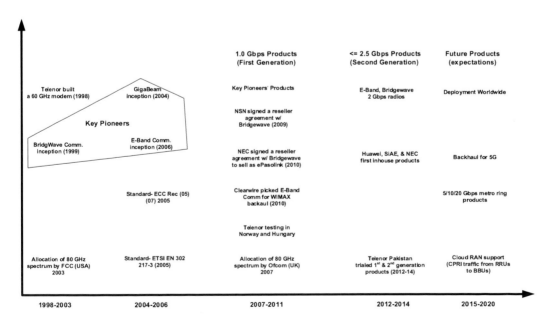

FIGURE 10.14   Evolution of e-band radios.

10.4   METRO TRANSPORT NETWORK

The question that needs to be answered here is how to migrate the current 3G/4G metro transport network to support 5G and/or how to create a new metro transport network for green field 5G deployment.

The connectivity beyond backhaul (and typically within a metropolitan area) toward the core transport network is normally in an optical fiber ring configuration as shown in Figure 10.15. The traffic from the backhaul can be dropped on an Ethernet based switch or on IP router which is then transported through metro transport network to core transport network.

In the majority of networks, this configuration has been and will be migrated from TDM based SDH (Synchronous Digital Hierarchy) to packet based technologies like MPLS-TP. The traditional metro networks based on SDH ADM (Add-Drop Multiplexer) products are interconnected in the form of physical rings using optical fiber cable. Services are protected in the ring using MSP/SNCP* (Multiplex Section Protection/Subnetwork Connection Protection) techniques and similar and better service survivability mechanisms have been proposed for MPLS-TP. This section will primarily look into the MPLS-TP standardization, challenges, and applications (utilization in 4G/5G scenario).

10.4.1   MPLS-TP

The MPLS-TP is an emerging packet transport technology that has been under research and investigation for many years. The technology was born from the efforts of the telecom industry that was looking to strengthen packet based transport networks. Its development has been shared between IETF and ITU-T where IETF brought its MPLS expertise while ITU-T added its significant knowledge of transport networks.

The MPLS-TP is designed to support packet transport services with a degree of predictability, similar to one that is available in existing circuit switched based transport networks. MPLS-TP networks offer layer-2 based services by re-use of the same MPLS label structure for packet forwarding yet avoiding the risks and vulnerabilities (like broadcast storms†) of traditional layer-2 networks.

MPLS-TP can be used in backhaul along with e-band microwave radios, and also can be deployed over OTN/DWDM (Optical Transport Network/Dense Wave Division Multiplexing) in the core transport network. The thinking behind the MPLS-TP network is to have a similar look and feel as the existing SDH/SONET (Synchronous Digital Hierarchy/Synchronous Optical Networking) network, but with better scalability. As a layer 2 connection oriented technology, it has continued to be deployed in mobile backhaul and metro/aggregation networks as an alternative to SDH/SONET.

10.4.1.1   Drivers and Objectives

MPLS-TP was developed primarily to address the shift in traffic from TDM to IP in mobile broadband networks. This transition toward all-IP is also increasing the uncertainty of the long-term future of SDH/SONET technologies. However, for years to come, operators have to support both circuit switched and packet switched traffic.

Thus, it was easily understood that the industry has to find a cost effective solution that could address current challenges, provides SDH/SONET like predictability, interoperability with widely deployed MPLS technology, and is futuristic. Furthermore, in the backhaul and metro/aggregation domains, there was a desire by operators to simplify transport networking to reduce CAPEX and OPEX. Finally, the solution came in the form of MPLS-TP [41].

---

* MSP is used to protect fibers between two adjacent nodes. SNCP based protection can be applied on a fiber path passing many nodes.
† A broadcast storm is the condition where messages are broadcast on a network that cause multiple hosts to respond simultaneously by broadcasting their own messages, which in turn, prompts further messages to be broadcast, and so on.

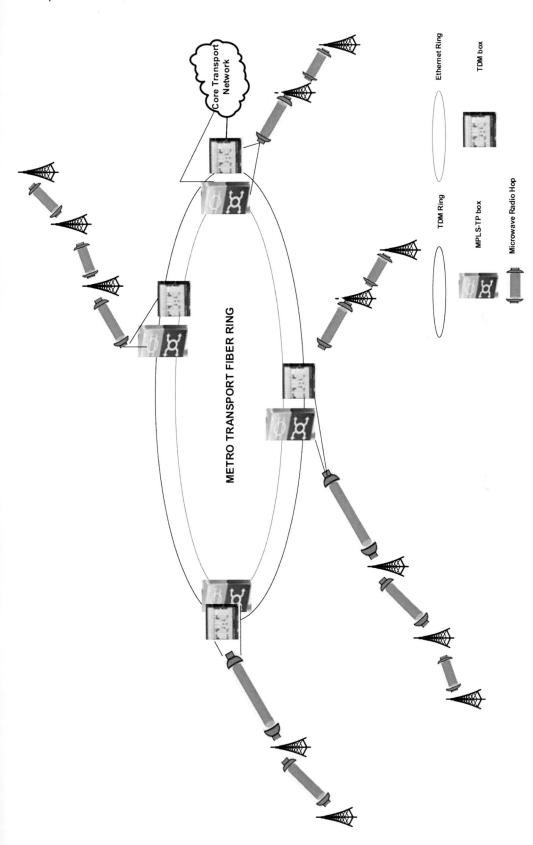

FIGURE 10.15 Metro transport network.

The primary objectives for the development of MPLS-TP as defined in RFC 5921 MPLS-TP Architecture Framework [42] are (1) to enable MPLS to be deployed in a transport network and operated in a similar manner to existing transport technologies; and (2) to enable MPLS to support packet transport services with a similar degree of predictability that is available in existing transport networks.

10.4.1.2 Standardization

The creation of MPLS-TP technology was a result of a joint standardization effort between IETF and ITU-T. The standardization work on MPLS-TP was started as T-MPLS (Transport-MPLS) at the ITU-T in 2005. ITU-T approved the key T-MPLS recommendations including G.8110.1 (Architecture of Transport MPLS layer network), G.8121 (Characteristics of Transport MPLS equipment functional blocks), G.8112 (Interfaces for the Transport MPLS hierarchy), and G.8131 (Linear protection switching for Transport MPLS networks) in 2006 and 2007.

The IETF was also working in parallel to extend the proven MPLS technology to be used in transport networks. As an Internet focused technology, the MPLS was designed on the principles of best effort routing and connectionless delivery of traffic. Given the deployments of MPLS in networks and the desire to align packet networking with more traditional transport operational KPIs (key performance indicators) like connection oriented, QoS, and so on, efforts started to take place in IETF.

After working for some years on separate paths, the two SDOs joined hands and decided to work together as prescribed in RFC 5317 [43]. The RFC 5317 "Joint Working Team report on MPLS architectural considerations for a Transport Profile" states that both parties work together, bring transport requirements into the IETF, and extend IETF MPLS forwarding, OAM, survivability, network management, and control plane protocols to meet those requirements through the IETF Standards Process. Later, T-MPLS was renamed MPLS-TP based on the agreement that was reached between the ITU-T and the IETF to produce a converged set of standards. The standardization of MPLS-TP for the most part has been completed. However, there are still some compromises needed to find a common ground on items like OAM, protection, and control plane which will be discussed in the next subsections.

10.4.1.3 MPLS-TP Concept

The MPLS-TP technology is both a subset and extension of MPLS[*] as shown in Figure 10.16. It builds upon the existing MPLS packet forwarding and MPLS-based pseudo wires but does not require MPLS control plane capabilities. It provides static provisioning via NMS (Network Management System) while the dynamic Generalized MPLS (GMPLS) control plane has been kept optional in the standard.

In MPLS-TP, some MPLS functions are turned off, namely LSP[†] (label switched path) merge, penultimate hop popping (PHP)[‡], and equal cost multi path (ECMP)[§]. Though these features have been successfully implemented with MPLS, they could disrupt the management of connection-oriented paths between end points in MPLS-TP. This possible disruption is due to the lack of full traceability of the said paths; for example, OAM packets must pass the through same path as the signal packets to exactly monitor the condition of the paths. In simple terms, MPLS-TP is a Layer 2, connection-oriented, and physical layer agnostic technology for packet switched transport networks.

10.4.1.4 Architecture

Figure 10.17 shows the architecture for an MPLS-TP network when single segment (single transport LSP) and multi-segment pseudo wires are used. This architecture is described in RFC 5921 [44]

---

* MPLS is layer 2.5 connectionless technology designed to carry all types of TDM and IP traffic.
† LSP is a path through an MPLS network, set up by a signaling protocol.
‡ PHP is a function performed by certain routers in an MPLS enabled network. It meant to remove the label one hop before its destination. In other words, it's a process where the outermost label of an MPLS tagged packet is removed by an LSR before the packet is passed to an adjacent Label Edge Router (LER).
§ IETF RFC 2992 defines Equal-Cost Multi-Path (ECMP) as a routing technique for routing packets along multiple paths of equal cost for a single destination.

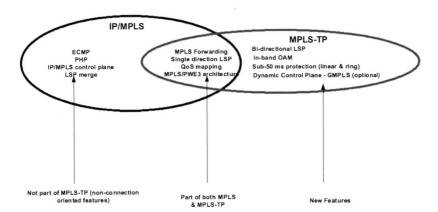

FIGURE 10.16 MPLS-TP overview.

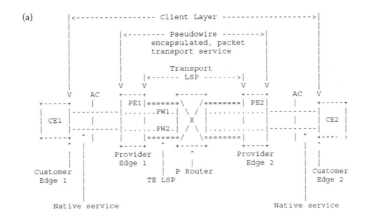

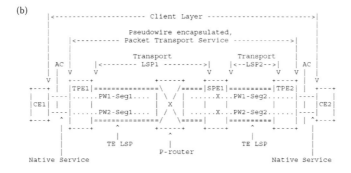

FIGURE 10.17 MPLS-TP architecture. (From RFC 59212010 A Framework for MPLS in Transport Networks, IETF, July [42].)

where the transport LSP is equivalent to the Packet Switched Network (PSN) tunnel. The client layer in this architecture is equivalent to the emulated service which is described in RFC 3985 Pseudo Wire Emulation Edge-to-Edge (PWE3) Architecture [45].

The key elements and functionalities of MPLS-TP are data and control planes, OAM function, and path protection mechanism.

Data Plane: A standard MPLS data plane consists of LSPs, sections, and pseudo wires* to provide packet transport service for client traffic. It primarily remains similar to MPLS to facilitate interoperability.

MPLS-TP forwarding plane uses the same standard MPLS forwarding architecture and is totally separated from the control plane. It employs bidirectional LSP which is defined by pairing the forward and backward directions to follow the same path: the same node and same links. In addition to that, it supports unidirectional point-to-point and point-to-multipoint LSPs.

The functions of LSP merge, and ECMP are discontinued in MPLS-TP. The amalgamation of two LSPs (going to the same destination) reduces the number of labels being used in the network. However, merging makes it rather difficult to differentiate between traffic coming from two different sources which were present before the merger. In IP/MPLS networks, different packets between a source destination pair can take different paths; particularly when multiple equal cost paths exist. However, this is in conflict with the concept of a circuit where all the traffic follows the same path; thus, ECMP was also discontinued.

Control Plane: The control plane is optional, and it can be static or dynamic.

OAM: The key was to have OAM capabilities that are similar to the OAM functionality of SONET/SDH. Its simpler yet stronger OAM capabilities using a separate G-Ach (Generic Associated Channel) allow for transport centric tasks like loop testing and circuit locking. The monitoring of faults and failures is performed using CC/CV messages (Connectivity Check/Connectivity Verification) functions.

Linear Protection: Linear protection switching mechanisms need to be capable of providing a recovery time of less than 50 ms similar to SONET/SDH. The linear protection allows speedy protection so that traffic following one path of the network can be switched to a backup path when the working path fails or falls below the required threshold or when approved human intervention takes place [46].

10.4.1.5   Challenges and Alternatives

This section highlights some of the key challenges that have been faced and alternatives to resolve these issues in the development of MPLS-TP.

10.4.1.5.1   OAM

One of the key features of MPLS-TP is existing transport network-like OAM functionality. From the very beginning of MPLS-TP development, there were two camps defining the OAM functions. The IETF was focusing on developing extensions to existing MPLS OAM tools and creating some new measurement and fault management tools. That includes extending Bidirectional Forwarding Detection (BFD) for proactive continuity check and connectivity verification and extending LSP Ping for on-demand connectivity verification and route tracing.

The ITU-T was focusing on an OAM mechanism that was later termed as ITU-T G.8113.1. The G.8113.1 standard draws upon key OAM elements of ITU-T Y.1731, which is used in carrier Ethernet networks. Until the very end, the two approaches remained apart and ultimately resulted in two different sets of OAM specifications. These two formulations, G.8113.1 [47] and G.8113.2 [48] from ITU-T and IETF, respectively, were approved at the WTSA in November 2012.

Recommendation G.8113.1 adheres to the OAM functions and mechanisms of ITU-T Recommendation G.8013 [49]. Recommendation G.8013 specifies mechanisms required to operate and maintain the network and service aspects of the Ethernet layer. It also specifies the Ethernet

---

* A section provides a means of identifying the type of payload it carries (a transport layer packet entity). Pseudo wire (PW) is an emulation of a native service (ATM, Frame Relay, Ethernet, SONET/SDH, etc.) over a packet switched network (IP, MPLS, etc.), commonly used for Ethernet over packet switched network.

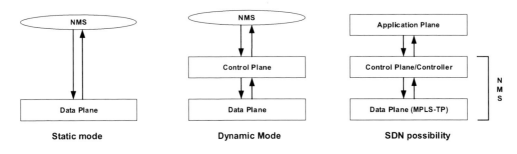

FIGURE 10.18 Static configuration or dynamic control plane and SDN.

OAM frame formats and syntax and semantics of OAM frame fields. When client nodes at both transmitting and receiving ends are based on Ethernet, the G.8113.1 will make the job of managing and monitoring MPLS-TX networks easier for operators due to reuse of OAM functions. In contrast, G.8113.2 gives priority to IP/MPLS router implementations and it depends on the IP layer's functions. G.8113.1, on the other hand, is independent of the IP layer.

G.8113.2 is built on pre-existing MPLS client diagnostic tools such as LSP Ping and BFD, and enhanced through new OAM protocols that can be carried in the MPLS ACh [46]. Overall, G.8113.2 tends to be more complex due to the addition of some functions such as four CV modes corresponding to G-ACh (Generic Ach), UDP (User Datagram Protocol), and IPv4, IPv6, and others. Within MPLS-TP, the ACh is extended to the G-ACh and a new label is introduced, G-ACh Alert Label (GAL), to identify packets on the G-Ach.

10.4.1.5.2 Control Plane

RFC 5654 [50] states that the MPLS-TP network must be able to operate without using a control plane. At the same time, it must also have the ability to set up static provisioning without the presence of any element of a control plane. RFC 5921 [44] and RFC 6373 [51] include support for both static configuration of the transport path (LSPs and PW) through NMS and dynamic provisioning via a control plane (Figure 10.18). This ability of supporting both static and dynamic attributes is one of the major strengths of MPLS-TP. The NMS based provisioning is sufficient for simple network scenarios; however, a control plane would be used as the networks evolve in complexity. The key motivation of using static configuration is to eliminate the cost associated with having the control plane functionality integrated into each node across an IP/MPLS based network.

The dynamic control, on the other hand, requires that all the routers along the path must have sufficient memory and capacity to support the variety of protocols stated in RFC 6373. The dynamic nature of the control plane will add complexity and cost. A switch to SDN would make introduction of a new feature easier as it does not have to go through the time-consuming standardization process. The SDN approach allows separation of a network's control and forwarding (data) planes to better optimize each one.

10.4.1.5.3 Protection

Protection, as the name suggests, refers to the availability of the network in case of faults. The mechanism for this network survivability is called APS or Automatic Protection Switching. APS involves reserving a channel/path (dedicated or shared) with the same capacity as the original channel/path which needs to be protected. The protection methods are categorized by topology (linear or ring), protection resource (dedicated or shared), direction (unidirectional or bidirectional), and/or operational types (revertive and nonrevertive). The APS mechanism was developed and later implemented in many SDH and Ethernet transport networks. For example, APS is generally specified by ITU-T in G.808.1, and also in G.841, G.873.1, and G.8031 for SDH, OTN (Optical Transport Network), and Ethernet networks, respectively.

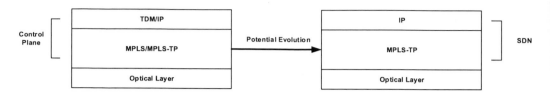

FIGURE 10.19   Evolution of MPLS-TP for 4G/5G.

Similar to the case of OAM standardization, the ITU-T and the IETF were on separate paths, however, unlike detachment between two SDOs on the OAM issue, they converge to a single solution for linear protection. The APS solution was initially considered in the IETF but failed to achieve a consensus in the MPLS working group [46]. However, ITU-T Recommendation G.8131 (2007) [52] was later revised as G.8131 (2014) [53] to allow for the solution which is also defined in IETF RFC 7271 [54]. This solution is called Automatic Protection Coordination (APC) in G.8131, and is called "Protection State Coordination (PSC) in "Automatic Protection Switching (APS) mode" in RFC 7271.

The "RFC 7271 MPLS-TP Linear Protection" for ITU-T defines two types of modes, namely PSC and APS, where a mode is a given set of capabilities. Both PSC and APS are single phased protocols which implies that each end point will notify its peer of a change in the operation path (switching to or from the protection path). Furthermore, this switching will take place without waiting for the acknowledgement.

Recommendation G.8131 (2007) was updated to allow for the differences between APS and PSC, which were both operational in nature, as well as those related to frame format. PSC changed some operational parameters including the priority levels of various protection switching triggers. For example, there is no EXER (Exercise) command in PSC which is used to test whether or not APS communication is taking place correctly. In PSC, it is assumed that a similar testing option shall be supported using lockout of protection (LO) or forced switch (FS), and the traffic cannot be recovered when a fault occurs during an exercise operation.

To reduce the said challenges, the ITU-T Recommendation prepared G.8131 (2014) which allows the co-existence of APS and PSC based frame formats. The unified solution APC through G.8131 (2014) is now available to solve the aforementioned deficiencies. This recommendation state that APC protocol is used to describe the means to coordinate the two ends of the protective domain via the exchange of a single message and represents the PSC protocol in APS mode as defined in IETF RFC 7271. The APC can be rolled out gradually in networks that already have PSC giving operators an option either to support PSC, APC or both according to their business plans.

10.4.1.6   Use of MPLS-TP in 4G/5G

The MPLS-TP network uses a layer architecture that completely separates the data plane from the control plane and can ease the introduction of SDN technology to any layer independently, for example, to layer 3 (IP) as shown in Figure 10.19. The separation of data and control planes and use of SDN are expected to be critical ingredients of 5G.

10.5   CORE TRANSPORT NETWORK

This segment of the transport network connects cities and countries mostly over long haul optical fiber. This connectivity can also be provided using leased lines that support STM1 or higher rates. The inter-connectivity to other operators within the country and to overseas operators with submarine cables can also fall under the core transport network.

Furthermore, the access to Internet or to the outside world for voice and data communications can be provided through cross border links. These links can be established through microwave radio, optical fiber, satellite, submarine cables or a combination thereof. In particular, the Internet

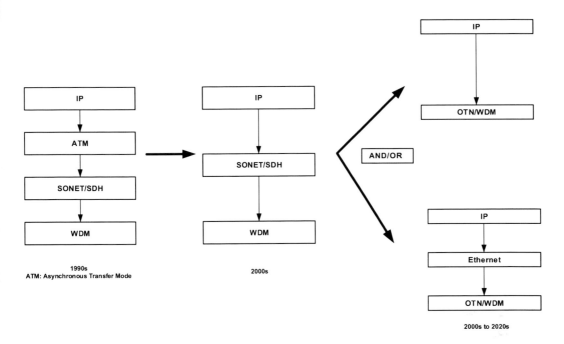

FIGURE 10.20   Service layering in core transport networks.

transit is a service allowing network traffic to cross a computer network connecting smaller ISPs (Internet Service Providers) to the larger Internet. The Internet connectivity is normally provided via interconnects' agreements, namely IP Transit and Peering.

10.5.1   SERVICE LAYERING IN TRANSPORT NETWORKS

The term layering is normally referred to as switching/routing technologies on top of the photonic layer. The common transmission layering structure that is used in the core transport network is IP/MPLS/SDH/DWDM. In the digital domain, it could mean the use of OTN (Optical Transport Network) switching (Layer 1), Ethernet switching (Layer 2), and IP routing (Layer 3) techniques. It is well established that there is a need for tighter integration between routing layer 3 and optical transport layer 1. However, the question that requires a thoughtful answer is whether the delivery of all types of services through routers is justifiable or Ethernet grooming/switching is required for 5G. The evolution of service layering is shown in Figure 10.20.

Depending on business and service needs, the two main and newer core transport network models deployed worldwide are [55]:

    a.   IP over DWDM networks with OTN OAM&P (operations, administration, maintenance, and provisioning).

    b.   IP/Ethernet/OTN/DWDM networks where transit traffic is mainly groomed/switched by OTN multiplexing/switching network elements.

Furthermore, the connectivity between distant cities is achieved through two complimentary platforms that work independently. The first is long-haul optical platform based on DWDM, ROADM* and probably OTN. It is responsible for switching wavelengths in the optical domain via ROADM and circuits in the electrical domain through the OTN cross-connect function. The second one is

---

* ROADM (Reconfigurable Optical Add-Drop Multiplexer) adds and drops wavelengths in the optical domain.

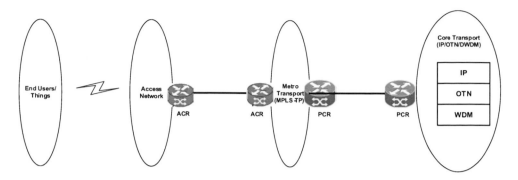

FIGURE 10.21   IP over DWDM.

the core packet platform, which is responsible for switching MPLS-labeled packets in the electrical domain through LSR* (Label Switching Router).

10.5.1.1   IP over DWDM

The traditional way to connect a router to a DWDM system is through transponders. The IP over DWDM concept is, however, based on the integration of DWDM or colored interfaces on IP routers. There is no role for SDH/SONET or Ethernet layers in this concept, thus this model reduces the overall TCO (total cost of ownership). The cost reduction is achieved by eliminating the gray† optics and transponder (element that sends and receives optical signals from a fiber) used in separate router and DWDM systems [55]. However, this architecture requires routers with much higher routing capacity for traffic aggregation and forwarding tasks [56].

This architecture normally corresponds to point-to-point and ring configurations of small distances, avoiding regeneration, where traffic grooming/switching at the intermediate nodes in not needed and when interworking between WDM interfaces in routers and legacy WDM is not present or is minimal. However, IP over DWDM can reach longer distances as proved by Sprint in 1998–99 [57].

To address some of these challenges, an alternative could be based on pluggable optics including OTN framing and FEC (forward error correction). OTN framing is used to ensure interoperability of integrated color interfaces while FEC is used to cover long distances. OTN, as defined in ITU-T Recommendation G.709, adds OAM&P functionality to optical carriers specifically in DWDM systems. FEC [58] also plays a key role in ensuring reliable transmission of information and has now become an essential technology for optical communication systems. The use of coherent detection [59] in fiber optics makes possible the use of soft-decision FEC. The 100G and 400G fiber optics systems use coherent detection, making soft-decision FEC an appropriate partner technology.

Figure 10.21 shows an IP over DWDM architecture with Aggregation Exchange Routers (ACR)/ Switches and Packet Core Routers (PCR). These routers/switches aggregate traffic from end users or things using 10/100 Gbps interfaces and drop it on a metro optical fiber ring which then transfers it toward the core IPoDWDM network. Any signal that needs transmission over DWDM can be preferably mapped on the OTN node and then transported over DWDM.

10.5.1.2   IP over Ethernet over DWDM

The second approach is based on the use of Ethernet technologies along with IP, OTN, and DWDM, that is, four distinct layers. In this model, the Ethernet layer based on MPLS-TP, PBB-TE‡, traditional Ethernet or carrier Ethernet supports L2 aggregation with standard OAM and protection mechanisms. A key capability of L2 aggregation is the support for statistical multiplexing that

---

* Such a switching capability router that supports MPLS is called a Label Switching Router or LSR.
† Gray or uncolored wavelength, typically 1550 nm.
‡ Provider Backbone Bridge Traffic Engineering (PBB-TE).

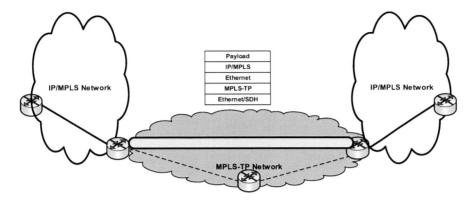

FIGURE 10.22   Overlay approach.

allows the bandwidth to be divided arbitrarily among users and uses the optical cable/medium only when there is data that needs to be sent. This is fundamentally different from the L1 aggregation (time or frequency multiplexing) approach where the number of users and data rates is fixed. This concentration of traffic at L2 allows traffic to be transferred to IP core routers with a few high-speed interfaces rather than over many lower capacity ones, and this is turn, simplifies OAM procedures. Furthermore, in this model, traffic is kept in its original form, that is, in packet form and all the grooming, aggregation, and switching is performed at L2 [55,56].

Though this multi-layer integration model seems more complicated than the IPoDWDM approach, it can multiplex both circuit and packet traffic, provide more enhanced OAM and protection schemes than SONET/SDH, and sub-lambda grooming for maximizing wavelength utilization through OTN. The WDM along with ROADM further allows transmission of client services over any wavelength and in any direction.

10.5.1.3   Interworking between MPLS and MPLS-TP

To provide end-to-end service, the preference still resides with IP/MPLS at the core level and MPLS-TP at the access/aggregation area. This requires interworking between the two technologies which can be achieved in two ways.

The first approach is called the 'overlay model,' which is used when the IP/MPLS network is carried over the MPLS-TP or the MPLS-TP network is carried over the IP/MPLS. As shown in Figure 10.22, the overlay is the service layer of one network (e.g., MPLS-TP) for the other network (e.g., IP/MPLS). After encapsulation at a network boundary node, client-layer data (including control plane data and transport plane data) is transparently transmitted to the corresponding service layer network boundary node via the service layer channel. The service layer network is articulated as a hop of the client layer network, that is, two boundary nodes in the service layer network are considered as adjacent nodes in the client layer network.

The second approach is called the 'peering model,' which is used when a customer network is carried partially over IP/MPLS subnetwork (e.g., via PW encapsulation) and partially over MPLS-TP subnetwork (a logical visible subdivision of an IP network). As shown in Figure 10.23, the two networks are peers in the model and independently process data in their own networks, while network boundary nodes map information between the two networks to transmit data. The details on interworking can be found in [60,61].

10.5.2   Integration of IP and Optical

The concept of IP-Optical integration has been around for nearly two decades. However, it has not taken off to the fullest extent primarily due to current investment in the traditional model that

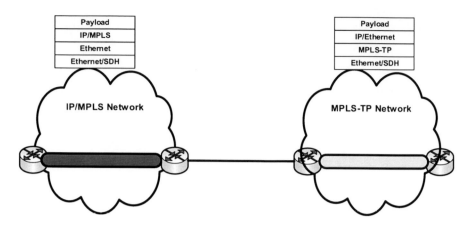

FIGURE 10.23 Peering approach.

runs into billions of dollars and the loose integration between the packet and optical transport organizations of the service providers.

Building a common network infrastructure, constructed with a mix of packet and DWDM layers, may be an ultimate goal for service providers, but the reality is that there are various ways to get to the target. Vendors and service providers have different views on the goal of IP and Optical Convergence (IPOC). Router vendors are more interested in putting optical transponders and OTN switching in a core router, while optical-centric vendors are pushing for the integration of packet switching to the optical network platform. The service provider community is also not on the same page and is taking a cautious approach that does not disrupt the packet and transport teams that are not accustomed to working in a converged network environment. Today, the service provider architectures are based largely on separate IP and optical network layers that are usually designed and operate independently. The problem with separate IP and optical transport networks is the unnecessary resource overhead it creates. Basically, each network has its own control and management mechanisms, leading to inefficient use of resources and increase in complexity [57,62,63].

While some steps have been made to integrate the two transport networks, a fully integrated solution that includes logical integration of all planes—data, control, and management—is still needed. This will eventually lead to the full scale convergence of IP and optical network layers. The IPOC offers significant operational gains as it permits the different layers of a network to operate in a single step. This, in turn, will lead to increasing service velocity, reducing time to market, optimizing resources utilization, decreasing total cost of ownership, and finally energizing 5G deployments.

10.5.2.1 IPOC Challenges and Enablers

The current SDN movement in the telecoms industry offers a much needed solution to attain the high levels of automation that modern networks demand for both IP and optical systems. SDN decouples control of the network from hardware to software and lets operators use a centralized control console to shape traffic prioritizing or deprioritizing/blocking packets at a granular level. However, a fully converged approach requires integration of data, control, and management planes, which is not completed yet (Figure 10.24).

10.5.2.1.1 Data Plane

The integration of packet and optical data planes is achieved by assimilating DWDM or colored interfaces into the IP routers. This model (i.e., IP over DWDM) makes good sense if traffic is all-IP and there is strong organizational integration between the packet and the transport organizations. Large and incumbent service providers will continue to handle a lot of SDH/SONET for a long time to come, and there is not always much organizational cohesion in such big companies.

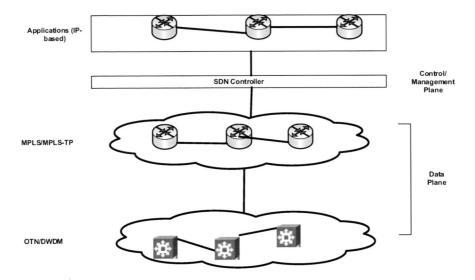

FIGURE 10.24   IP-Optical integrated approach.

In addition to its advantages, this model also has some shortcomings. The approach requires a number of different router line cards to allow for many kinds of optics such as 10G, 40G, 100G, and 400G covering distances from 10s to 1000s km. To reduce the development cost for producing this distinct number of cards, pluggable optical modules can be used. These modules come in several variants to allow for the different data rates and distances. At the same time, it may be kept in mind that the development in packet forwarding and fiber optics usually does not happen at the same time, dropping the productivity of cards as those cannot take advantage of the most recent technological advancements [63].

Along the same lines, data plane integration can run into vendor-lock. The physical layer interfaces serve as demarcation points between vendors. If there are fewer interfaces, it could result in less modularity. Thus, the industry may need to find a balance between data plane integration and realistic modularity/interoperability when different vendors are present in a network. The black-link approach defined in ITU-T G.698.2 Recommendations [64] may come handy; it removes transponders and provides optical interfaces Ss and Rs*. Every path from Ss to its corresponding Rs must comply with the parameter values of the application code.

10.5.2.1.2   Multi-Layer Control Plane

A multi-layer control plane will allow better coordination between an optical transport layer and its client layer (i.e., IP) in this case. This approach will reduce TCO by optimizing bandwidth resource utilization.

A control plane can be implemented in many ways in an optical-IP converged architecture. For example, the GMPLS protocol suite enables automated provisioning for services that use resources in several network layers, and it is also the first standard control plane for wavelength switched optical networks (WSONs)†. Moreover, an integrated multi-layer control plane provides path selection in the case of a network failure and signaling capabilities through software agility. The SDN which has been standardized improves scalability and simplifies operation. SDN decouples control of the network from hardware to software and enables operators to use a centralized console for traffic management [65]. The control plane is all together separated from the data/forwarding plane for all

---

* Ss is a single-channel reference point at the DWDM network element tributary input while Rs is a single-channel reference point at the DWDM network element tributary output.

† WSON is a telecommunications network that consists of two planes, namely data and control.

switching technologies and standardized interfaces. The controller in SDN has a wider scope than the ASON*/GMPLS control plane and it needs to support additional functions such as NFV[†] [66].

10.5.2.1.3 Management Plane

The management plane from L0-L2 is considerably more robust than L3 (IP layer). The IP layer has been traditionally managed via CLI (command line interface) functions. Service creation also usually requires access to two different network management systems. The target is to develop a single multi-layer integrated provisioning system. SDN, which is under development, is designed to provision and configure paths and services in converged networks. However, a comprehensive management system is required that can address the true convergence of IPOC and at the same time be upgradable to support 5G.

10.6 CONCLUSION

A mix of transport technologies will be used to carry 5G traffic. Backhaul is expected to be a major bottleneck that will be required to support a dense HetNet 5G environment. Small cells and traditional BTSs can use lower frequency bands microwave radios, but for the majority, e-band radios and/or fiber will be necessary. MPLS-TP and true IP-Optical converged architecture with SDN is perhaps needed for 5G networks.

PROBLEMS

1. Define the key domains of transport networks?
2. Reproduce LTE and 5G transport capacity requirements (Table 9.1), assuming 40 and 120 MHz channel sizes, respectively, and compare the results with the said table?
3. What are the minimum and maximum channel bandwidths of e-band radios? Discuss the most suitable channel sizes for the deployment of e-band radios?
4. Define the technique for the licensing of e-band radio links and provide at least two examples.
5. State the key differences between 1st and 2nd generation e-band radio products?
6. What are the key propagation challenges at such e-band frequencies?
7. A line-of-sight microwave radio link has a time availability of 99.99%. What is its unavailability in percentage and in minutes?
8. What will be the free space loss at 70 GHz for 2 km link? Is it more than the loss at 80 GHz?
9. If a vertically polarized, 80 GHz link located in Islamabad, Pakistan, was required to have 4 lines availability over a distance of 2 km, what would be the loss in signal strength due to rain per kilometer? Islamabad lies in the L rain region?
10. What will be the availability of an 83.5 GHz microwave radio link operating over a distance of 1.5 km? The transmit power is 20 dBm, antenna gains of 45 dBi, transmitter/receiver losses (cable, connecters, etc.) of 1.2 dB, miscellaneous losses of 0 dB, and a path loss of 134.32 dB. The receiver sensitivity is -64 dBm @ 10^{-6} bit error rate with 16-QAM modulation for supporting 1 Gbps traffic in a 250 channel bandwidth?
11. Describe the key reasons e-band radios are needed by the telecom sector of Pakistan?
12. What were the key IETF objectives for developing MPLS-TP?
13. What is MPLS-TP and how it is different from MPLS?
14. Describe the key elements and functions of MPLS-TP?
15. Describe the MPLS and MPLS-TP interworking approaches?
16. Define the key differences between the two MPLS-TP OAM standards?
17. What are the key options to manage and control the operations of MPLS-TP?

---

* ASON (Automatically Switched Optical Network) is a control plane for SDH and OTN layers.
[†] NFV is a concept for virtualization network services that are currently being carried out by proprietary/dedicated hardware.

18. What is automatic protection switching and list some networks where it has been implemented.
19. What are Automatic Protection Coordination and Protection State Coordination in APS?
20. What are two key services layering models that exist in core transport networks?
21. Define the IP over DWDM concept?
22. Define IP/Ethernet/OTN/DWDM networks?
23. Define IP and Optical convergence?
24. Describe the role of SDN for transport networks?

REFERENCES

1. Asif, S. et al. 2009. An Evaluation of Future Mobile Networks Backhaul Options. *5th Int'l. Conference Wireless and Mobile Communications*, Cannes/La Bocca, France, Aug. 23–29, 2009, pp. 146–51.
2. NGMN 2011. A White Paper by the NGMN Alliance—Guidelines for LTE Backhaul Traffic Estimation.
3. Sharma, S. 2014. *Integrated Backhaul Management for Ultra-Dense Network Deployment* (Master Thesis Report), KTH Royal Institute of Technology.
4. Basit, S.A. 2009. *Dimensioning of LTE Network Description of Models and Tool, Coverage and Capacity Estimation of 3GPP Long Term Evolution Radio Interface* (Master Thesis), Helsinki University of Technology.
5. Allen, J., Chevalier, F. and Bora, B. 2014. Mobile Backhaul Market: Phase 1 Report (Report for Vodafone), Analysys Mason.
6. Croy, P. 2011. *LTE Backhaul Requirements: A Reality Check*. Aviat Networks, Inc, California, USA.
7. Penttinen, J.T.J. 2011. *The LTE/SAE Deployment Handbook*, Wiley, UK.
8. Penttinen, J.T.J. 2015. *The Telecommunications Handbook: Engineering Guidelines for Fixed, Mobile and Satellite Systems*, Wiley, UK.
9. Wiltse, J.C. 1984. History of Millimeter and Submillimeter Waves. *IEEE Transactions on Microwave Theory and Techniques*, 32(9):1118–1127.
10. Rosenblum, E.S. 1961. Atmospheric Absorption of 10–400 KMCPS Radiation: Summary and Bibliography to 1961. *Microwave Journal*, 4:91–96, Mar. 1961.
11. ECC Recommendation (05)07 2009. Radio Frequency Channel Arrangements for Fixed Service Systems Operating in the Bands 71–76 and 81–86 GHz, ECC with CEPT.
12. ETSI EN 302 217-3 (V1.3.1) 2009. Fixed Radio Systems; Characteristics and Requirements for Point-to-Point Equipment and Antennas; Part 3: Equipment Operating in Frequency Bands where Both Frequency Coordinated or Uncoordinated Deployment Might be Applied; Harmonized EN Covering the Essential Requirements of Article 3.2 of the R&TTE Directive, ETSI.
13. NEC. iPasolink EX. http://www.nec.com/en/global/prod/nw/pasolink/products/ipasolinkEX.html
14. Proakis, J.G. and Salehi, M. 1994. *Communications Systems Engineering*. Prentice Hall, Englewood Cliffs, New Jersey, USA.
15. Wells, J. 2010. *Light Licensing*. E-Band Communications Corporation, California, USA.
16. Lehpamer, H. 2002. *Transmission Systems Design Handbook for Wireless Networks*. Artech House, Inc., Norwood, MA, USA.
17. Asif, S.Z. 2015. E-band Microwave Radios for Mobile Backhaul. *International Journal Wireless and Microwave Technologies*, 2015(4):37–46.
18. Adhikari, P. 2008. *L1104-WP: Understanding Millimeter Wave Wireless Communication*. Loea Corporation, Hawaii, USA.
19. Johnson, E. 2013. *Mobile Data Backhaul: The Need for E-Band*. Sky Light Research, Arizona, USA.
20. Bridgewave. 80 GHz Wireless. http://www.bridgewave.com/products/80ghz.cfm
21. E-Band Communications, LCC. 4G Evolution Series. http://www.e-band.com/ProductOverview
22. Huawei. E-band Microwave. http://www.huawei.com/en/solutions/broader-smarter/hw-196711.htm
23. NEC. iPasolink EX. http://www.nec.com/en/global/prod/nw/pasolink/products/ipasolinkEX.html
24. SiAE. ALFOplus80 series. https://www.siaemic.com/index.php/products-services/telecomsystems/microwave-product-portfolio
25. ITU-R 2012. Recommendation ITU-R P.530-14 Propagation Data and Prediction Methods Required for the Design of Terrestrial Line-of-Sight Systems.
26. ITU-R 2005. Recommendation ITU-R P.838-3 Specific Attenuation Model for Rain for Use in Prediction Methods.

27. 4Gon Solutions 2013. Path Loss, Link Budget & System Operating. http://www.4gon.co.uk/solutions/technical_path_loss_link_budget_som.php
28. LightPointe 2010. Millimeter-Wave (MMW) Radio Transmission: Atmospheric Propagation, Link Budget and System Availability.
29. ITU-R 2001. Recommendation ITU-R P.676-5 Attenuation by Atmospheric Gases.
30. CCIR (ITU-R) 1978. Attenuation and Scattering by Precipitation and Other Atmospheric Particles, Report 721, in Vol. V, Propagation in Non-Ionized Media, Recommendations and Reports of the CCIR, 1978, pp. 107–115.
31. Recommendation ITU-R PN.837-1 1994. Characteristics of Precipitation for Propagation Modeling (Question ITU-R 201/3).
32. Freeman, R.L. 2006. *Radio System Design for Telecommunications*, Third Edition. Wiley Inc., UK.
33. Tjelta, T., Tran, D.V. and Tanem, T. 2012. Radiowave Propagation and Deployments Aspects of Gigabit Capacity Radio Links Operating at 70/80 GHz. *IEEE-APS Topical Conference on Antennas and Propagation in Wireless Communications (APWC)*, Cape Town, South Africa, September 2–7, 2012, pp. 412–415.
34. Tjelta, T. 2009. Measured Attenuation Data and Predictions for a Gigabit Radio Link in the 80 GHz Band. *3rd European Conference on Antennas and Propagation*, Berlin, Germany, March 23–27, 2009, pp. 657–661.
35. Asif, S. 2011. *Next Generation Mobile Communications Ecosystem: Technology Management for Mobile Communications*. Wiley Inc., UK.
36. Faisal, N. et al. 2013. Third Successive Active Monsoon over Pakistan—An Analysis and Diagnostic Study of Monsoon 2012. *Pakistan Journal of Meteorology*, 9(18):73–84.
37. IETF RFC 2544 1999. Benchmarking Methodology for Network Interconnect Devices.
38. Pakistan Telecom Authority. Telecom Indicators. http://www.pta.gov.pk/index.php?option=com_content&task=view&id=269&Itemid=658
39. GSMA 2016. Country Overview: Pakistan—A digital future.
40. Kowalke, M. 2014. Microwave Transmission Evolves Thanks to New Ring Topology Standard, Next Generation Communications. http://next-generation-communications.tmcnet.com/topics/service-provider/articles/382733-microwave-transmission-evolves-thanks-new-ring-topology-standard.htm
41. Hubbard, S. 2011. *MPLS-TP in Next-generation Transport Networks*, Heavy Reading, New York, USA.
42. RFC 5921 2010. A Framework for MPLS in Transport Networks, IETF, July.
43. RFC 5317 2009. Joint Working Team (JWT) Report on MPLS Architectural Considerations for a Transport Profile, IETF, February.
44. RFC 5921 2010. A Framework for MPLS in Transport Networks, MPLS Working Group, July.
45. RFC 3985 2005. Pseudo Wire Emulation Edge-to-Edge (PWE3) Architecture, March.
46. King, D. et al. 2014. MPLS-TP Linear Protection for ITU-T and IETF. *IEEE Communications Magazine*, 52(12):16–21.
47. ITU-T G.8113.1/Y.1372.1 2012. Operations, Administration and Maintenance Mechanism for MPLS-TP in Packet Transport Networks, November.
48. ITU-T G.8113.2/Y.1372.2 2015. Operations, Administration—Transport and Maintenance Mechanisms for MPLS-TP Networks using the Tools Defined for MPLS, August.
49. ITU-T G.8013/Y.1731 2015. OAM Functions and Mechanisms for Ethernet based Networks, August.
50. RFC 5654 2009. MPLS-TP Requirements, Network Working Group, September.
51. RFC 6373 2011. MPLS Transport Profile (MPLS-TP) Control Plane Framework, September.
52. ITU-T G.8131/Y.1382 2007. Linear Protection Switching for Transport MPLS (T-MPLS) Networks, February.
53. ITU-T G.8131/Y.1382 2014. Linear Protection Switching for MPLS Transport Profile, July.
54. RFC 7271 2014. MPLS Transport Profile (MPLS-TP) Linear Protection to Match the Operational Expectations of Synchronous Digital Hierarchy, Optical Transport Network, and Ethernet Transport Network Operators, June.
55. Rios, F.J. et al. 2014. IP and Optical Convergence: Use Cases and Technical Requirements, White Paper.
56. Tzanakaki, A. et al. 2014. MS103 (MJ1.1.1) White Paper—Future Network Architectures, Document Code: GN3PLUS14-976-35.
57. Asif, S.Z. et al. 1999. The Optical Internet: A New Network Architecture. *15th Annual National Fiber Optics Engineers Conference*, Chicago, IL, USA, Sept. 26–30, 1999.
58. Huawei 2015. Soft-Decision FEC: Key to High-Performance 100G Transmission—FEC Ensures Reliable Transmission. http://www.huawei.com/ilink/en/solutions/broader-smarter/morematerial-b/HW_112021
59. ACG Research 2012. Why Service Providers Should Consider IPoDWDM for 100G and Beyond.
60. ZTE 2013. Technical Whitepaper on IP MPLS and MPLS-TP Interoperability.

61. Azizi, M., Benaini, R. and Mamoun, M.B. 2013. Key Requirements for Interworking between MPLS-TP Network and IP/MPLS Network. *International Journal of Engineering and Technology*, 5(4):3351–3358.
62. Cisco 2013. Cisco Open Network Environment: IP and Optical Convergence.
63. Ericsson 2014. IP-Optical Convergence: A Complete Solution, Ericsson Review.
64. ITU-T G.698.2 2009. Amplified Multichannel Dense Wavelength Division Multiplexing Applications with Single Channel Optical Interfaces, November.
65. FierceTelecom 2013. Converging the Optical Core to IP.
66. Optical Interworking Forum 2013. OIF Carrier WG Requirements on Transport Networks in SDN Architectures Transport SDN, September.

11 Core Network and Operational Support System

Core network is the segment that connects the RAN with operators' in-house application servers, IP multimedia subsystem or the Internet. It is comprised of both circuit switched and packet switched elements. Historically, it supported only circuit switched (CS) services, but later with the advent of 3G it started to support packet switched (PS) services as well. LTE and 4G systems only require PS support and with the passage of time, core networks are expected to only provide for IP/Ethernet services.

Figure 11.1 describes the evolution of core networks from 2G to 5G based on 3GPP standards. The initial 2G systems supported CS only with key elements like MSC and SMSC, and common elements HLR, VLR, EIR, and SGW*. To support data services, SGSN and GGSN were introduced in 3GPP Release-99 which was the first release on UMTS (3G). Release-4 (2001) divided the MSC into two functional elements, namely, an MSC server for signalling and a media gateway function as the user plane to reduce operational challenges. Later, Release-5 (2002) brought IMS, which was primarily developed for the mobile 3G devices communicating over IP with embedded SIP clients. In Release-6 (2005), a new functional node, that is, BM-SC (the Broadcast Multicast-Service Center), was added to support MBMS. Release-7 (2007) introduced the concept of direct tunneling which enables a split between the control plane and user plane toward the packet core networks.

LTE (Release 8) brought EPC that was only designed to support PS services including elements like MME to manage user equipment mobility and identity and Gateway (Serving and Packet) for packet routing and connecting to external networks, respectively. Rel-9, Rel-10 (LTE-Advanced/4G), Rel-11, and Rel-12 have not brought any fundamental architectural changes in EPC. Rel-13 has introduced the concept of Dedicated Core Network (DCN) along with network slicing which will be explained later in the chapter. The specifications on the 5G core network are expected to be finalized in Rel-15 (2018) and Rel-16 (2020). For 5G, we may see EPC or core network going in the Cloud supported by technologies such as SDN and NFV, addressing all types of IP based services.

The chapter provides a short overview of EPC, the evolution of EPC, and the evolution of IMS, whereas the details on core networks can be found in [1]. Additionally, this chapter also briefly presents insights on 5G core network, CDN (Content Delivery Network), and LTE and 5G OSS/BSS (Operational/Business Support Systems).

11.1 EVOLVED PACKET CORE

3GPP Rel-8 produced EPC to support LTE and only packet switched traffic [2,3]. The EPC consists of two main entities, namely MME and Gateway as illustrated in Figure 11.2. MME supports the control plane functionalities like SSGN, and it is separated from the node that performs bearer-plane functionality, that is, Gateway (GW). Key functions of EPC are as follows:

The *MME* main function is to manage UE mobility and UE identity. It also performs authentication and authorization; idle-mode UE tracking and reachability; security negotiations; and NAS (Non Access Stratum)[†] signaling. It is connected to EUTRAN via S1-MME interface.

---

* Abbreviations are spelled out in the glossary.

[†] NAS is a functional layer in the UMTS and LTE wireless telecom protocol stacks between the core network and user equipment. This layer is used to manage the establishment of communication sessions and for maintaining continuous communications with the user equipment as it moves [4].

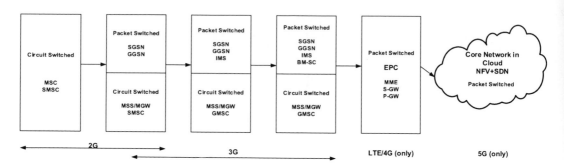

FIGURE 11.1   Core network evolution.

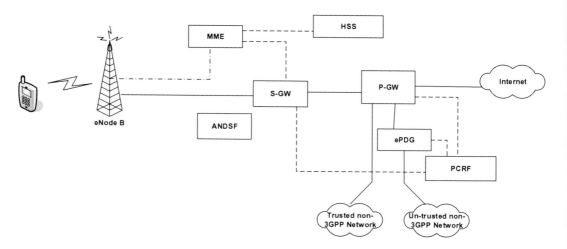

FIGURE 11.2   Evolved packet core.

The *S-GW* connects to EUTRAN via S1-U interface. For each UE associated with the EPS, at a given point of time, there is one single S-GW. The S-GW routes and forwards user data packets, while also acting as the mobility anchor for the user plane during inter-eNodeB handovers and as the anchor for mobility between inter-3GPPP technologies. For idle state UEs, the S-GW terminates the downlink data path and triggers paging when downlink data arrives for the UE. It manages and stores UE contexts, for example, parameters of the IP bearer service and network internal routing information. It also performs replication of the user traffic in case of lawful interception.

The *Packet Gateway (PGW)* serves as the entering point in EPS (Evolved Packet System). It is connected to external PDNs, operators' IMS, and non-IMS IP services, and provides access for trusted and nontrusted non-3GPP IP networks. It acts as the anchor for mobility between 3GPP and non-3GPP access systems. The PGW performs policy enforcement, packet filtering (by, e.g., deep packet inspection*) for each user, charging support, and lawful interception. It also performs transport level packet marketing in the uplink and downlink.

EPC also consists of the following elements to support authentication and mobility management [4].

---

* Deep Packet Inspection enables advanced security functions as well as Internet data mining, eavesdropping, and censorship.

The *HSS (Home Subscriber Server)* is a central database that contains user related and subscription related information. The functions of the HSS include functionalities such as mobility management, call and session establishment support, user authentication, and access authorization. The HSS is based on pre-Rel-4 Home Location Register (HLR) and Authentication Center (AuC).

The *ANDSF (Access Network Discovery and Selection Function)* provides information to the UE about connectivity to other 3GPP and also to non-3GPP access networks (such as WiFi). The purpose of the ANDSF is to assist the UE to discover the access networks in their vicinity and to provide rules (policies) to prioritize and manage connections to these networks.

The *ePDG's (Evolved Packet Data Gateway)* main function is to secure the data transmission with a UE connected to the EPC over an untrusted non-3GPP access. For this purpose, the ePDG acts as a termination node of IPsec* tunnels established with the UE.

The *Policy and Charging Rules Function (PCRF)* is an optional component which can be supported even when IMS is not supported. It determines policy rules in a multimedia network in real time. It also supports service data flow detection, policy enforcement, and flow-based charging.

11.1.1 KEY CORE NETWORK FEATURES

Some key features of core networks from Rel-8 onwards are briefly described in this section [5–15].

11.1.1.1 SMS in MME

SMS (Short Message Service) in MME was introduced to support networks that do not have a 3GPP complaint MSC and also do not support MAP (Mobile Application Part).

Another option is having SMS over IMS, however, this would require SIP clients in devices and not all devices (for example, machine type devices and dongles) have such capability. Additionally, it is for inbound roamers whose home networks do not support IMS, cannot be offered SMS over IMS, and thus will need support for SMS over NAS. To address such challenges, 3GPP introduced a new architecture in TS 23.272 [16] for supporting SMS services in EPC which is shown in Figure 11.3. This feature can be enabled or disabled in the MME via simple reconfiguration. From the UE perspective, it remains transparent whether SMS in MME or SMS over SGs interface (between MME and MSC) is offered by the network. The HSS can support both a Diameter protocol S6c and

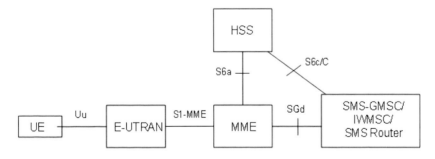

FIGURE 11.3 SMS in MME architecture. (From 3GPP TS 23.272 (V11.4.0) 2013 Circuit Switched (CS) Fallback in Evolved Packet System (EPS); Stage 2. Technical Specification (Release 11), Technical Specification Group Services and System Aspects, 3GPP, March [16].)

---

* Internet Protocol Security (IPsec) is a protocol suite for securing IP communications by authenticating and encrypting each IP packet of a communication session [4].

a MAP C interface to SMS entities that have not evolved to S6c. Otherwise, SMS entities can support both interfaces to obtain routing information from HSS's that have not evolved to S6c.

11.1.1.2   QoS Control Based on Subscriber Spending Limits

A Policy Charging Control (PCC) architectural enhancement called QoS-SSL (Quality of Service - Subscriber Spending Limit) has been specified by 3GPP. QoS-SSL allows monitoring of subscriber usage in relation to a spending limit and permits action to be taken when the limit is reached. Such spending limits may be by user or by application.

The feature of QoS-SSL provides an operator with the ability to deny access to a user to a particular service if the user has reached his/her allocated spending limit within a certain time period. Furthermore, the said feature could modify QoS if such a threshold (spending limit) has been reached, providing operators additional means to shape a user's traffic. Such authority allows operators to deny monopolization of a network resource at any one time by a user. Users also have the opportunity to purchase additional credit to increase their spending limit [9–10,12,17].

11.1.1.3   User Data Convergence

The User Data Convergence (UDC) as described in 3GPP TS 22.101 [18] supports a layered architecture separating the data from the application logic. The user data is stored in a logically unique repository allowing access from core and service layer entities, named application front-ends. Furthermore, network elements and functionalities shall have the flexibility to access profile data remotely and without having to store them locally on a permanent basis.

11.1.1.4   Enhanced Nodes Restoration for EPC

Various components of EPC are designed to be highly reliable, but occasional restart or failure due to various reasons cannot be avoided. 3GPP technical report TR 23.857 [19] specifies certain serious operational errors if the MME, S4-SGSN*, SGW or PGW fails along with the recovery mechanisms. For example, a user may not be able to receive any IMS terminating call following an MME failure, and thus certain enhancements are needed to provide service resiliency.

Work on this item was started in Rel-10, but the release mainly specified restoration procedures for MME/S4-SGSN failure. Certain enhanced EPC node restoration procedures are specified in Rel-10 and Rel-11 to restore CS and PS services upon occurrence of various failures in EPC. Rel-12 specified enhanced restoration procedures defining EPS behavior and enabling restoration of the eMBMS service when possible in order to minimize impact on end user service.

11.1.1.5   Lawful Interception in the 3GPP Rel-13

Studies are ongoing on how to extend 3GPP Legal Interception (LI) service to accommodate Rel-13 service enhancements and extensions. The said service will be further extended as needed to meet national LI requirements. The national LI requirements will apply to a portion of such Rel-13 enhancements including but not limited to EPS enhancements, IMS VoIP (voice over IP), mobile terminated call, proximity service, and so on.

The enhancements to specifications TS 33.106, TS 33.107 and TS 33.108 will address LI service requirements, LI architecture, LI functions, and the HI2 (Intercepted Related Information) and the HI3 (Content) interfaces to delivery to the law enforcement monitoring facilities [20].

11.2   EVOLUTION OF IMS

IMS is designed to be the architectural framework for enabling various multimedia services and required signalling. It is an access agnostic system, introduced in 3GPP Release-5.

---

* S4: This provides related control and mobility support between GPRS core and the 3GPP anchor function of a S-GW and is based on Gn reference points as defined between SGSN and GGSN.

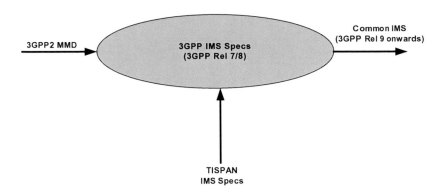

FIGURE 11.4 Common IMS concept. (From Asif, S. 2011 *Next Generation Mobile Communications Ecosystem: Technology Management for Mobile Communications.* Wiley Inc., UK [1].)

IMS was originally designed by 3GPP as a part of the vision for evolving mobile networks beyond GSM. Its original formulation (3GPP Rel-5) represented an approach to delivering "Internet services" over GPRS. This vision was later included by 3GPP2 and TISPAN in order to include networks other than GPRS, such as Wireless LAN, CDMA2000, and fixed line.

It was agreed in 2007 that 3GPP would take overall responsibility of IMS specifications to avoid further fragmentation of IMS specifications. A common IMS understanding was envisioned to address the interoperability between the different IMS standards and future IMS-enabled networks. Common IMS, as shown in Figure 11.4, is expected to bring economies of scale and reductions in capital and operational costs. The ETSI TISPAN IMS specifications were moved to 3GPP in 2007 while 3GPP2 MMD requirement specifications were merged with Release 8. 3GPP maintains one common set of IMS specifications starting from Rel-9. The Rel-9 specifications as a whole were frozen in December 2009 [1].

The following subsections will primarily look into the evolution of IMS as defined from Rel-9 on and will only list key features [5–15,20]. The details on IMS from 3GPP, 3GPP2, and WiMAX perspectives including architectures can be found in [1].

11.2.1 IMS Service Continuity and Centralized Services

These two functionalities allow IMS to control the service delivery as well as enable services to progress over HetNets. There is a wide degree of commonality between these two areas (e.g., they are both provided by the same application server and are functionalities that can be invoked as part of the same service) which expedites its delivery to the market. Though the performance management measurements for these functions were defined in Rel-11, work has been ongoing since Rel-8.

11.2.1.1 IMS Service Continuity

3GPP has been evolving aspects of IMS service continuity in its several releases to support the multimedia session continuity experience. Rel-7 developed the definition of Voice Call Continuity (VCC) and Rel-8 built on this to define Service Continuity (SC) and VCC for SRVCC. Rel-9 and onward releases have added further enhancements to this functionality.

In Rel-8, SC allows a user's entire session to be continued seamlessly as the user's device moves from one access network to another. In Rel-9, this functionality has been enhanced to allow components of a user's session to be transferred to, and retrieved from, different devices belonging to the user. For example, a video call or video stream in progress on a mobile device could be transferred to a laptop or even a large-screen TV (assuming both support IMS), for an enhanced user experience and then, if necessary, retrieved to the mobile device. As well as transferring existing media, the user can add or remove media associated with a session on multiple devices, all controlled from a single device. These devices may be on different 3GPP or non-3GPP access networks.

Rel-11 gives operator the ability to assign an additional MSISDN, in addition to the original MSISDN, to a subscriber with a packet switched subscription. If the additional MSISDN is available, it can be used for correlation of circuit switched and IMS in voice call and service continuity as well as IMS Centralized Service.

11.2.1.2   IMS Centralized Services

The IMS centralized services feature was developed in Rel-8 and provides voice services and service control via IMS mechanisms and enablers, while providing voice media bearers via CS access. Users, therefore, subscribe to IMS services and can receive those services regardless of whether the voice media is carried over PS access or CS access. Within the limitations of the CS access capabilities, the user has the same experience with the services.

The services are controlled via a channel that is provided either by IMS (via PS access, if supported simultaneously with CS access) or through interworking of legacy CS signaling into IMS by the MSC server. The latter capability allows support of legacy user devices, but cannot provide new, richer voice services to the user.

Rel-9 has enhanced this functionality to add support of video media. Also added is an optional service control channel from the user's device to IMS that is transparent to the MSC server. This avoids the need to update legacy CS networks and allows new services to be developed but cannot support legacy user devices.

A new feature in Rel-11 gives the operator the ability to assign an additional Mobile Subscriber Integrated Services Digital Network-Number (MSISDN), in addition to the original MSISDN, to a subscriber with a PS subscription. When the additional MSISDN is available, it is used for correlation of CS and IMS in voice call and service continuity as well as IMS centralized service. This improves the ability to give simple and flexible implementations to perform IN (Intelligent Network) type services in the PS environment. The implementation results in the possibility to conditionally include an additional MSISDN field in the location update procedure from the HSS to the MME/SGSN [10].

11.2.2   eMPS FOR IMS

The Multimedia Priority Service or MPS allows certain security and emergency preparedness users to make priority calls and/or establish sessions using the public networks during network congestion conditions. Such users are likely government-authorized personnel, emergency management officials, and/or other authorized users. Rel-9 based MPS only allowed IMS-based voice call origination/termination between such users with priority. To extend the process to circuit switched users, Rel-10 specified mechanisms for the priority handling of IMS based multimedia services (voice, video, etc.), PS data, and CS-voice (for CS-fallback) with regard to LTE/EPC which is called eMPS. Rel-11 provided a mechanism to handle SRVCC for an IMS based priority call established in LTE in order for the network to successfully hand over the call from LTE to CS in GERAN/UTRAN or 1xCS [9].

The mechanism to handle SRVCC for an IMS-based priority voice or voice plus video session established in LTE is to reuse the priority handling mechanisms that were already defined for UMTS and GSM/EDGE in TS 25.413 [21] and TS 48.008 [22], respectively. In simple terms, when a user has an on-going IMS voice call and the call is prioritized using eMPS, and when the call is handed over from LTE to circuit switched GERAN/UTRAN or 1xCS using SRVCC procedure, the procedure can be prioritized depending on regulatory requirements/operator policy even under network congestion situations.

11.2.3 IMS, IETF—Protocol Alignment

The ongoing work since Rel-9 is about maintaining alignment of the development of SIP used in IMS with that defined in IETF. The scope of this work is protocol alignment, and those capabilities that may lead to new or enhanced IMS applications that are not part of this process.

11.2.4 UICC Access to IMS Specifications

A work item was introduced in Rel-10 to specify necessary mechanisms in UICC (Universal Integrated Circuit Card) and UE to make use of IMS functionality implemented in UE. It will consider both presence and absence of ISIM (IP Multimedia Services Identity Module) in UICC.

IMS allow operators to develop new value-added applications as well as to enhance their existing solutions. These IMS-based applications may be located in the UE whereas additional IMS-based applications could reside and be executed in the UICC. This will lead to new opportunities and allow, for example, the development of operator specific IMS-based applications that require a high level of security and portability.

11.2.5 IMS Emergency Calls

The support for IMS emergency calls over GPRS and HSPA was introduced in Rel-7 including support for PSAP (Public Safety Answering Point) call back and to handle supplementary services accordingly.

There has also been a need to support IMS emergency sessions over LTE, and other wireless packet switched technologies (e.g., WLAN and CDMA2000). Additionally, support for IMS emergency sessions with other media on UTRAN and E-UTRAN is also required.

This enhancement has been added to support IMS emergency sessions that allow the UE to use other media and communication types (real time video, file transfer, etc.) than voice during an IMS emergency session. This occurs when the network supports IMS voice emergency calls and the UE also supports other media or communication types.

When a UE with an active IMS emergency session moves out of IMS voice coverage, the voice call continuity is supported by the UE and network and it then becomes a CS emergency call. Other media will be dropped when a UE with an active IMS emergency session moves out of IMS voice coverage, regardless of whether or not there is an active voice session. The deployment of IMS emergency calls is dependent on local regulatory requirements.

11.3 5G CORE NETWORK

Core networks traditionally have been designed as a single architecture addressing a range of requirements and supporting backward compatibility. This one size fits all approach has been successful in keeping the costs down to a reasonable level and by supporting legacy circuit switched and today's packet switched functionalities.

This core network, however, is rigid in the sense that it is not flexible enough to accommodate the customized and variable connectivity needs of individual users and businesses that are expected in the future. However, with virtualization, NFV, SDN, and network slicing, it is possible to make core networks more flexible and scalable. Thus, the next generation core network is expected to exist in a cloud-based environment with a high degree of virtualization and software-based networking. Such flexibility is needed to support a variety of access networks such as 3G, LTE, 4G, WiFi, and tomorrow's 5G.

11.3.1 Components of Core Network/High Level Architecture

The current EPC will further evolve to support virtualization and network slicing to become NGC applicable for 5G networks.

Network slicing is often termed as logical instantiation of a network possibly due to virtualization technologies [23]. The concept is seen as the natural extension/evolution of the current network sharing methodologies [24]. Network slicing is one of the promising techniques that will likely exist in both radio access and core networks. It allows multiple logical networks to be created on top of a common physical infrastructure. Either DCN (Dedicated Core Network) or a combination of NFV and SDN can be used as a technology to enable network slicing along with orchestration and analytics [25,26]. DCN or Décor as defined in 3GPP TS 23.401 [27] is a feature that enables an operator to deploy multiple logical mobile core networks connected to the same RAT or multiple RATs (e.g. GERAN, UTRAN, E-UTRAN, WB-E-UTRAN and NB-IoT). A DCN consists of one or more MME/SGSN and it may be comprised of one or more SGW/PGW/PCRF. This feature enables subscribers to be allocated to and served by a DCN based on subscription information (e.g., "UE Usage Type").

With 5G, a single terminal can use multiple services with different characteristics almost simultaneously. In such cases, a network slice can be created for each service, requiring all such slices to coordinate control for that particular single terminal. These slices can be mapped to respective radio and core network slices to provide end-to-end connectivity. The methodology is currently being specified for selecting radio/core networks particularities for supporting slicing in existing as well as in future 5G systems.

Control and User Planes' Separation: The separation of control and user planes is one of the key principles of 5G core network architecture. 3GPP started a study in TR 23.714 [28] on user/control planes' separation involving core network elements such as P-GW, Traffic Detection Function (TDF), and so on. This separation allows independent scaling of each plane and migration toward cloud-based architecture. For example, the control plane can be placed in a centralized location with complex hardware and processing capabilities. On the other hand, the user plane can be distributed to a larger number of local sites making reachability from the perspective of a user easier. A good example of this will be content caching in local sites instead of securing it from the main server sitting thousands of miles away. This separation is the fundamental concept of SDN and having SDN will make core networks more flexible.

Further details on network slicing, NFV, and SDN can be found in Chapter 5 5G Concepts.

11.4 CONTENT DELIVERY NETWORK

CDNs have been developed and evolved to meet the growing needs of efficient delivery of content over the Internet. A CDN uses a widely distributed network of cache servers (surrogate servers) to spread content closer to the edge of telecommunication networks that would otherwise be concentrated in a few centralized locations. A CDN enables the delivery of replica content from multiple highly distributed sites and reduces the issues of congestions, latencies, and failures. However, as the Internet, which is an amalgamation of loosely-coordinated heterogonous autonomous networks, works mostly in a best effort fashion, a number of the bottlenecks are outside the control of any given entity including CDNs [29–31].

11.4.1 CDN Platform

A basic CDN consists of several surrogate servers, a transport (delivery) system, data and analysis system, content routing system, and operations center as shown in Figure 11.5. CDN can use different models for communications, where each model lists its own challenges within the context of a highly distributed Internet platform. For all such models, a CDN provider faces key challenges of scale

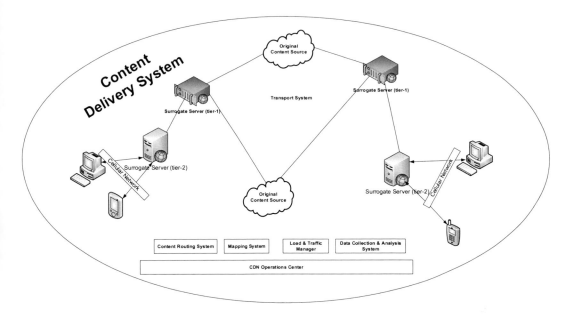

FIGURE 11.5 CDN block diagram.

(communicating with 1000s of machines) and reliability (providing the best connectivity to the rest of the Internet). A brief overview of CDN components is as follows:

Surrogate servers are the delivery servers which are the original servers responsible for forwarding the copies of the original content to the concerned user. These make the delivery of content faster and more effective. Such servers are typically deployed throughout the CDN in a hierarchical fashion [32].

Transport system is a collection of network elements called content distributors supporting the delivery of publishers' content from the original server to one or more surrogate servers. These servers may also transport content signals that contain information such as validation, integrity, and expiration of the content. The CDN uses these content signals to maintain the integrity and consistency of the content in its surrogate servers. These can interact with the request routing system to inform about the content availability in different surrogate servers and with the accounting system to inform the content distribution activity so that it later can measure the volume of content distribution [30].

Content routing system routes the user to the appropriate surrogate server that can serve the required content. It can be said that CRS is the first that the user encounters when he/she places the request for content. The selection of the surrogate server is made on the basis of certain parameters such as its service availability, geographical proximity to the user, and capacity. The routing can be performed in multiple ways including the popular DNS (Domain Name System) based mechanism or less popular ones such as transport layer or application layer based.

Data collection and analysis system includes the logging server and analytics server. The logs are used for generating historical reports and customer bills whereas the analytics server allows near real-time monitoring and generating reports on service utilization and service availability.

CDN operations center like other operational management units is responsible for the management and provisioning of network and services. It is responsible for the provisioning of services such as live streaming, on-demand streaming, and so on, and configuring devices such as routers, switches, and surrogate servers.

Additional components such as the *mapping system* use historic and real-time data to inquire about the health of both the CDN and the Internet at large. This monitoring assists in the creation of maps to direct traffic in an effective manner. Another component, that is, *load and traffic manager,* is used for load balancing and fault-tolerance. It monitors Internet performance and collects traffic info and feeds this data to the mapping system [23].

11.4.2 CDN Delivery Model

Market research reports show that operators face enormous increase in IP traffic and in particular for video-on-demand. Equipment and Network Infrastructure vendors such as Cisco, Ericsson, and Alcatel-Lucent have varying estimates of video traffic growth; however, even the lowest prediction from Alcatel-Lucent suggests a 12-fold rise by 2020. The Alcatel-Lucent report predicts that viewing hours will rise from 4.8 to 7.0 hours per day driven by OTT (over the top) video, which will increase from 11% of the traffic to 48% [31].

CDN delivery mechanisms come handy in addressing the video traffic growth.

The three key categories of CDN delivery mechanism, each with its advantages and disadvantages are peer-to-peer, cache based, and Enterprise CDN [33].

With *peer-to-peer* mode, content is delivered to users from a central server. However, after its delivery, the user devices themselves become point-of-presence (POPs) and can deliver to other users. This platform is economical, but poses issues of security, content updation, and copyrights.

Private CDNs like *Enterprise CDN* operate within a corporation and deliver the same data at the same time to many users (i.e., operates in a multi cast manner). This mode is economical in terms of bandwidth resources and delivering, training, and corporate communications to an internal audience. At the same time, it has shortcomings when comes to delivery of video-on-demand/unicast traffic.

The *cache based* approach which is the most common one is for delivering content in both unicast and multicast modes.

Furthermore, the new VoD/unicast traffic demand represents a far larger share of overall traffic at the IP service edge than traffic at subscriber links. Unlike multicast delivery techniques which leverage a single video server stream to reach hundreds of users, unicast delivery requires a separate stream for each user, even when users are consuming the same content. This leads to congestion for those networks that use a centralized IP edge, based on legacy broadband gateways or broadband regional access server routers. It also raises transport and peering costs.

The two approaches to address VoD/unicast and, in general, for high demand content delivery are to either use an operator's home-grown CDN setup or outsource the content delivery to a CDN specialist.

11.4.3 CDN in Mobile Networks

This approach gives operators more control over how the content requests are routed across their networks and takes into account their own topologies and efficiencies. In this approach, the distribution infrastructure is placed into an operator's network, which could be either active (revenue impact) or passive (no revenue impact) CDN.

The active CDN will allow operators to gain revenue from content service providers by placing the required infrastructure in their network. Active CDN means that the required CDN network element can be placed in the cellular mobile operator's RAN or in the packet core as shown in Figure 11.6. Active CDN can improve end users' QoS and reduce mobile network investment costs, as well as create new active revenue streams in particular coupled to mobile network QoS (policy controller). Passive CDN is more like transparent caching where the CDN network element is getting

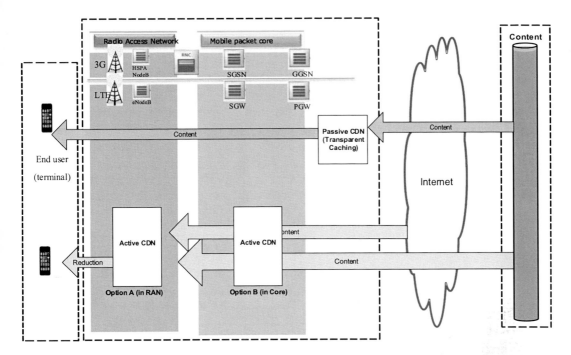

FIGURE 11.6   CDN in mobile networks.

traffic from external sources that are terminating at GGSN. This type of caching helps in reducing IP transit and peering expenses.

However, a consensus is building in the industry that the managed CDN services are more practical for operators. The operators can partner with CDN specialists for three key reasons, namely traffic offload, network performance, and revenue. Another reason in favor of this approach is the capacity crunch that is shifting the telco network design from a centralized multicast topology to a decentralized IP edge plus cache design. This approach will help operators to monetize on high value content through agreements with CSPs to offer improved QoE (Quality of Experience) through utilizing mobile network QoS.

11.4.4   CDN Use Cases

CDN improves the performance of networks by moving content closer to the end user. Some cases where CDN can make a difference are as follows [34]:

- As an Application Optimization Engine, CDN can optimize the amount of data being sent while increasing the utilization of available user bandwidth. Moreover, as CDNs connect the user to the geographically closest server, the net effect is in latency reduction introduced by the network. Such engines handle optimization at various networking levels to speed up the delivery of content to all types of users.
- As a file replicator, particularly when large data files are required at multiple geographies, CDNs can replicate data from a corporate headquarters in one region to one or more regional headquarters in a timely manner.
- As an ultimate bandwidth source, CDNs can be used to mitigate the effect of high volume, bursty, and ill-defined traffic growth, and keep the application performance at an acceptable level.

11.5   OPERATIONS SUPPORT SYSTEM

OSS/BSS provides OAM&P functions for all the different elements of mobile networks. The current OSS/BSS standards have been designed to handle a multitude of interfaces, fixed and mobile networks, mixed vendor environment, and to some extent rapid technological changes [1].

Until the mid-1990s, mobile telecommunications were solely run on postpaid subscriptions with few services making the job easier for OSS/BSS systems. Then from the mid-1990s until the early 2000s, there was a surge of prepaid subscriptions all around the world. This was a major shift requiring new thinking in the area of OSS/BSS since this necessitates the implementation of real time and online processes [35].

The inception of data with GPRS and 3G in the early 2000s was a ground-breaking development for mobile telecommunications. This caused the next shift for OSS/BSS since the services included data in addition to voice and SMS. Furthermore, toward the end of the last decade, the communications were not only meant for people, but also for devices and this trend is continuing today [35].

Frameworx was TM Forum's* answer to service providers, suppliers, and in general to the industry for their above-mentioned OSS/BSS challenges. The details on Frameworx and its components can be found in [1,36].

11.5.1   OSS/BSS FOR LTE

The E-UTRAN/EPC, collectively named EPS, uses the existing management reference model elaborating the operating systems interfacing with other systems within the single PLMN (Public Land Mobile Network) organizations or different organizations specified in TS 32.101 [37]. The TR 32.816 [38] on the other hand is a report on management of E-UTRAN and EPC. The EPS management reference model is a combination of the existing management reference model and the use case specific management reference models. The use cases, for example, can be "establishment of new eNodeBs in the network," "coverage and capacity optimization," and so on [1].

Figure 11.7 shows a 3GPP and EPS architecture showing management interfaces and management areas. The following interfaces are highlighted in Figure 11.7:

- Itf-T between a terminal and an NE (Network Element) manager.
- Itf-B and Itf-R between UTRAN and an NE manager.
- Itf-G1 between GSM NSS and NE manager.
- Itf-G2 between GSM BSS and NE manager.
- Itf-S between E-UTRAN, EPC, and domain managers.
- Itf-G3 between GPRS NEs and an NE manager.
- Itf-N between E-UTRAN, EPC Domain Managers (DM), and NMS.

11.5.2   OSS/BSS FOR 5G

In the next decade, 5G will exist with legacy 2G, 3G, and 4G data networks along with numerous IoT communications. The multi vendor, multi technology infrastructure will be augmented with SDN/NFV based network infrastructure. This could result in another shift in the working of OSS/BSS systems [40]. The ETSI NFV MANO specification [41] was the first step in support of this expected shift. As stated in this specification, NFV-MANO alone cannot deliver all the NFV business benefits; it needs to integrate and interwork with other management entities for this purpose such as OSS, BSS, and so on, using interfaces offered by those entities and offering its own interfaces to be used by external entities. However, just extending existing OSS/BSS models to account for virtualization will

---

* TM Forum is the premier nonprofit global industry association providing for OSS/BSS needs.

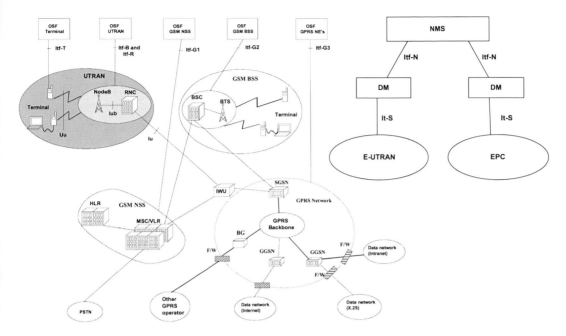

FIGURE 11.7 3GPP and EPS management architecture. (From Asif, S. 2011 *Next Generation Mobile Communications Ecosystem: Technology Management for Mobile Communications.* Wiley Inc., UK; 3GPP TS 32.102 V13.0.0 2016 Telecommunication Management; Architecture. Technical Specification (Release 13), Technical Specification Group Services and System Aspects, 3GPP, 3GPP, December [1,39].)

not be sufficient as it would not entertain value added services provided by NFV. Thus, evolutions of both NFV-MANO and OSS/BSS need to be coordinated to address rapid development [41].

11.6 SUMMARY

This chapter provided information on the prevalent core network, that is, EPC. EPC, defined in 3GPP Rel-8, was articulated first to support LTE and then LTE-Advanced (4G) access technologies. The overall EPC architecture has not been changed since Rel-8 and no significant change was made until Rel-14 which was frozen in June 2017. The current Rel-15 and Rel-16 are expected to thrash out the architecture and features of 5G core network which are expected to be frozen in 2018 and 2020, respectively.

A high-level background was also provided on IMS including its evolution. Certain key features of both core network and IMS were also presented. Information on LTE and 5G OSS/BSS was also presented. CDN was another topic that was discussed in the chapter. The cellular operators are likely moving toward managed CDN rather than home-grown CDN technology.

And, finally, for 5G, perhaps moving core network in the Cloud along with SDN/NFV is the answer.

PROBLEMS

1. Which types of services can current core networks handle?
2. What is the function of ANDSF?
3. Why was SMS in MME introduced?
4. Describe QoS-SSL?
5. What is UDC?
6. What are IMS and common IMS?
7. Define IMS service continuity?

8. Define IMS centralized services?
9. Define MPS and eMPS and their benefits?
10. How are emergency calls are supported in IMS?
11. What are the expected characteristics of the 5G core network?
12. Define content delivery network?
13. Briefly define the key components of CDN?
14. Describe CDN delivery models?
15. Discuss the way forward for providing CDN in mobile networks?
16. What is the basic purpose of having operations/business support systems?
17. Define Frameworx?
18. What is the next shift in the area of OSS/BSS?

GLOSSARY

3GPP   Third Generation Partnership Project
eMBMS   Evolved Multimedia Broadcast Multicast Services
EIR   Equipment Identity Register
EPC   Evolved Packet Core
EPS   Evolved Packet System
E-UTRAN   Evolved Universal Terrestrial Radio Access
GERAN   GSM EDGE Radio Access Network
GGSN   Gateway GPRS Support Node
GPRS   General Packet Radio Service
HLR   Home Location Register
HSPA   High Speed Packet Access
IETF   Internet Engineering Task Force
ISIM   IP Multimedia Services Identity Module
LTE   Long Term Evolution
MSISDN   Mobile Subscriber Integrated Services Digital Network-Number
NB-IoT   Narrow-Band Internet of Things
NFV   Network Functions Virtualization
PCRF   Policy Control and Charging Rules Function
PCC   Policy Charging Control
PDN   Packet Data Network
QoS   Quality of Service
SDN   Software Defined Networking
SGW   Service Gateway
SGSN   Serving GPRS Support Node
SMSC   Short Message Service Center
SSL   Secure Socket Layer
UE   User Equipment
UTRAN   Universal Terrestrial Radio Access Network
VLR   Visitor Location Register
WB-E-UTRAN   Wideband E-UTRAN

REFERENCES

1. Asif, S. 2011. *Next Generation Mobile Communications Ecosystem: Technology Management for Mobile Communications*. Wiley Inc., UK.
2. ETSI Mobile Competence Centre 2011. Overview of 3GPP Release 8, V0.2.4.
3. ETSI Mobile Competence Centre 2014. Overview of 3GPP Release 8, V0.3.2.

4. Wikipedia. http://en.wikipedia.org/wiki/Main_Page.
5. ETSI Mobile Competence Centre 2013. Overview of 3GPP Release 9, V0.2.9.
6. ETSI Mobile Competence Centre 2014. Overview of 3GPP Release 9, V0.3.3.
7. ETSI Mobile Competence Centre 2013. Overview of 3GPP Release 10, V0.1.8.
8. ETSI Mobile Competence Centre 2014. Overview of 3GPP Release 10, V0.2.1.
9. ETSI Mobile Competence Centre 2013. Overview of 3GPP Release 11, V0.1.4.
10. ETSI Mobile Competence Centre 2014. Overview of 3GPP Release 11, V0.1.9.
11. ETSI Mobile Competence Centre 2013. Overview of 3GPP Release 12, V0.0.8.
12. ETSI Mobile Competence Centre 2014. Overview of 3GPP Release 12, V0.1.3.
13. ETSI Mobile Competence Centre 2013. Overview of 3GPP Release 13, V0.0.1.
14. ETSI Mobile Competence Centre 2014. Overview of 3GPP Release 13, V0.0.6.
15. Americas 2012. 4G Mobile Broadband Evolution: Release 10, Release 11 and Beyond—HSPA+, SAE/ LTE and LTE-Advanced.
16. 3GPP TS 23.272 (V11.4.0) 2013. Circuit Switched (CS) Fallback in Evolved Packet System (EPS); Stage 2. Technical Specification (Release 11), Technical Specification Group Services and System Aspects, 3GPP, March.
17. 3GPP TS 23.203 V12.0.0 2013. Policy and Charging Control Architecture. Technical Specification (Release 12), Technical Specification Group Services and System Aspects, 3GPP, March.
18. 3GPP TS 22.101 V12.4.0 2013. Service Aspects; Service Principles. Technical Specification (Release 12), Technical Specification Group Services and System Aspects, 3GPP, March.
19. 3GPP TS 23.857 V11.0.0 2012. Study of Evolved Packet Core (EPC) Nodes Restoration. Technical Specification (Release 11), Technical Specification Group Core Network and Terminals, 3GPP, December.
20. 3GPP TSG CT Meeting #69 RP-151569 2015. Release 13 Analytical View Version Sept. 9th 2015.
21. 3GPP TS 25.413 V11.3.0 2013. UTRAN Iu Interface Radio Access Network Application Part (RANAP) Signaling. Technical Specification (Release 11), Technical Specification Group Radio Access Network, 3GPP, March.
22. 3GPP TS 48.008 V11.4.0 2013. Mobile Switching Centre—Base Station System (MSC-BSS) Interface; Layer 3 Specification. Technical Specification (Release 11), Technical Specification Group GSM/EDGE Radio Access Network, 3GPP, March.
23. Ericsson 2014. Ericsson White Paper Uen 284-23-3248: Network Functions Virtualization and Software Management.
24. Samdanis, K. et al. 2017. *From Network Sharing to Multi-tenancy: The 5G Network Slice Broker*, arXiv:1605.01201. Accepted for publication in IEEE Communications Magazine.
25. Shimojo, T. et al. 2016. Future Core Network for the 5G Era. *Technology Reports, NTT Docomo Technical Journal*, 17(4):50–59.
26. Marek, S. 2016. Dynamic Network Slicing Is the 'Secret Sauce' Behind 5G. https://www.sdxcentral.com/ articles/news/dynamic-network-slicing-secret-sauce-behind-5g/2016/08/
27. 3GPP TS 23.401 V14.0.0 2016. General Packet Radio Service (GPRS) Enhancements for Evolved Universal Terrestrial Radio Access Network (E-UTRAN) Access. Technical Specification (Release 14), Technical Specification Group Services and System Aspects, 3GPP, June.
28. 3GPP TR 23.714 V14.0.0 2016. Study on Control and User Plane Separation of EPC Nodes. Technical Report (Release 14), Technical Specification Group Services and System Aspects, 3GPP, June.
29. Erik, N., Ramesh, K.S. and Jennifer, S. 2010. The Akamai Network: A Platform for High-Performance Internet Applications. *ACM Digital Library Newsletter ACM SIGOPS Operating Systems Review*, 44(3):2–19.
30. Md. HK, Eric GM, Gholamali CS (20) Request-Routing Trends and Techniques in Content Distribution Network. Available at http://faculty.pucit.edu.pk/fawaz/cs595/lectures/10.1.1.91.5053.pdf
31. Ian, G. 2013. CDNs Rapidly Becoming Key to Operator Content Strategies. European Communications (Tektronix Communications Special Promotion).
32. The Tech Corner 2012. Components of a Content Delivery Network (CDN). http://www.thetech. in/2012/10/components-of-content-delivery-network.html
33. Ari, B. and Robert, K. 2010. The Content Delivery Network (CDN): Delivering the Ultimate Web Experience, Bell, June.
34. Don, M. 2012. Building a CDN with F5, F5 Networks, Inc.
35. Fourie, J. 2012. OSS/BSS Explained, Part 1: It Used to be Simple; Now a Massive Transformation Is Required. EBR #2 2012. https://www.ericsson.com/res/thecompany/docs/publications/business-review/2012/issue2/oss-bss_explained.pdf
36. TM Forum, Frameworx. https://www.tmforum.org/tm-forum-frameworx/

37. 3GPP TS 32.101 V8.2.0 2008. Telecommunication Management; Principles and High Level Requirements. Technical Specification (Release 8), Technical Specification Group Services and System Aspects, 3GPP, June.

38. 3GPP TR 32.816 V1.9.0 2008. Telecommunication Management; Study on Management of Evolved Universal Terrestrial Radio Access Network (E-UTRAN) and Evolved Packet Core (EPC). Technical Report (Release 8), Technical Specification Group Services and System Aspects, 3GPP, July.

39. 3GPP TS 32.102 V13.0.0 2016. Telecommunication Management; Architecture. Technical Specification (Release 13), Technical Specification Group Services and System Aspects, 3GPP, 3GPP, December.

40. Americas 2015. NFV and SDN Networks, November.

41. ETSI GS NFV-MAN 001 V1.1.1 2014. Network Functions Virtualisation (NFV); Management and Orchestration. ETSI, December.

12 Connected Devices

Today and tomorrow is all about smart and savvy connected devices and applications. The term device from the perspective of mobile networks is rather broad and can include feature phones, smart phones, tablets, wearables, and even sensors. This chapter describes the possibilities for addressing the following device relevant requirements of 5G as identified by ITU-R in Rec. ITU-R M.2083-0 [1].

1. *Very Long Battery Life*: particularly for mMTC. For mMTC, battery life may need to be more than 10 years.
2. *Connection Density*: up to 1 million devices per km^2 fulfilling a specific quality of service.
3. *Diverse Set of Performance Requirements*: such as energy efficiency, media consumption capabilities, eMBB, and URLLC.

The first section describes the architectural elements of smart phones while the succeeding sections will look into three components, namely batteries, processors, and antennas. It is the understanding that these three components will require significant attention to meet the demands of future communications. In addition, a section is dedicated to D2D communications while the last section briefly describes the role of SDOs for the various aspects of connected devices.

12.1 SMART PHONE COMPONENTS

Smart phones consist of multiple hardware components and software, that is, architectural elements. These elements work in collaboration to achieve the required level of performance and user experience. The key elements include mobile software platform, radio frequency transceiver, baseband processor, application processor, signal processing section (vocoder, logic control), user interface, power supply, antenna, and accessories as shown in Figure 12.1 [2–3].

The mobile software platform provides an environment for the development and execution of applications. It usually consists of the operating system, middleware, and application programming interface.

The RF unit converts the baseband information suitable for communications between base stations/small cells and wireless devices. It consists of transceiver (transmitter and receiver chains), antennas assemblies, and RF power amplifiers. The antenna converts radio waves into electrical power and vice versa. The power amplifier, as one of the most important components of a mobile phone, still remains a standalone component.

A baseband processor manages all the radio functions while an application processor manages applications' operations in a mobile phone. The signal processing section conducts operations on the information and controls the internal operations of the devices (i.e., logic). To process the digital signals, devices use either digital signal processors (DSPs) or application integrated circuits (ASICs). Speech coding, channel coding, data processing, and SIM functionalities exist in this section.

The user interface enables the user to originate and respond to calls, messages, and data communications. This interface at a minimum consists of audio input (microphone) and output (speaker), display unit, touch screen, external memory slots, sensors keypad, and accessory connectors to allow battery chargers and other devices to be connected to the mobile phone.

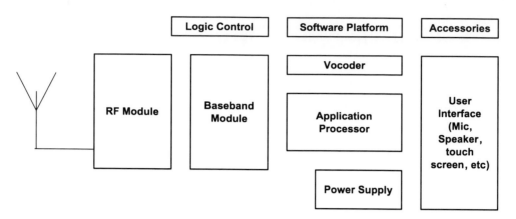

FIGURE 12.1 Components of a smart phone. (From Asif, S. 2011. *Next Generation Mobile Communications Ecosystem: Technology Management for Mobile Communications*. Wiley Inc., UK [2].)

The power supply (normally rechargeable batteries) provides the resource necessary to operate the devices. The details on smart phone components can be found in [2].

12.1.1 COMPONENTS—WHAT'S NEXT?

Enhancements may have to be made in most of the components to achieve the targets of 5G. Some elements will require considerable upgrades and some may need out of the box thinking. Additionally, the targets may end up replacing a current element with a new one and/or totally eliminating an element altogether. All these thoughts have to be explored to meet the goal.

This section will primarily look into power supply (batteries), processors, and antennas that will require reasonable improvements to meet the targets of 5G communications.

12.2 BATTERIES

The battery is the largest but one of the most underdeveloped components of devices and, as a rule, is the most unappreciated technology in use today. These hardly make the list of important innovations and technologies that are part of consumers' daily lives like smart phones, though these technologies heavily rely on them (i.e., batteries).

The requirement to remain always-on and the enormous usage of social networks and data applications are taking a toll on the battery life*. As far as 5G is concerned, the devices will not just be connected to one network but to hundreds of other devices. It will further require rethinking in order to improve energy storage capacity, battery life, overall performance, and perhaps the sizes of batteries as well.

12.2.1 RECHARGEABLE BATTERY

Batteries are devices that convert stored chemical energy into useful electrical energy. Each battery consists of three parts, that is, a negative electrode (anode) that holds charged ions, a positive electrode (cathode) that holds discharged ions, and an electrolyte that separates these terminals (electrodes). The electrolyte is a chemical medium that allows the flow of electrical charge between the cathode and anode.

A rechargeable battery or storage battery is a battery that can be recharged and reused many times through reversible electrochemical reactions. Some common combinations of chemicals that are used

---

* Battery life is how long a device can work on a single charge (capacity) of a rechargeable battery.

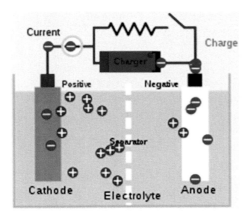

FIGURE 12.2 Charging of a rechargeable battery. (From Electopedia 2005. http://www.mpoweruk.com/life.htm [4].)

in batteries include lead–acid, nickel cadmium (NiCd), nickel metal hydride (NiMH), lithium ion (Li ion), and lithium ion polymer (Li ion polymer). Figure 12.2 shows the recharging process of a battery.

The importance of rechargeable batteries cannot be ignored as these are extensively used in everyday life, from portable devices such as phones, to light vehicles such as motorized wheelchairs, and uninterruptible power supplies. The lead–acid batteries are the most common SLI (Starting, Lighting, and Ignition) batteries that are used in the automotive industry. NiCd batteries are used in cordless and wireless telephones but are being phased out (including NiMH) due to a combination of cost and environmental factors [4]. The most common rechargeable batteries for mobile phones and wireless devices are of lithium ion and lithium polymer. These are in immense use as they are one of the most energetic rechargeable and lightest batteries available today.

12.2.1.1 Battery Life

The usefulness of a rechargeable battery is normally defined in terms of its battery life. *Battery life* refers to how long a device can work on a single charge (capacity) of a rechargeable battery. Along the same lines, the *battery lifetime* refers to the number of charge/discharge cycles until the end of its useful life, that is, until it degrades irreversibly and cannot hold a useful charge.

The capacity is how much charge a battery can hold, often measured in units of mAh (milli-amp-hour). It is determined by the hours of service of 1 volt times the discharge rate. In general, batteries with larger mAh ratings will last longer than those with smaller ratings, assuming that the two are subjected to the same patterns of usage.

Battery life can be calculated by:

$$hours = \frac{mAh}{mA}$$

12.2.1.2 Common Rechargeable Batteries

The lithium polymer (LiPo) is a more technologically advanced version of a lithium ion (Li ion) battery. Its electrolyte is held in a solid polymer film and not in an organic solvent which is directly bonded to the lithium electrode. Unlike lithium ion cylindrical cells, which have a rigid metal case, pouch cells of LiPo have a flexible, foil-type (polymer laminate) case. Since individual pouch cells have no strong metal casing, by themselves these are lighter than equivalent cylindrical cells of lithium ion batteries [4–7].

Liquid electrolytes in lithium ion batteries consist of lithium salts, such as $LiPF_6$, $LiBF_4$ or $LiClO_4$,* in an organic solvent, such as ethylene carbonate, dimethyl carbonate, and diethyl carbonate. A

---

* Lithium hexafluorophosphate (inorganic compound), Lithium tetrafluoroborate (chemical compound), and Lithium Perchlorate (inorganic compound).

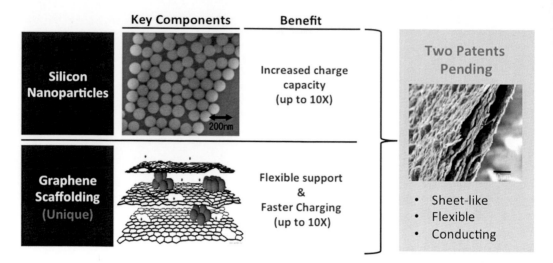

FIGURE 12.3 SiNode systems technology. (From SiNode Systems 2013. Superior Anode Technology for Next Generation Li-Ion Batteries. http://sinodesystems.com/ [9].)

liquid electrolyte acts as a carrier between the cathode and the anode when current flows through an external circuit. Graphene is used for anodes while cobalt oxide is commonly used for cathodes. A LiPo battery is constructed with either $LiCoO_2$ (lithium cobalt oxide) or $LiMn2O_4$ (lithium manganate) as the positive electrode (cathode) and with Li or carbon Li intercalation* compound as the negative electrode (anode). A nonconducting polymer electrolyte such as polyethylene oxide is used as a separator between the two electrodes.

For a given battery mass, an LiPo holds more charge due to simple casing. However, the LiPo lifespan is slightly less than that of a lithium ion battery, but the difference is decreasing over time. It is arguable that a LiPo battery is safer when disposed of as the electrolyte is solid and less likely to leak out into the environment, but that makes the polymer less green than the liquid electrolyte.

12.2.1.3 Improvements in Rechargeable Battery

Though progress has been made to make better rechargeable batteries, there is still room for improvement. Numerous studies and approaches are underway to make improvements in the battery life. In this section, three such approaches have been considered.

12.2.1.3.1 Approach A—Better Way of Building Li Ion Batteries

One way of improvement involves building a Li ion battery using a piece of graphene drilled with tiny holes. This work focuses on the use of a composite of silicon nanoparticles and graphene for anode. By creating a porous structure in graphene, movement of electrons between anode and cathode can be sped up and silicon can be stabilized. This development is carried by SiNode Systems, a startup spun out of Northwestern University, U.S.

SiNode technology utilizes a composite material of silicon nanoparticles and graphene in a layered structure. In this approach, as shown in Figure 12.3, the graphene scaffolds the composite material which considerably reduces the rapid performance degradation associated with silicon-based anode materials [8–9].

The company claims that the capacity of their anodes is approximately 3200 mAh/g, which is almost ten times that of current graphite anodes (370 m/Ah/g). Higher energy density anodes will

---

* To insert between or among existing elements or layers; in chemistry, intercalation is the reversible inclusion or insertion of a molecule (or ion) into compounds with layered structures.

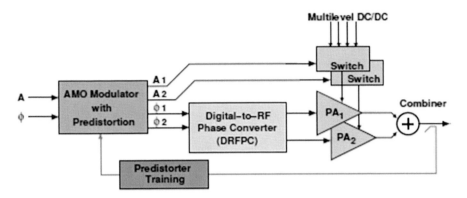

FIGURE 12.4 Asymmetric multilevel outphasing (AMO) architecture. (From Godoy, P.A. et al. 2009 Asymmetric Multilevel Outphasing Architecture for Multi-Standard Transmitters. 2009. *IEEE Radio Frequency Integrated Circuits Symposium*, Boston, MA, USA, June 7–9, 2009, pp. 237–240 [10].)

allow smaller battery form factors and extended battery life. However, the success of this technology depends on the availability of a cost-effective manufacturing process [8].

12.2.1.3.2 Approach B—Same Li Ion Batteries but with Better Amplifier Technology

Power amplifiers are necessary pieces of hardware but are not efficient as compared to their use in mobile communications. Power amplfiers, while turning electricity into radio signals, end up eating battery life. An approach called AMO (Asymmetric Multilevel Outphasing), described in [10] by two MIT electrical engineering professors, increases the efficiencies of the amplifiers which in turn is expected to improve the battery life [11,12]. Thus, this concept envisions the use of the same old batteries, but with better amplifier technology called AMO (Figure 12.4).

AMO, the alternative amplifier technology, is currently being pursued by Eta Devices, a startup cofounded by the respective MIT professors. AMO provides the ability to select the voltage that minimizes power consumption from among a set of voltages that are sent across the transistor [11]. The incoming predistorted amplitude and phase are divided into two pairs of amplitude and phase commands. The DRFC (Digital to RF Phase Converter) performs phase modulation by embedding the phase component of the AMO mapper output into an RF carrier. The time delay mismatch between the two paths is maintained within <1 nanosecond using a time aligner between the AMO modulator and amplitude switch [10].

AMO modulation breaks a complex vector into two vectors such that the sum of the two vectors constructs the original complex vector with the minimum out-phasing angle as shown in Figure 12.5. The complex vector itself represents a baseband constellation point whereas the two vectors are the baseband representation of the two PA outputs. As compared to LINC and multilevel-LINC* (ML-linear amplification using nonlinear components) techniques [13], AMO allows independent changes in the supply voltage for each of the two PAs resulting in smaller out-phasing angles. This ultimately results in higher efficiency even in relatively high PAPR conditions [10].

12.2.1.3.3 Approach C—Metal Air Batteries

Energy storage is a critical element of portable electronics like smart phones. The success of smart devices that the world has witnessed so far would have not been possible without the high-quality energy storage/Li ion batteries [14]. It may be recalled at this stage that energy density is the amount of energy stored in a given system or region of space per unit volume measured in MJ/L (megajoules per liter) and if it is per unit mass, then it is called specific energy measured in MJ/kg (megajoules per kilogram).

---

* LINC is a technique whereby a linear modulation signal is converted into two constant envelope signals.

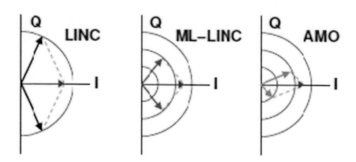

FIGURE 12.5 Signal component vector diagram for LINC, ML-LINC, and AMO. The smallest outphasing angle is achieved with AMO. (From Godoy, P.A. et al. 2009. Asymmetric Multilevel Outphasing Architecture for Multi-Standard Transmitters. *2009 IEEE Radio Frequency Integrated Circuits Symposium*, Boston, MA, USA, June 7–9, 2009, pp. 237–240 [10].)

Currently, we get less than 20% of the theoretical volumetric capacity out of a battery so there is room for improvement [15]. Thus, to overcome this challenge, metal air batteries have resurfaced. A metal air electrochemical cell is an electrochemical cell that consists of a metal anode, an air cathode, and an aqueous electrolyte. A critical component of all metal air batteries (which have been under investigation since the 1970s) is the air cathode that leads to a high theoretical energy density. Several metal/air couples have been evaluated for use in rechargeable batteries, however, none has been commercialized until recently. The most known couple is Li air, while zinc air is also in the news [15–16]. Metal air batteries have exceptionally high theoretical specific energy including oxygen weight. For example, Li air has 5200 Wh/kg whereas Zn air's specific energy is 1090 Wh/kg (watt-hour/kilogram). Compared to Li air, the specific energy of Li ion batteries is around 150 Wh/kg [16].

Along with advantages, there are numerous challenges, such as enablement of reversible cycling of metal air cells with high energy efficiency and system design challenges to take advantage of the intrinsically high energy [15–19]. The foremost important task requires the identification of a stable electrolyte for Li air batteries. Another problem facing nonaqueous Li/O_2 is the quality of the oxygen supply. Nonaqueous Li/O_2 electrochemistry is predominantly sensitive to contamination by CO_2 (carbon dioxide) and H_2O (water). The use of highly purified O_2 (oxygen) tanks may also have an impact as such tanks can decrease the energy density and specific energy of practical Li air batteries. Li air cells often discharge just below 3 V, but charge at 4 V or higher, with round trip energy efficiency below 70%. Thus, improvement in the overall charge/discharge process is required [16].

ZAF Energy Systems claimed to have developed a zinc air prototype that produces two times the energy of a Li ion battery at a third of the cost, with applications ranging from powering cell phones to airplanes. The company claims that they used a fuel technology developed at Lawrence-Livermore National Labs under a license. The fuel cell was mechanically rechargeable, but ZAF found a way to make it electronically rechargeable. The electrolyte in the ZAF battery is rechargeable at 500 charges which results in extended battery life. Rechargeable zinc air batteries also face challenges similar to other metal air batteries [20].

Though it is too early to say which battery advancement option would be able to fulfill the demands of 5G networks and IoT (Internet of Things), the air seems to be flowing in the direction of metal air batteries.

12.3   DEVICE PROCESSORS

A device processor is the component that controls almost everything the device does or is supposed to do. Though a battery is responsible for powering the unit, a processor, on the other hand, works as both the brain and brawn, controlling many functionalities of the device [21]. The key elements that need to be considered for processors are the speed, manufacturing process, and number of cores.

The speed of a processor is measured in gigahertz (GHz). In 2013, a typical smart phone processor could have anything from 0.8 GHz to 2 GHz. It was just four years ago that a 1 GHz processor (central processing units) in a smart phone was considered a big deal, but by today's standards, 1 GHz is rather slow [23]. The processing speed (clock speed) is reaching 3.0–3.5 GHz which is good for high performance applications.

Another performance element is the semiconductor manufacturing process which is measured in nanometers. For example, a 45-nm process will carry as many as 410 million transistors [22]. Furthermore, advancements from the 28 nm process to a 20 nm is expected to deliver a 30% increase in clock speed while at the same time a 25% reduction in heat [23]. As forecast by ITRS (International Technology Roadmap for Semiconductors) the manufacturing process is expected to reach 5 nm by 2020 making it more energy efficient.

Besides single core processors, multicore processors are also commonly found in smart phones. Multicore processors help in multitasking, meaning a device with such functionality is capable of performing more tasks at once. The best way to envision multicore processors is like an octopus; the more cores or arms the device has, the more tasks can be accomplished simultaneously and thus this makes it more efficient.

To be amazingly fast as required for 5G communications, processor speed also needs to be increased. High speed processors are good for high performance applications like 3D rendering (translating a 3D scene into a 2D image) and gaming. Such applications can eat up a majority of a CPU's (central processing unit) power making multitasking difficult. Thus, in order to serve process intensive applications and at high speeds, devices may need both multicore and high clock speed enabled energy efficient processors [24].

12.3.1 Processor Types

Most smart phones consist of two processors that are application and baseband processors while in the low-end feature phones, the general purpose processor is often used as the only processor. The key reason for separation is not to compromise the performance of application, particularly in mid to high end devices.

The wireless phone market is dominated by the ARM (Advanced RISC* Machine) 32-bit processor architecture[†] which is present in more than 95% of smart phones. ARM develops instruction sets and processor core architecture, but does not manufacture processors. The architecture describes the rules for how the microprocessor will behave, but without constraining or specifying how it will be built. The microprocessor architecture defines the processor's instruction set, the programmer's model, and how the processor interfaces with its closest memory resources. The low power characteristics, high data bandwidth, and media processing capability coupled with a high performance core make the ARM micro-architecture particularly suited for networking, wireless, and consumer applications [25–26].

A baseband processor manages all the radio functions (signal modulation, encoding, radio frequency shifting, etc.). A baseband processor consists of two parts, namely a modem to modulate and demodulate the radio signal and a protocol stack unit. The protocol stack unit handles the communication between base station and mobile terminal by establishing connections, managing radio resources, handling errors, and packetizing incoming and outgoing data [27].

The application processor, often called the mobile processor, on the other hand, manages the rest of the functions of the smart phones. The application processor runs the majority of the applications for the user and at the same time communicates with the baseband processor. The communications between the two processors either takes place via serial link or a high speed multiplexed bus as shown in Figure 12.6.

---

* Reduced Instruction Set Computer.
[†] 64-bit architecture has become part of high end devices.

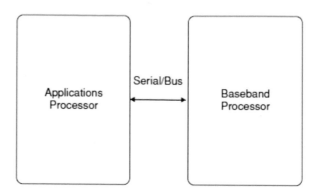

FIGURE 12.6 Smartphone 2-chip solution. (From Asif, S. 2011. *Next Generation Mobile Communications Ecosystem: Technology Management for Mobile Communications*. Wiley Inc., UK [2].)

12.3.2 KEY IMPROVEMENT AREAS

The focus of this subsection will be in those areas where improvement can help in achieving the targets of 5G. Namely, the discussion will be on multicore, manufacturing process technology or process technology, and high clock speed.

12.3.2.1 Manufacturing Process Technology

The microprocessors consist of billions of transistors which are glued together via a process called semiconductor device fabrication. The process technology (or technology node) is measured in nanometers (nm). For example, in the recent past, the 45 nm node referred to the smallest structure size of a single transistor, which is approximately 1,000 times thinner than the diameter of a human hair. Recently, for instance, Qualcomm's Snapdragon 820 mobile processor uses 14 nm low process technology. Thus, the advancements are ongoing to reduce the node size in order to achieve higher efficiencies.

Going to a better process technology (22 nm, 16 nm, and then 10 nm) will improve the efficiency of the mobile processors. A chip using 14 nm semiconductor fabrication would mean it can be about half the size of current 28 nm smartphone chips. This means less power consumption, less heat, and naturally more space in the device [28].

The semiconductor industry uses the concept of die shrink. The term die shrink (sometimes optical shrink or process shrink) refers to taking an existing design and scaling it down for use with a smaller lambda*, without making major design changes. This shrinking process creates a similar circuitry using an advanced fabrication process. The result is lower cost and more space on the same piece of silicon.

In CPU fabrications, a die shrink follows an advance to a lithographic† node which is defined by ITRS (Table 12.1) [29]. However, for GPUs which are designed by fabless companies, the die shrink involves shrinking the die on a node not defined by the ITRS and is referred to as a half node. For example, the 40 nm node is called a stopgap half node between the main ITRS nodes of 45 nm and 32 nm. This is a stopgap measurement between two ITRS-defined lithographic nodes, before further shrinking to the lower ITRS-defined node, which helps in saving R&D costs. The decision to perform die shrinks to either full nodes or half nodes primarily rests with the foundry (for example, Global Foundries, U.S.) and not with the integrated circuit designer (such as Texas Instruments).

---

* In the area of semiconductors, all dimensions are sized in lambda.
† Photolithography or lithography is a process used to create each layer of a chip and used to apply patterns onto a silicon wafer.

TABLE 12.1
IRTS versus Stopgap Nodes

Main IRTS Node	Stopgap Half Node
250 nm	220 nm
180 nm	150 nm
130 nm	110 nm
90 nm	80 nm
65 nm	55 nm
45 nm	40 nm
32 nm	28 nm
22 nm	20 nm
16 nm	14 nm
10 nm	8 nm
7 nm	6 nm
5 nm	4 nm

12.3.2.2 Multicore and Higher Clock Speed

Two smart phones having the same clock speed does not mean that they will have a similar performance. This is because different chipset manufacturers take different approaches when it comes to designing the processors. A multicore processor is a single computing component with two or more independent central processing units (called "cores"). Smart phones are now commonly available with dual core and quad core processors. A key feature of multicore design from the perspective of energy is load balancing. Load balancing enables specific tasks to be prioritized and specific cores to be used for differentiating tasks. This keeps energy consumption as low as possible providing better battery life than single core processors [30]. Another advantage of having the multicores on the same die is allowing the signals to travel shorter distances, and thus have less signal degradation. These higher quality signals allow more data to be sent in a given time period as these can be shorter and do not need to be repeated as often [29].

As the processing power of mobile processors increases so does the power limitation and overheating challenges. Given the tiny footprint of the processor in devices and the lack of a fan, the heat dissipation problem could become acute. Thus, a user will not get as much use from all the cores if the device itself is too hot to hold. Therefore, multicore offers a possible solution where individual cores are specialized for a particular processing requirement, for example, video, thereby reducing power drain and heat generation [30,31].

To run applications, a processor needs to continually complete calculations and if the smart phone has a higher clock speed then these calculations can be done faster which will make the applications run faster as well.

12.3.3 SAMPLE LTE-ADVANCED PROCESSOR

LTE-Advanced (4G) has specified stringent requirements on latency. For example, ARM Holdings, which is the world's dominant mobile phone processor manufacturer, has developed low latency RAM (random access memory) in its Cortex-R processors to address this challenge. Low latency RAM is an area of memory that is used to hold critical software and data such as Interrupt Service Routines. ISR can be executed immediately without waiting for main AXI (Advanced Extensible Interface) bus transactions to finish and/or for the ISR to be fetched into level 1 cache (or primary cache used for temporary storage of instructions and data organized in blocks of 32 bytes) [32]. Very fast interrupt response is necessary in order to maintain reliable LTE-Advanced communications [25]. Figure 12.7 shows an abridged representation of an LTE-Advanced user device [32].

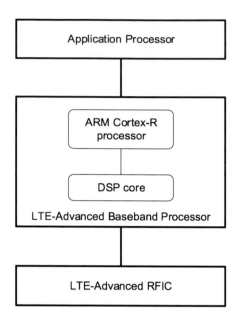

FIGURE 12.7   LTE-Advanced user device.

An LTE-Advanced modem consists of receive and transmit signal processing chains that serve air-interface through an RF transceiver integrated circuit. According to 3GPP specifications, the signal processing is divided into layers with layer 1 providing all the low-level signal functions such as forward error correction, interleaving, constellation-modulation, MIMO, and so on, and maintaining connection with the base station. The layer 1 functions can be handled through a DSP core while its control and management functions are implemented on an ARM processor. Layer 2 (MAC, PDCP, and RLC) and layer 3 (RRC) processing can be performed with the same ARM Cortex-R5/Cortex-R7 processors. The baseband processor interfaces to the applications processor which is running the operating system such as Android.

This section shows a number of components that need to be evaluated to design future 5G devices. In the end, the right mix of multicores (quad to hexa and beyond), manufacturing process (16–5 nm), and clock speed (2.0–3.5 GHz and beyond) is needed to get the job done.

12.4   MASSIVE MIMO

MIMO in simple terms is a smart antenna concept that utilizes more than one antenna both at the base stations and at terminals. MIMO have been explored and used primarily to increase the capacity of cellular and broadband systems. Various forms of MIMO have been deployed in LTE and 4G networks. A 2 × 2 MIMO, which refers to two transmit antennas at the base station and two receive antennas at the device, is commonly used in LTE networks. The 4 × 4 and 8 × 8 configurations which result in having four/eight transmit antennas at the base station and four/eight at the receiver are also getting attention [33].

When it comes to 5G, it is known by other names such as large-scale antenna systems, very large MIMO, hyper MIMO, and full dimension (FD-)MIMO (3GPP term). In 5G, MIMO may be comprised of tens of antennas at both ends, thus the term massive MIMO is commonly used. This concept was first proposed by Thomas L. Marzetta in 2010-11 [34] stating that the only impediment in massive MIMO is intercell interference due to pilot contamination [35]. This concept is applicable to all types of cells (macro, micro, and small) making networks denser than ever before.

Determining the right number of antennas per terminal and base station along with a detection scheme is a difficult task. It is also important to understand the spectral, power, and energy efficiencies of large scale MIMOs. For example, authors in [36] showed that matched filter requires more antennas than an MMSE detection scheme to achieve the same level of performance. Similarly, authors in [37] showed that power efficiency varies whether the receiver has perfect channel-state information (CSI) or whether it has only an imperfect CSI that is derived from uplink pilots. The radiated power of the terminals can be made inversely proportional to M (array of antennas at base station) while maintaining spatial multiplexing gains. For imperfect CSI, the power can only be made inversely proportional to the square root of M [37].

12.4.1 3GPP PERSPECTIVE

The 3GPP specifications for LTE include support for various configurations, that is, 4×4 MIMO, 8×2 MIMO, and multi-user MIMO (MU-MIMO) on the downlink and 1×4 on the uplink [38]. For future networks such as 5G, 3GPP has completed a study [39] on FD-MIMO to understand the performance benefit of standard enhancements targeting a two-dimensional antenna array operation with 8 or more transceiver units (TXRUs) per transmission point. The two-dimensional antenna arrays may use up to 64 antennas at the base station. The use of cmWave and mmWave bands with their short wavelengths will facilitate FD-MIMO. FD-MIMO simultaneously supports both elevation (in high rises) and azimuth (on ground) beamforming.

FD-MIMO is a promising technology that can significantly improve cellular capacity. However, a number of challenges [40] need attention before it becomes part of commercial networks. Economies of scale are required to build cost efficient radio frequency chains and down/up converters. This cost efficiency may result in impairments in hardware. Coherent signal processing needs to be performed fast and in real time given the vast amount of baseband data that needs to be processed. Finally, incorporation of tens of antennas given the tiny space in smart phones will be a daunting challenge.

Compared to base stations, it is relatively difficult to increase the number of antennas in smart phones. However, the jury is still out to determine the most feasible energy efficient ratio of antennas at both sides.

12.5 D2D COMMUNICATIONS

The demand for data is continuously on the rise which is constantly pushing the capacity-coverage envelop of 3G/4G mobile networks [41]. Therefore, the industry is seeking new paradigms to revolutionize the traditional communication methods of cellular networks. D2D communication is one of the paradigms that appears to be a promising concept for addressing such phenomenal data demands in the next generation cellular technologies.

D2D communications is a technology of the near future. It is established as a communication link between two devices, either directly using the ISM (Industrial, Scientific, Medical) band or through a cellular network via its licensed spectrum for exchanging and sharing information. Its use in cellular networks will not only enhance network performance, but also improve user experience with emerging applications such as content distribution and location aware advertisement [41]. However, details around how and when to activate D2D mode, how to manage QoS, and how to schedule resource and manage power are still sketchy.

A number of use cases for D2D have been introduced in the literature such as such as multicasting [42,43], peer-to-peer communication [44], video dissemination [45–48], machine-to-machine (M2M) communication [49], cellular offloading [50], social networks [51, mobile cloud computing [52], public safety communications, and so on. Figure 12.8 shows a representation of such use cases of D2D communications in cellular networks.

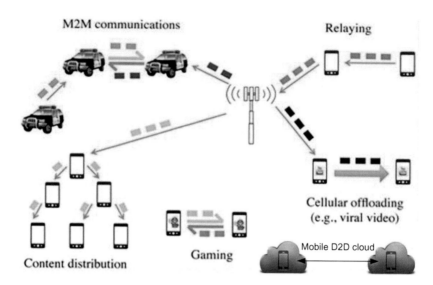

FIGURE 12.8 Representative use cases of D2D communications in cellular networks. (From Asadi, A., Wang, Q. and Mancuso, V. 2014. *A Survey on Device-to-Device Communication in Cellular Networks.* Cornell University, arXiv: 1310.0720v6 [cs.GT], New York, USA [41].)

One of the first attempts to implement D2D communication in a cellular network was made by Qualcomm's FlashLinQ. Its PHY/MAC network architecture that allows cellular devices to discover other FlashLinQ devices and communicate on a peer-to-peer basis. FlashLinQ, by using OFDM/OFDMA technologies and distributed scheduling, creates an efficient method for timing synchronization, peer discovery, and link management in D2D-enabled cellular networks [53]. However, it has not become part of any cellular network so far.

At this stage, it is important to identify the difference between M2M and D2D modes of communication. M2M is a method of data communications between machines with or without human interaction, without as such any distance limitations. A machine could be anything besides a user terminal (TV, refrigerator, etc.). D2D, on the other hand, is restricted to handheld user terminals at least for now and is distance limited [41].

12.5.1 Taxonomy

The D2D can be categorized into two types, namely inband D2D and outband D2D. Figure 12.9 shows the breakup of these two categories. During inband D2D communications, only the licensed cellular spectrum can be used while in outband D2D communications, the licensed cellular spectrum is dedicated for cellular services while the unlicensed ISM band is used for D2D.

The *underlaid inband D2D* proposes the use of the cellular spectrum for both D2D and cellular communications, that is, radio resources are shared. The *overlay inband D2D* on other hand proposes to dedicate part of the cellular resources only to D2D communications. The underlaid inband D2D can improve the spectrum efficiency of cellular networks by reusing spectrum resources. The overlaid inband D2D on the other hand improves the same by allocating dedicated cellular resources to D2D users. However, the former may suffer from interference issues while the latter one faces the hurdle of proper resource allocation. The motivation for choosing inband communication is usually to have high control over the cellular licensed spectrum.

The D2D communications which exploits the use of the unlicensed ISM band is called *outband D2D* communications. This form of D2D take place without establishing communications with the cellular network and thus it reduces interference and increases the capacity of the network. The coordination for radio resources can take place either through the base station which is controlled or could be autonomous

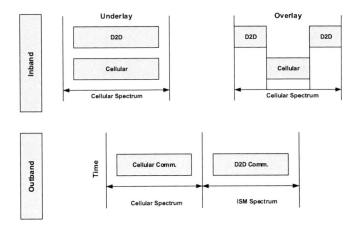

FIGURE 12.9 Schematic representation of overlaid inband, underlaid inband, and outband D2D. (From Asadi, A., Wang, Q. and Mancuso, V. 2014. *A Survey on Device-to-Device Communication in Cellular Networks.* Cornell University, arXiv: 1310.0720v6 [cs.GT], New York, USA [41].)

if it is controlled by users themselves. This use of the ISM band requires the use of an extra interface supporting wireless technologies such as WiFi Direct [54], ZigBee [55] or Bluetooth [56]. Thus, the use of unlicensed wireless technology brings some sense of insecurity to the D2D picture [57].

12.5.2 D2D Standardization

D2D communication was first proposed in year 2000 [58] to enable multihop relays in cellular networks. Later, roughly after a decade, the feasibility of D2D communications was studied by the 3GPP for LTE and 4G (LTE-Advanced) networks. 3GPP published a study on D2D communications under the heading of TR 22.803 Feasibility study for Proximity Services (ProSe) [59] in June 2013. The study identifies two types of ProSe for next generation networks. These two data paths are envisioned to use the licensed spectrum (or inband D2D as referred to in the literature).

(1) *Direct mode* data path (Figure 12.10) where two devices exchange packet data without involvement of any network element. The operator can move the data path (user plane) off the access and core networks onto direct links between the UEs, and (2) *locally-routed* data path (Figure 12.11) where D2D user equipment (UE) exchanges the data locally by relaying through the controlling node (for example, eNodeB).

There are few choices for the control path in ProSe communication. If two UEs are served by the same eNB, the system can decide to perform ProSe communication using control information exchanged between UE, eNB, and EPC. In this scenario, the UEs can also exchange control signaling directly with each other to minimize signaling modification. In the case when two UEs are served by different eNBs, the system can decide to perform ProSe communication using control information exchanged between UE, eNB, and EPC. In this scenario, the eNBs can coordinate with each other or with EPC for radio resource management. Similar to the former case, the UEs can also communicate directly with each other to exchange control signaling.

The 3GPP also defines other aspects of ProSe communication, such as ProSe direct discovery, roaming, support for public safety service, support for WLAN direct communication, and so on. Details of these aspects can be found in [60]. Based on the above-mentioned schemes, the 3GPP proposes tens of use cases [61], such as ProSe-enabled UEs discovering other ProSe-enabled UEs (which can be used for social networking), supporting a large number of UEs in a dense environment (which can be used for city parking services), establishing ProSe-assisted WLAN direct communications (which can be used for cellular traffic offloading), and so on.

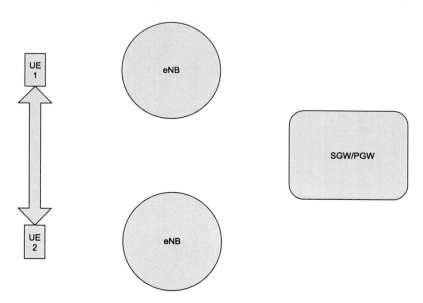

FIGURE 12.10   The "direct mode" data path in the EPS for communication between two UEs. (From TR 22.803 (V12.2.0) 2013. Feasibility Study for Proximity Services (ProSe). Technical Report (Release 12), Technical Specification Services and System Aspects, 3GPP, June [59].)

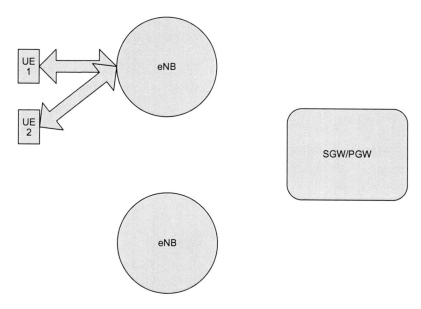

FIGURE 12.11   A "locally-routed" data path in the EPS for communication between two UEs when UEs are served by the same eNBs. (From TR 22.803 (V12.2.0) 2013. Feasibility Study for Proximity Services (ProSe). Technical Report (Release 12), Technical Specification Services and System Aspects, 3GPP, June [59].)

12.5.3   TECHNICAL CONSIDERATIONS

A key technical component of D2D is mode selection that selects the cellular or direct communication mode for a D2D pair based on factors such as the current resource situation, traffic load, and interference level. The fundamental question is when and how to enable D2D mode. In [61], the authors have proposed four modes of communication, namely cellular, optional D2D, pathloss based D2D, and force D2D. That is,

- Cellular: All devices are in cellular mode and no D2D communication mode is available.
- Optional D2D: Mode selection vector for all devices is searched in order to minimize the first norm of the device powers combined over two time slots and to fulfill eigenvalue* criteria.
- Path loss D2D: D2D mode is selected if path loss between communicating devices is less than minimum of path losses between each device and base station.
- Force D2D: This D2D mode is always selected for locally communicating devices (i.e., devices that have a peer).

It was shown in [61] that the distance plays an important role in mode selection. If the distance is small, that is, 5 m or so, all four modes have a comparable performance. As the distance increases, the performance of pathloss based D2D and force D2D decreases as these methods do not consider interference situations for the most part.

In underlaid inband D2D that reuses the spectrum resources of the primary system (for example, cellular), *resource allocation of D2D pairs* is an important task in order to keep the performance of heterogeneous macrocell-smallcell networks at a satisfactory level [62,63]. The task is challenging since the receivers in this system suffer from three tier interference, that is, macro-to-device, smallcell-to-device, and device-to-device *interference* [64]. Another challenge along the same lines is to guarantee the *fairness* among D2D UE when scheduling.

In [65], the authors have proposed a joint scheduling and resource allocation scheme to address these challenges. The Stackleburg game framework was used by them to improve the performance of D2D communication in a HetNet. The analysis showed that the spectral efficiency improvements for macrocell and smallcell are in the order of 2–3 bps/Hz.

12.5.4 APPLICATIONS

The wireless operators have started to include the D2D functionality in their cellular networks. This is largely because this mode of communications was only envisioned in the past as a tool to reduce the cost of local service provision, which is fractional according to the operators' current market statistics [66]. However, this concept is expected to change as presented by some of the use cases in this subsection.

- The variations in user distribution cause *inhomogeneous traffic load* among the networks having different coverage sizes in the HetNet. 5G will have HetNets in abundance; thus, D2D communications may be used as proposed in [67] to address this inhomogeneity.
- D2D communications will increase the traffic of *mobile social networks* and that in turn jacks up the revenue. In [51], the authors have shown that common profile matching (CPM) is a key application of mobile D2D social networks. CPM refers to the scenario where a group of smartphone users meet in a small region (such as a ball room) and these users are interested in identifying the common attributes among them from their personal profiles efficiently via short-range (such as D2D) communications. For example, a group of strangers may want to find common hobbies, friends or countries they visited before, and a group of students may want to know the common courses they have taken.
- *Physical layer coding* can be added to D2D communications to increase the spectral and energy efficiencies of the network [68].
- In [52], the authors have explored D2D communications for *mobile cloud computing*. For such a scenario, the question is how users can detect and utilize the computing resources on other mobile devices. The authors have suggested that the mobility patterns of users have significant impact on the performance of the mobile cloud.

---

* Eigenvalues are a special set of scalars associated with a linear system of equations (i.e., a matrix equation).

In conclusion, the network operator controlled D2D (underlaid inband, overlaid inband, and controlled outband) should enable the operators to provide better user experience and make more profits accordingly. However, for the direct path case, network operators will be in the chicken or the egg situation. At one end, they will have room to offer more services, on the other they will lose hard earned revenue. Finally, the operators still face several challenges in providing such a D2D solution that can address the above two contradicting objectives simultaneously [66].

12.6   ROLE OF SDOS FOR CONNECTED DEVICES

It is important to note that there is no central SDO at the international level that takes care of the majority of the standardization activities for smart phones. For example, 3GPP, ETSI, and OMA play some specific roles, but they do not develop specifications for power supply and processors which have been addressed in this chapter.

- *3GPP* primarily looks into terminal capabilities (such as execution environments) and develops specifications for interfaces (logical and physical) and for smart card (SIM) applications.
- *ETSI* mainly creates specifications for a smart card platform on which other bodies can base their system-specific applications to achieve compatibility between all applications resident on the smart card.
- *OMA* (Open Mobile Alliance) is the leading standards body with several working groups and committees looking into various concepts and issues of mobile devices. Their focus is on architecture, messaging, content delivery, device management, and interoperability of location services.

For batteries, a number of organizations are involved in developing standards such as Battery Council International, IEEE—SA, TIA, IEC, and so on. For processors, regional Semiconductors Associations play the most important role via ITRS. The ITRS (International Technology Roadmap for Semiconductors) is a set of documents produced by a group of semiconductor industry experts. These experts are representative of the sponsoring organizations which include the Semiconductor Industry Associations of the U.S., Europe, Japan, South Korea, and Taiwan.

12.7   SUMMARY

This chapter provided an in-depth look on meeting the key requirements for 5G devices. Batteries, processors, and antennas were the three main topics discussed. Various options were described to improve the battery life including lithium polymer which is a promising alternative to lithium ion batteries. Metal air batteries were also suggested, however, these are facing a number of challenges as described. Next, advancements in manufacturing process technology, multicore technology for specialized functions, and better clock speed were presented for mobile processors. Massive MIMO technique was also highlighted to determine the optimum number of antennas for base stations and devices. D2D communications was also investigated to assist the industry in fulfilling the data demands of the future. The chapter concluded by providing a note on the role of SDOs on the various domains of connected devices.

PROBLEMS

1. What is the main purpose of the RF unit of smart phones?
2. What are the key differences between Li ion and Li poly batteries?
3. Describe two techniques to improve the battery performance?

4. What will the battery life of a Li ion battery that has a capacity of 220 mAh and a discharge rate of 4 mA?
5. What are the key benefits of a multicore processor?
6. Define the differences between baseband and application processors?
7. Define MIMO and Massive MIMO?
8. What are the key challenges associated with FD-MIMO?
9. What is D2D communications and list some of the use cases for it?
10. Define inband D2D and outband D2D types?
11. What are the Proximity Services as defined by 3GPP?

REFERENCES

1. ITU-R 2015. Recommendation ITU-R M.2083-0 IMT Vision—Framework and Overall Objectives of the Future Development of IMT for 2020 and Beyond.
2. Asif, S. 2011. *Next Generation Mobile Communications Ecosystem: Technology Management for Mobile Communications.* Wiley Inc., UK.
3. Lawrence, H., Levine, R. and Kikta, R. 2002. *3G Wireless Demystified.* McGraw-Hill, New York, NY, USA.
4. Electopedia 2005. http://www.mpoweruk.com/life.htm.
5. MIT School of Engineering 2012. http://engineering.mit.edu/ask/how-does-battery-work
6. Albright, G., Edie, J. and Al-Hallaj, S. 2012. *A Comparison of Lead-Acid to Lithium-ion in Stationary Storage Applications,* AllCell Technologies LLC, Illinois, USA.
7. Yahoo 2008. What's the Difference between Lithium-Ion and Lithium-Polymer Batteries? http://answers.yahoo.com/question/index?qid=20080504170740AAevnfI.
8. Wang, U. 2013. *A Better Way to Build a Lithium Ion Battery? Bloomberg Businessweek Technology.* http://www.businessweek.com/articles/2013-04-30/a-better-anode-gives-batteries-a-boost.
9. SiNode Systems 2013. Superior Anode Technology for Next Generation Li-Ion batteries. http://sinodesystems.com/.
10. Godoy, P.A. et al. 2009. Asymmetric Multilevel Outphasing Architecture for Multi-Standard Transmitters. *2009 IEEE Radio Frequency Integrated Circuits Symposium,* Boston, MA, USA, June 7–9, 2009, pp. 237–240.
11. Breeden, II, J. 2012. Emerging Techn Smart-Phone Battery Life could Double—Without Better Batteries. GCN. http://gcn.com/blogs/emerging-tech/2012/11/smart-phone-battery-life-could-double-without-better-batteries.aspx.
12. Phys.org 2012. Ex-MIT Company Rethinks Power-Feasting Amplifiers. http://phys.org/news/2012-11-ex-mit-company-rethinks-power-feasting-amplifiers.html
13. Chen, Y.J. et al. 2007. Multilevel LINC System Design for Wireless Transmitters. *2007 Int'l Symposium on VLSI-DAT,* Hsinchu, Taiwan, April 25–27, 2007, pp. 1–4.
14. Fletcher, S. 2011. *Bottled Lightening: Superbatteries, Electric Cars, and the New Lithium Economy.* Hill and Wang, New York, USA.
15. Whittingam, M.S. 2012. History, Evolution, and Future Status of Energy Storage. *Proceedings of the IEEE* v.100 1518-1534.
16. NPC 2012. Topic Paper #17. *Advanced batteries: Beyond Li-Ion. Working Document of the NPC Future Transportation Study.*
17. Linden, D. and Reddy, T.B. 2002. *Handbook of Batteries.* McGraw-Hill, USA.
18. Argonne National Laboratory 2013. Researchers Tackle New Lithium Battery Challenge. Nature Communications. http://www.rdmag.com/news/2013/09/researchers-tackle-new-lithium-battery-challenge
19. Cleantechnica 2013. Zinc-Air Battery Company Claims Novel Electrolyte will Do the Trick (CT exclusive). http://cleantechnica.com/2013/09/09/zinc-air-battery-company-claims-novel-electrolyte-will-do-the-trick-ct-exclusive/
20. Energizer 2009. Rechargeable Batteries and Chargers: Frequently Asked Questions. http://data.energizer.com/PDFs/Rechargeable_FAQ.pdf
21. Smartphone Processors 101: A Quick Guide 2013. http://www.phones4u.co.uk/community/smartphone-processors-101-a-quick-guide/
22. Tom's Hardware 2007, Semiconductor Production 101. http://www.tomshardware.com/reviews/semiconductor-production-101,1590.html

23. Gantt, C. 2011. ARM will Boost Mobile Processor Clock Speeds to the 3 GHz Range by 2014. http://www.tweaktown.com/news/31644/arm-will-boost-mobile-processor-clock-speeds-to-the-3ghz-range-by-2014/index.html

24. Yahoo 2011. http://answers.yahoo.com/question/index?qid=20110413115614AALo3XX

25. Georgescu, D. 2003. Evolution of Mobile Processors. *2003 IEEE Conference on Communications, Computers and Signal Processing*, Vol. 2, Victoria, Canada, August 28–30, 2003, pp. 638–641.

26. Khoushanfar, F. et al. 2000. Processors for Mobile Applications. *2000 IEEE International Conference on Computer Design*, Austin, USA, September 17–20, 2000, pp. 603–608.

27. Turner, C. 2011. *New Cortex™ – R Processors for LTE and 4G Mobile Broadband*. ARM Ltd, Cambridge, UK.

28. CellularNews 2013. Samsung Working on More Powerful Smartphone Chip. http://www.cellular-news.com/story/Handsets/62485.php

29. Wikipedia. https://en.wikipedia.org/

30. Goodacre, J. 2011. The Evolution of the Microprocessor—From Single CPUs to Many Core Devices. newelectronics. http://www.newelectronics.co.uk/electronics-technology/the-evolution-of-the-microprocessor-from-single-cpus-to-many-core-devices/35556/

31. Chitkara, R. 2012. *Mobile Technologies Index, Application Processors: Driving the Next Wave of Innovation*.

32. Maidment, D., Turner, C. and Bergman, E. 2012. *Designing an Efficient LTE-Advanced Modem Architecture with ARM® Cortex™-R7 MPCore™ and CEVA XC4000 Processors*. ARM Limited and CEVA, Cambridge, UK and Hoofddorp, Netherlands.

33. Fitchard, K. 2013. T-Mobile's Plan to Supercharge LTE: A Whole Lot of Antennas. http://gigaom.com/2013/06/04/t-mobiles-plan-to-supercharge-lte-a-whole-lot-of-antennas/

34. Marzetta, T.L. 2010. Noncooperative Cellular Wireless with Unlimited Numbers of Base Station Antennas. *IEEE Transactions on Wireless Communications*, 9(11):3590–3600.

35. Marzetta, T.L. and Ashikhmin, A. 2010. *Beyond LTE: Hundreds of Base Station Antennas!*

36. Hoydis, J., Brink, S.T. and Debbah, M. 2011. *Massive MIMO: How Many Antennas Do We Need?* Cornell University, arXiv:1107.1709v2 [cs.IT], New York, USA.

37. Marzetta, T.L., Larsson G.E. and Ngo, Q.H. 2011. Uplink Power Efficiency of Multiuser MIMO with Very Large Antenna Arrays. *49th Annual Allerton Conference on Communication, Control, and Computing*, Allerton House, Monticello, Illinois, USA, September 28–30, 2011, pp. 1272–1279.

38. Rysavy Research/5G Americas 2016. Mobile Broadband Transformation LTE to 5G.

39. TR 36.897 (V13.0.0) 2015. Study on Elevation Beamforming/Full-Dimension (FD) Multiple Input Multiple Output (MIMO) for LTE. Technical Report (Release 13), Technical Specification Group Radio Access Network, 3GPP, June.

40. NuRAN Wireless/Nutaq 2015. Massive MIMO. https://www.nutaq.com/blog/massive-mimo-%E2%80%93-part-1-introduction-theory-implementation

41. Asadi, A., Wang, Q. and Mancuso, V. 2014. *A Survey on Device-to-Device Communication in Cellular Networks*. Cornell University, arXiv: 1310.0720v6 [cs.GT], New York, USA.

42. Du, J. et al. 2012. A Compressed HARQ Feedback for Device-to-Device Multicast Communications. *IEEE Vehicular Technology Conference*, Quebec City, Quebec, Canada, September 3–6, 2012, pp. 1–5.

43. Zhou, B., Hu, H., Huang, S-Q. and Chen, H-H. 2013. Intra-Cluster Device-to-Device Relay Algorithm with Optimal Resource Utilization. *IEEE Transactions on Vehicular Technology*, 62(5):2315–2326.

44. Lei, L., Zhong, Z., Lin, C. and Shen, X. 2012. Operator Controlled Device-to-Device Communications in LTE-advanced Networks. *IEEE Wireless Communications*, 19(3):96–104.

45. Doppler, K., Rinne, M., Wijting, C., Ribeiro, C. and Hugl, K. 2009. Device-to-Device Communication as an Underlay to LTE-Advanced Networks. *IEEE Communications Magazine*, 47(12):42–49.

46. Golrezaei, N., Molisch, A.F. and Dimakis, A.G. 2012. Base-Station Assisted Device-to-Device Communications for High-Throughput Wireless Video Networks. *IEEE International Conference on Communications*, Ottawa, Canada, June 10–15, 2012, pp. 7077–7081.

47. Golrezaei, N., Molisch, A.F. and Dimakis, A.G. 2012. Device-to-Device Collaboration Through Distributed Storage. *IEEE Global Communications Conference (GLOBECOM)*, Anaheim, CA, USA, December 3–7, 2012, pp. 2397–2402.

48. Li, J.C., Lei, M. and Gao, F. 2012. Device-to-Device (D2D) Communication in MU-MIMO Cellular Networks. *IEEE Global Communications Conference (GLOBECOM)*, Anaheim, CA, USA, December 3–7, 2012, pp. 3583–3587.

49. Pratas, N.K. and Popovski, P. 2013. *Low-Rate Machine-Type Communication via Wireless Device-to-Device (D2D) Links*. Cornell University, arXiv:1305.6783, New York, USA.

50. Bao, X., Lee, U., Rimac, I. and Choudhury, R.R. 2010. Data Spotting: Offloading Cellular Traffic via Managed Device-to-Device data Transfer at Data Spots. *ACM SIGMOBILE Mobile Computing and Communications Review*, 14(3):37–39.

51. Chen, Y.-A., Lin, W.-H. and Tseng, Y.-C. 2014. On Common Profile Matching Among Multiparty Users in Mobile D2D Social Networks. *IEEE Wireless Communications and Networking Conference*, Istanbul, Turkey, April 6–9, 2014.

52. Li, Y., Sun L. and Wang, W. *Exploring device-to-device communication for mobile cloud computing.* IEEE International Conference on Communications, Sydney, NSW, Australia, June 10–14, 2014.

53. Wu, X. et al. 2010. FlashLinQ: A Synchronous Distributed Scheduler for Peer-to-Peer ad Hoc Networks. *48th Annual Allerton Conference on Communication, Control, and Computing*, Allerton House, Monticello, Illinois, USA, Sep 29—Oct 01, 2010, pp. 514–521.

54. Alliance, W. 2010. *Wi-Fi Peer-to-Peer (P2P) Specification v1.1.* Wi-Fi Alliance Specification, vol. 1, pp. 1–159.

55. Alliance, Z. 2006. *Zigbee Specification.* Document 053474r06, Version, vol. 1.

56. Bluetooth, S. 2001. *Bluetooth Specification Version 1.1.* http://www.bluetooth.com.

57. Asadi, A., Wang, Q. and Mancuso, V. 2013. *A Survey on Device-to-Device Communication in Cellular Networks.* Cornell University, arXiv:1310.0720v2 [cs.GT], New York, USA.

58. Lin, Y.-D. and Hsu, Y.-C. 2000. Multihop Cellular: A New Architecture for Wireless Communications. *IEEE INFOCOM Nineteenth Annual Joint Conference of the IEEE Computer and Communications Societies*, March 26–30, 2000, pp. 1273–1282.

59. TR 22.803 (V12.2.0) 2013. Feasibility Study for Proximity Services (ProSe). Technical Report (Release 12), Technical Specification Services and System Aspects, 3GPP, June.

60. TR 23.703 (V12.2.0) 2014. Study on Architecture Enhancements to Support Proximity-Based Services (ProSe). Technical Report (Release 12), Technical Specification Services and System Aspects, 3GPP, February.

61. Koskela, T. et al. 2010. Device-to-Device Communication in Cellular Network—Performance Analysis of Optimum and Practical Communication Mode Selection. *IEEE Wireless Communications and Networking Conference (WCNC)*, Sydney, NSW, April 18–21, 2010, pp. 1–6.

62. Das, D. and Ramaswamy, V. 2009. Co-Channel Femtocell-Macrocell Deployments-Access Control. *IEEE 70th Vehicular Technology Conference Fall (VTC '09)*, Anchorage, Alaska, USA, September 20–23, 2009, pp. 1–6.

63. Liu, P. et al. 2012. Distributed Cooperative Admission and Power Control for Device-to-Device Links with QoS Protection in Cognitive Heterogeneous Network. *7th International ICST Conference on Communications and Networking in China (CHINACOM '12)*, Kunming, China, August 8–10, 2012, pp. 712–716.

64. Chandrasekhar, V., Andrews, J.G. and Gatherer, A. 2008. Femtocell Networks: A Survey. *IEEE Communications Magazine*, 46(9):59–67.

65. He, Y., Wang, F. and Wu, J. 2014. Resource Management for Device-to-Device Communications in Heterogeneous Networks using Stackelberg Game. *Hindawi Publishing Corporation International Journal of Antennas and Propagation* 2014, 1–10 http://dx.doi.org/10.1155/2014/395731.

66. Lei, L. and Zhong, Z. 2012. Operator Controlled Device-to-Device Communications in LTE-Advanced Networks. *IEEE Wireless Communications*, 19(3):96–104.

67. Liu, J. et al. 2014. An Efficient Traffic Detouring Method by Using Device-to-Device Communication Technologies in Heterogeneous Network. *IEEE Wireless Communications and Networking Conference (WCNC 2014)*, Istanbul, Turkey, April 6–9, 2014, pp: Not Available.

68. Fodor, G., Pradini, A. and Gattami, A. *On Applying Network Coding in Network Assisted Device-to-Device Communications.* 20th European Wireless Conference, Barcelona, Spain, May 14–16, 2014.

13 Mobile Applications

Applications and platforms to support them are essential for the success of the cellular/broadband industry. 5G may bring the killer application, however, until that time, the focus is still on improving value added services.

Yesterday's view on VAS (Value Added Service), which included SMS and MMS services, is no longer valid. Applications like m-money, mHealth, and IoT (Internet of Things) have become or will become important revenue generation segments of the cellular/broadband industry. These services can be termed advanced VAS. A newer concept, IoT, has emerged and it has become one of the industry's buzz words which will perhaps shape the future of the Internet.

Millions of applications are currently available from simply taking photos to conducting financial transactions. The main crux of this chapter is on the three advanced VASs namely, mobile financial services, mobile health, and IoT. It provides details on the standardization, typical applications, and challenges of these three services*. Before diving into these services, background information on SDPs, IMS, and cloud computing is provided.

13.1 DELIVERY PLATFORMS

Looking back at the history of the delivery of telecommunications services, it can be seen that the journey started with the concept of feature node which later became an intelligent network. The intelligent network was defined by the ITU-T back in the 1980s. An SCE (Service Creation Environment) was created in intelligent network enabling operators to create new communication services independently without having to change it in the exchange. From there, the industry has come a long way and has developed various SDPs and IMS. SDPs and IMS have been developed with a common goal in mind, that is, to develop and deliver any type of application level service via a standardized delivery mechanism [2].

In many ways, SDP can be seen as a precursor to IMS. The SDP arose from the mobile operators' need to deliver 2G services more quickly and cost effectively across a complex infrastructure consisting of proprietary network elements and end user devices, while IMS was first standardized by 3GPP in 2002 in its Release-5 to meet the needs of GSM/UMTS operators seeking to deploy IP applications in new 3G networks [1–3].

This section will briefly touch upon these two topics along with cloud computing from a telecommunications' perspective and not an IT one.

13.1.1 SDPs

An SDP is a service delivery architecture (infrastructure) that allows defining, developing, and deploying circuit switched communication, packet switched, and content and broadcast services [2–4]. In the network, it is sandwiched between enablers (core network elements like messaging gateway, content server, etc.) and applications supporting wired and wireless access networks (Figure 13.1).

The emergence of application stores to create, host, and deliver applications for devices such as Apple's iPhones and Google's Android based smart phones, has made SDP an enabling platform

---

* For detailed information on other advanced VAS such as Mobile TV, Machine-to-Machine, and Location based Services, readers can refer to [1].

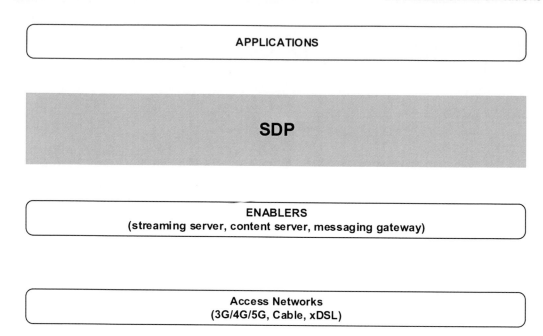

FIGURE 13.1   SDP in the network.

for Communication Service Providers (CSPs) to generate revenue from data services. Using SDPs, CSPs expose their network assets to both the internal and external development communities, through which they can manage the lifecycles of thousands of applications. The business objective of implementing the SDP is to enable rapid development and deployment of new converged multimedia services, starting from basic POTS (plain old telephone service) phone services to complex audio/video conferencing for multiplayer games [5].

However, there is no single standardized definition of SDP and various standard bodies have provided specifications on the bits and pieces of the puzzle. OSA (Open Service Access) Parlay, OMA Service Environment (OSE), and TMF (TeleManagement Forum) Service Delivery Framework (SDF) are the key organizations working on SDP. OSA Parlay is part of the 3GPP IMS and exposes enablers such as presence and call control to applications. The OSE is an architecture that provides a common structure and rule set for specifying enablers, for example, presence, location, and so on. The TMF SDF (Service Delivery Framework) definition provides the terminology and concepts needed to reference the various components involved, such as applications and enablers, network and service exposure, and orchestration. The TMF uses these in the definition of the processes and entities needed for managing the SDF components. And, as there is no consensus, operators have been deploying vendor specific solutions to meet their needs [2–5].

13.1.1.1   SDP Architecture

Due to the absence of a single standardized architecture, each manufacturer has used its own knowledge and expertise in the development of an SDP platform. However, it can be safely said that an SDP consists of five main blocks, namely service creation environment, service execution environment, service oriented architecture environment, OSS/BSS Gateways, and common service support functions as shown in Figure 13.2 [2,4,6].

SCE: SCE is used to rapidly create new services and improve and customize existing services. This (typically graphical) interface offers developers an easy way to identify and combine the appropriate predefined software code building blocks required for new service creation.

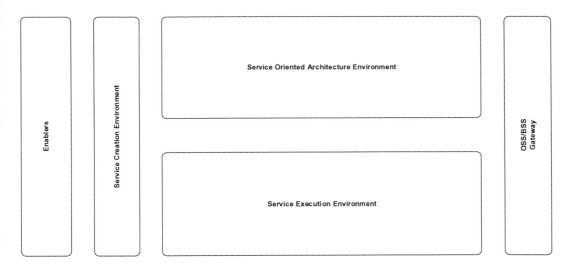

FIGURE 13.2 SDP architecture block diagram.

Service Execution Environment (SEE): The SEE is used to execute the services developed in SCE. Along with SCE, SEE enables rapid development and deployment of commercial scale applications.

Service Oriented Architecture (SOA) Environment: SOA is an evolution of distributing computing and modular programming enabling the creation and execution of business processes. The goal of SOA is to reduce the cost of developing a new application to almost zero. SOA principles are important for successful implementation of SDP as these allow complex service logic to be easily transferable across devices and interfaces. Using SOA and open APIs (application programming interfaces), SDP eases the process of integrating new devices in the system.

OSS/BSS Gateway: The gateway provides access to the service provider's OSS/BSS systems. The integration with OSS and BSS systems is critical for provisioning, billing, and monitoring operations. This integration is also important for reallocating network resources instantly in the event of hardware failure. For example, once a subscriber selects a new service, provisioning functionality adds that subscriber's usage statistics to the service provider's system-wide management view and initiates required billing.

Common Service Support Functions (Enablers): These are basic software elements which are used to construct a service. These could include online portal, online charging, presence and location, identity management, device management, and so on. The SCE presents the enablers to the developers in an uncomplicated and normally visual format. This makes it easy to develop services without requiring hardcore expertise of the underlying code.

13.1.2 IMS

IMS defines a standard framework for the deployment of next generation IP-based applications and services. SDP can be considered as a precursor to IMS. IMS is designed to be access independent so that services can be provided over IP connectivity across wireline and wireless networks. IMS is independent of access technology thus it can support wired cable, DSL, and wireless 3G, 4G, and future 5G systems (Figure 13.3).

IMS is a signaling system that can be used to provide various services. It only implements signaling procedures and application-common functions, but does not offer services. It uses IETF

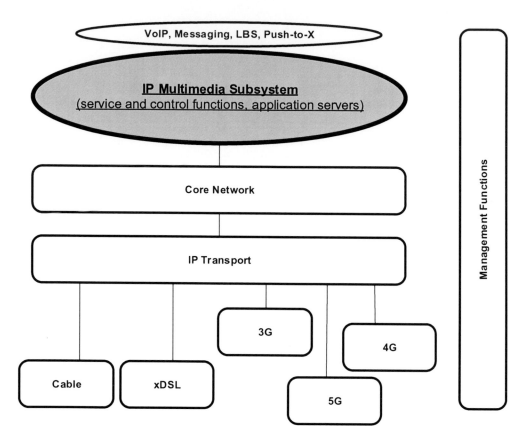

FIGURE 13.3 IMS concept. (From Asif, S. 2011. *Next Generation Mobile Communications Ecosystem: Technology Management for Mobile Communications.* Wiley Inc., UK [1].)

SIP for establishing, terminating, and modifying multimedia sessions within an IP network. It supports the migration of Internet applications (like VoIP, video conferencing, messaging, etc.) to the mobile environment and offers enhanced service control capabilities. This is done by having a horizontal control layer that isolates the access network from the service layer. The services by themselves need not have their own control functions, as these functions are provided by the common horizontal control layer.

13.1.2.1 Historical Background

IMS was originally designed by 3GPP as part of its vision of evolving mobile networks beyond GSM. Its original formulation (3GPP Rel-5) represented an approach to delivering "Internet services" over GPRS. This vision was later augmented by 3GPP in its following releases and was also incorporated by 3GPP2 and TISPAN in their respective applications to include networks other than GPRS, such as WLAN, CDMA2000, and fixed line.

To avoid fragmentation of IMS specifications, it was decided in 2007 that 3GPP would take overall responsibility of IMS. 3GPP started to maintain one common set of IMS specifications starting from its Release-9. Release-9 specifications were frozen in December 2009.

13.1.2.2 IMS Architecture

At a high level, there are three layers in the IMS architecture, namely transport, session control, and application. The layered IMS model allows each layer to be deployed independently of the other layers, providing flexibility required by the next generation networks [1].

- The transport (media) layer transports media streams between subscribers and between subscribers and IMS media (announcements) generating and processing functions.
- The session or control layer controls access to services and provides operator control over subscriptions and services.
- The application layer consists of application servers (presence servers, messaging servers, etc).

The key IMS entities include a signaling routing entity called Common Session Control Function (CSCF), databases, namely Home Subscriber Server (HSS) and Subscription Locator Function (SLF), and a number of supporting and inter-working functions like Media Resource Function (MRF), and so on [7–9]. The service requirements of the 3GPP IMS are specified in 3GPP TS 22.228 [10] while the technical details are in 3GPP TS 23.228 [11]. Details of IMS can be found in [1].

13.1.3   SDP, IMS, and Cloud Computing

SDP and IMS have both been developed with a common goal of liberation of application-level services from silo delivery mechanisms and application-specific boxes (servers). It can be said that SDP and IMS are complementary to each other where SDP acts as the service layer for creating applications that sit on top of IMS as shown in Figure 13.4. SDP and IMS both are key components for the delivery of IP-based applications [2,3]. Details on this aspect can be found in [12].

As services and applications are moving in the Cloud, it is important to evaluate its potential for telecommunications in this context. In simple terms, cloud computing means storing, accessing, and processing data/applications over the servers hosted on the Internet instead of using the hard drive of a personal computer. Cloud computing enables capabilities and offerings on demand that can support internal as well as external customers. Cloud computing lacks a signaling control mechanism, QoS, beyond best effort, and a billing mechanism, which can be addressed by IMS [13,14].

Cloud computing also needs to keep up with the abundance of smart phones in the networks. Mobile cloud computing tends to move the computing power and data storage away from smart phones and into the cloud, bringing applications to the end users in a quick fashion.

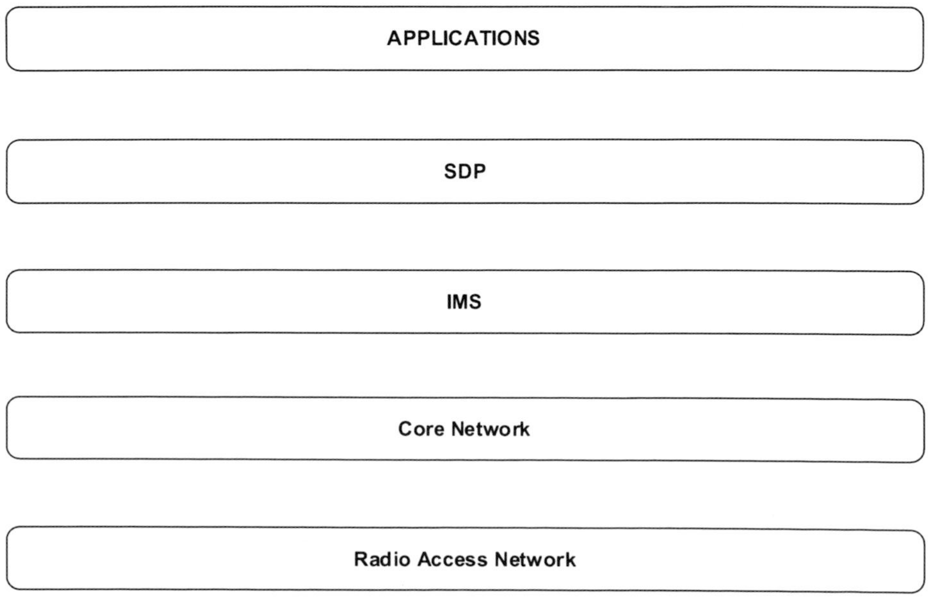

APPLICATIONS

SDP

IMS

Core Network

Radio Access Network

FIGURE 13.4   SDP and IMS.

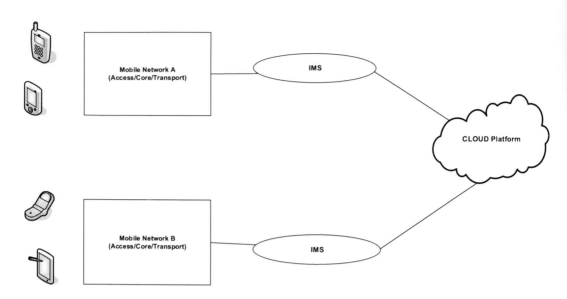

FIGURE 13.5 Mobile cloud computing architecture.

The general architecture of the mobile cloud computing is shown in Figure 13.5 where multiple mobile networks can get access to the same cloud for getting applications for their respective users. Various mobile applications like m-commerce, mHealth, and m-gaming, can use the advantages of this architecture. For example, m-gaming can completely offload the game engine, requiring a large computing resource (e.g., graphic rendering) of the server in the cloud, and gamers only interact with the screen interface on their devices [15].

13.2 ADVANCED VALUE-ADDED SERVICES

VAS is one of those areas where telecom operators are continuously looking to bring new items to the table to increase profitability. VAS includes everything from the basic voice mail and call forwarding services to the complex mobile financial and location-based services and the horizon is continuously expanding. With 5G, user experience and the reliability of existing VAS including the complex ones will definitely improve. The focus of this section is on two such key revenue generating complex services, that is, mobile money and mHealth and on the emerging trend that is IoT.

13.3 MOBILE FINANCIAL SERVICES

The form of money has evolved several times in human history starting from barter trade, to coin, to paper, then plastic, and now phones. Mobile financial services, m-money, and m-commerce are interchangeably used referring to the use of mobile devices (primarily phones, iPads, and tablets) to conduct financial transactions [16].

The term m-commerce was first coined in 1997 by Kevin Duffey. Also, during 1997, the first trial of m-commerce took place in Finland with the installation of two mobile phones-enabled Coca Cola vending machines. These vending machines accepted the payment through SMS. It was also in 1997 that the first mobile phone-based banking service was launched by Merita Bank of Finland, also using SMS [5].

13.3.1 Is There a Need?

According to the World Bank, financial inclusion means that individuals and businesses have access to useful and affordable financial products to meet their needs [17].

Globally, more than 2 billion adults do not have a formal bank account, most of them in developing economies. In the developing countries, mobile money has gotten much more attention due to the presence of a larger unbanked population. For example, around 85% of the adult population of Pakistan does not a have bank account. Contrary to this, in the developed world, where most people already have a bank account, mobile banking gives them another channel (not a necessity) to conduct financial transactions.

In poor economies, people rely more on informal risk-sharing monetary arrangements with friends and family members than on formal arrangements with financial institutions. The apparent obstacles are—bank accounts are too expensive, banks are too far away (especially in rural areas), personal documentation is lacking, and people do not trust banks. Additionally, as banks and interest (profit on the cash, interest on loans) go hand-in-hand, people in some countries do not want to take this (banking) route due to religious reasons.

Realizing this need and opportunity, the telecom and banking industries have been producing concrete projects since the last decade. According to GSMA's annual report on the subject, there were 277 live mobile money services available in 92 countries (including two-thirds of low and middle-income countries) in 2016 to assist the unbanked population [18].

13.3.2 MOBILE PAYMENT

Mobile payments refer to P2P and C2B/B2C (consumer-to-business/business-to-consumer) transactions for physical goods and services that are made using a mobile phone. It also includes P2P mobile money transfers (both domestic and international) where customers use their mobile devices to send, receive or transfer money electronically.

There are two types of mobile payments, namely remote payment and proximity payment. Figure 13.6 shows the various forms of mobile payment services that are available, differentiated by technology, location, transaction value, and cash handling function. Remote mobile payments and proximity mobile payments differ by virtue of the location of the mobile device and the merchant, as well as the nature of the transaction.

13.3.3 REMOTE PAYMENT

A remote mobile payment is one where consumers use mobile phones to conduct transactions without interacting directly with the merchant/physical POS (point of sale). Remote payments make use of SMS, Wireless Application Protocol (WAP), browser or a mobile application. These are ideal for P2P, B2C, and C2B transactions such as domestic remittances, bill payments, and so on [16,19,20].

13.3.3.1 Mobile Money Transfers

One widely used type of remote payment is mobile money transfer. Mobile money transfers mainly refer to the transfer of money from one individual to other using mobile phones. It requires senders to give the money along with a fee to a remittance center. The remittance center then transfers the money electronically through the phone service provider to the recipient's phone. The person receiving the money gets a text message advising of the transfer. The recipient can go to any licensed outlet, including a retail store or restaurant, to get the money. The recipient may have to pay a fee to collect the money [20].

Business Models: Different business models have emerged to transfer mobile money. The adoption depends on a number of factors including regulatory regime, culture, and population size. The three prevalent models include bank centric model, mobile operator-led (MNO-centric) or non-bank-based, and partnership based model.

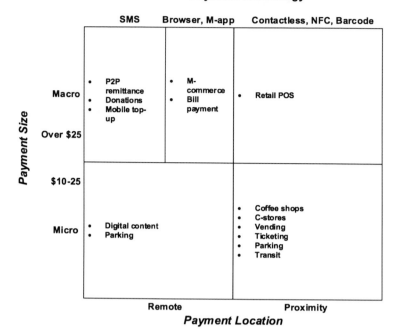

FIGURE 13.6 Mobile payment spectrum. (From Smart Card Alliance Payments Council 2011. Mobile Payments and NFC Landscape: A US Perspective. Publication Number: PC-11002 [19].)

- MNO-centric models are more prevalent in developing economies due to their inadequate financial services infrastructure. Within this model, the MNO is practically the owner of the entire value chain whereas the bank merely acts as a funds storage entity. MNOs manage the distribution/agent network which is their key strength.
- Bank-centric models are likely suitable in countries where there is a good level of financial infrastructure and regulations for such transactions. The bank is the owner of the value chain and provides this service just as a new channel in addition to existing services. The MNO is mainly responsible for provision of the telecommunications facility for such transfers.
- Partnership model is likely prevalent in developed economies since it requires tight integration between telecom and financial systems/infrastructures. Banks and MNOs both contribute to the value chain where the former provides the financial management while the latter manages the distribution network [20,22].

Examples: Some examples of mobile money transfer systems in developing nations are shown in Table 13.1.

In a nutshell, MNOs are good at branding campaigns, product development, and have a vast distributing/agent network. Banks, on the other hand, are good at safeguarding the funds. The success, however, depends on how much the telecom and banking regulators are allowing the MNOs and banks to practice their expertise.

TABLE 13.1

Examples of MFS in Emerging Economies[a]

M-money Application	Countries Implemented	Main Features	Technology
Easypaisa (bank-led model)	Pakistan	• Pay utility bills • Make P2P transfers • Increase air time credits • Save money • Pay for goods and services	USSD, Internet
M-PESA (MNO-led model)	Kenya, Tanzania, South Africa and Afghanistan	• P2P transfers • Pay school fees • Pay electricity bills • Pay for goods and services	STK, USSD
T-Cash	Haiti	• Receive salary • Make P2P transfers • Pay bills	USSD
Globe GCash	Philippines	• Pay utility bills • Make P2P transfers • Use as a mobile wallet • Increase air time credits • Pay for goods and services	SMS, STK
Airtel money	India and 14 African countries including Uganda, Tanzania, and Kenya	• Make P2P transfers • Pay for goods and services • Bill payments	USSD
MTN mobile money (MNO-led model)	Africa, including Uganda, Ghana, Cameroon, Ivory Coast, Rwanda and Benin	• P2P transfers • Buy air time • Check balances • Pay utility bills	STK, USSD
EKO	India	• Make P2P transfers • Bill payments • Loan payments	USSD
WIZZIT	South Africa	• P2P transfers • Buy air time • Check balances • View statements • Pay electricity	USSD

Source: ITU-T Technology Watch Report 2013. The Mobile Money Revolution Part 2: Financial Inclusion Enabler [20].

[a]   USSD (Unstructured Supplementary Service Data); SKT (SIM Toolkit); MTN: Mobile Telephone Network; M-PESA: M for mobile, pesa is Swahili for money..

13.3.4   PROXIMITY PAYMENT

With proximity payments, the mobile device interacts with a physical POS or with a mobile POS device to obtain a consumer's payment information to complete the transaction. Mobile POS is defined as the use of a mobile device (smart phone, tablet) for electronic payment, inventory management, and so on. It is based on technologies such as bar codes, magnetic stripe readers or contactless NFC-enabled (near field communication) devices. B2C/C2B transactions are the key form of proximity payments where mobile phones act as debit/credit cards or POS terminals [16,20,21].

13.3.4.1 Proximity Payment Types

Some key options for proximity are provided in this section [19,20].

Mobile Wallet: A common form of mobile money is mobile wallet which is an electronic account held on a mobile device. It is a software application that is loaded onto a mobile phone to manage payments made from the mobile device. It is a menu on the device which provides access to different payment instruments and payment account information. This could be either based on a tap and go approach (i.e., the user taps his/her mobile device for payment instead of the credit card) or on built-in NFC wireless technology, in much the same way as a credit card. There are numerous examples of mobile wallets including Google Wallet which works with an NFC-enabled device. YES-pay International Limited and YES-wallet.com Limited also offer a cloud-based payment platform using NFC. However, this approach requires software upgrades at contactless POS terminals and low-cost availability of NFC-enabled devices, particularly in developing nations.

Stickers, microSDs, and similar approaches (e.g., key fobs, watches): These allow consumers to transmit payment credentials to a contactless POS terminal. Some of these options such as microSDs can be integrated with a mobile whereas fobs can be affixed to a handset. These solutions are essentially identical to the use of a contactless card in terms of data personalization, security, and functionality. A number of challenges are associated, such as stickers are primarily from a single issuer providing less flexibility to the users and nonintegrated solutions add complexity to the supply chain.

Bar Codes: The famous bars are also applicable for proximity payments. A 2D bar code or QR (quick response) code is displayed on the screen of the mobile device and read by an optical scanner at POS or vice versa. Starbucks is using 2D bar code technology in its stores.

13.3.5 PLAYERS

MFS (Mobile Financial Service) depends heavily on the active collaboration between several players. The key players that are involved in m-commerce are financial institutions (banks, credit card companies), mobile network operators, government (legislative and regulator), device manufactures, and end users. They not only play a critical role, but also contribute to the overall value chain and at the same time have some expectation from MFS as shown in Table 13.2. At the same time, it is important to note that there is no single global standard that addresses the payment process associated with MFS [1].

13.3.6 CHALLENGES

MFS does not come without some critical challenges which need to be understood and resolved [16,20]:

Security: There are primarily two main security concerns - one involving security theft at the user end and the other is end-to-end (consumer to network to recipient) security. Security theft is mainly addressed by having both the correct device and correct password to conduct the transaction. The absence of one piece will not allow execution of the transaction. The second issue is rather complex as all the main technologies (SMS, STK, USSD, and WAP) have their own security issues. WAP is comparatively better and commonly used by banks adding mobile as another channel for users to access their accounts.

Interoperability: This challenge appears when more than one MNO is involved in the execution. Some countries make it a mandatory requirement and others do not. For example, Bank of Ghana requires a 'many-to-many' relationship model where all banks and telecom operators would collaborate and allow transactions across their networks. As a result, each of the MNOs with branchless banking services has signed up at least three partner banks. Contrary to this, Pakistan has not made it compulsory and thus operators have little incentive to make their services interoperable.

TABLE 13.2

MFS Players (List Is Not Exhaustive)

Players	Contributions	Expectations
End users (customers)	• Monetary support	• Fast and secure transactions • Personalize service • Minimal learning curve • No hidden costs • Being able to pay "anywhere," "anytime," and in any currency
Merchant	• Merchandise support • Merchant base	• Fast and secure transactions • Minimal investment cost • Real time status of transactions
Mobile network operator	• Customer base • MFS enabled network	• Potential to add value to existing services • Increase customer loyalty • New revenue channel • Increase average revenue per user • Minimal investment cost
Financial institutions	• Customer base • Merchant base • Secure payment infrastructure	• Increase customer loyalty • New revenue channel • Minimal investment cost
Device manufacturers	• MFS enabled devices	• Open, interoperable, widely-used standards • Low cost of new technologies and features that need to be integrated
Franchisee	• Agent role/middleman	• Increase customer loyalty • Additional revenue channel
Government/regulator	• Legislations/policies • Monitoring infrastructure	• Economic uplift for residents • Taxable income

Source: Asif, S. 2011. *Next Generation Mobile Communications Ecosystem: Technology Management for Mobile Communications.* Wiley Inc., UK [1].

Standardization: There are no as such common technical standards for MFS which is a challenge for all the involved key players. The MFS arena is crowded with tens of forums and SDOs and what is missing is an "active collaboration" among the players. Some of these forums are operator driven (ETSI, UMTS), bank driven (Mobey forum), device manufacturer driven (Mobile Electronic Transactions), technology driven (3GPP, OMA, ITU-T), and cross industry driven (Mobile Payment Forum).

Regulatory: Mobile money encompasses the regulatory space of both telecommunications and banking and therefore requires partnership and collaboration between both sectors in order to mitigate the risks. The regulatory regime is different from country to country and this is a challenge depending on the business model that has been adopted. In some countries, the telecommunication regulation regime prevents mobile operators from entering into nontelecommunication activities and thus forbids them from launching MFS activities of their own. For example, in South Africa and also in Pakistan, a mobile operator has to partner with a bank in order to operate mobile money transfer services. MNOs can be at a disadvantage compared to banks when it comes to taxation of revenue and financial services. For example, telecom is not considered an industry in Pakistan thus it bears heavy taxes on revenue. In some countries, value added tax is not levied on financial services such as person–to–person transfers operated by mobile operators, whereas it can apply to telecommunication services.

Bitcoin: This is a phenomenon that may revolutionize MFS in the future if allowed by the major powers of the world in its true sense. However, Bitcoin is, by its nature, an unregulated currency with no central bank. Thus, if one loses the phone and/or the private keys associated with Bitcoin wallet, one will lose not only the actual currency (USD, GBP, etc.), but also the Bitcoins.

13.3.7   CASE STUDY—EASYPAISA

Telenor Pakistan's Easypaisa mobile money service has been a successful story. A recipient of international awards, it was the first mobile financial service launched in the country in October 2009.

Easypaisa is also the country's largest mobile money service and the third largest in the world, serving over 6 million customers every month. It has over 25,000 agents in 750 cities and towns across the country.

Telenor Pakistan is a 100% owned subsidiary of Telenor Group. It is the second largest mobile operator out of five in Pakistan, with over 35 million subscribers and a market share of 26% in 2013.

13.3.7.1   Background

Pakistan is the sixth most populous country of the world having 182 million inhabitants in 2013. Around two-thirds of the population lives in rural areas with insufficient banking infrastructure. In addition, more than 40% of the population lives below the poverty line. Keeping these hardships in mind, State Bank of Pakistan introduced guidelines for branchless banking in coordination with the Ministry of Information Technology and Telecommunication which also issued a similar directive (both were issued in 2008).

Both regulations called for a bank-led model, which meant that only commercial banks and microfinance banks with an existing banking license were eligible to apply for a branchless banking license. The regulation also specified that MNOs could operate as "super agents" on behalf of a bank, providing marketing and distribution in addition to participating in product development [23].

In spite of a regulation that specified a bank-led model, Telenor decided to move forward with developing mobile financial services. Given the regulatory requirements and Telenor Pakistan's lack of experience in financial services, they saw majority ownership in a microfinance bank as the optimal way to enter the mobile money market.

Thus, in late 2008, Telenor acquired 51% of Tameer Microfinance Bank. Tameer Microfinance Bank is a fully licensed microfinance bank licensed by the State Bank of Pakistan since 2005. It maintains a network of ATMs (Automated Teller Machines) and branches across Pakistan. This acquisition of the bank provided Telenor Pakistan with a mobile banking license and paved the way for it to offer mobile financial services [24]. In partnership with Tameer bank, Easypaisa was launched in October 2009.

According to [25,26], the household mobile money market is dominated by Telenor Easypaisa with an approximately 88% share at the end of 2013 followed by a private bank UBL (United Bank Limited) with 6%; the other mobile money products share the rest of the market (Table 13.3).

13.3.7.2   What Is Easypaisa?

As stated, Easypaisa is the mobile money solution of Telenor Pakistan. It can be used by any cell phone user regardless of whether he/she is a subscriber of Telenor. With Easypaisa, customers can conduct multiple sorts of financial transactions, namely pay bills, send/receive money within Pakistan, purchase airtime (easyload*) for their mobile phones, give donations, and so on. Easypaisa also offers ATM cards, Traveler checks, and allows interbank fund transfers and funds transfer

---

* Easyload allows users to charge their prepaid accounts in small denominations (as low as 20 rupees, approximately U.S.D 0.20), and pay postpaid bills.

TABLE 13.3
Mobile Money Market Share in Pakistan

Operator/Bank	Service Name	Launch Date	Market Share (January 01, 2014; Approximately)
Telenor Pakistan	Easypaisa	October 2009	88%
United Bank Limited	Omni	April 2010	6%
Operators			
Mobilink	Mobicash	November 2012	6%
Zong	Timepey	December 2012	
Ufone	Upaisa	September 2013	
Warid	Mobile paisa	January 2014	
Other banks	--	--	

from over 80 countries. Its services can also be used by people who do not have a cell phone via an authorized agent.

13.3.7.3 Go to Market Strategy

The hardest part in the beginning was to decide on the choice of business model. There were two choices, that is, mobile account or over-the-counter (OTC) service [23,24].

Mobile Account versus OTC: The mobile account is an electronic wallet on the customer's phone, usually run on USSD (Unstructured Supplementary Service Data) or STK (SIM Application Toolkit). The key feature of the wallet is stored value—customers visit agents to cash in, converting cash to digital currency, after which they can trigger transactions from anywhere. However, it comes with a baggage of challenges; first, in 2009, the Telenor Pakistan's market share was only 22%, meaning it would need to lure 40 million non-Telenor Pakistan subscribers as well as those with no mobile subscription, all of whom were potential mobile money customers. Second, regulations for mobile account registration mandated comprehensive Know-Your-Customer (KYC) procedures, which were cost prohibitive and time consuming. Registering for a mobile account required a photo, a copy of the customer's CNIC (Computerized National Identity Card), and a signed account opening form. In addition, the regulation specified that a biometric[*] fingerprint of the customer had to be obtained. Registration points, therefore, required an Internet-enabled device (computer or smartphone) to take the photo, scan the CNIC, and upload the signed form to the back office for processing. Finally, the CNIC card had to be verified with the government bureau NADRA (National Database and Registration Authority) to confirm the customer's identity, which created an additional cost. In total, it costs between U.S. $1.50 and $2.00 per account opening and the investment in registration equipment for each point of service was roughly U.S. $150. The Easypaisa team was concerned that these registration requirements would be too costly for the business model and would present a major barrier to customer adoption. Thus, it was decided to launch with OTC in 2009, whereby all the transactions were agent assisted and no registration was required. The OTC model was not explicitly listed in the regulation; a special acceptance letter from State Bank was obtained to operate the service. Customers who wanted to pay bills or send money simply went to any Easypaisa agent, presented their government issued CNIC, and handed over cash to the agent who performed the transaction. The customer did not have to register and did not need a mobile account. In 2010, the "Easypaisa M-Wallet" was launched with money transfer and bill payment services.

---

[*] A temporary exception from the biometrics requirements was obtained from State Bank. However, it was unclear how long that exception would last. Easypaisa's optimal registration device was a mobile handset, but as it lacked the biometric functionality, Easypaisa delayed investing in handsets. Updated regulation issued in 2012 removed the biometric requirement.

Agent Selection: The key to success lies with the selection of agents and their location and building trust with the society. During Easypaisa launch, the company employed over 2500 agents which is a staggering number. By the end of the first year (2010), there were 8000 agents, and three years after launch, that is, by 2012, there were 20,000. The company carefully selected the number and location of these agents in order to link the disparate and disconnected Pakistani population; a problem which was augmented by the challenging geography of Pakistan. The roll out of the offering was also supported with a targeted marketing campaign to aid consumer understanding. The marketing campaign was designed to maximize outreach, included advertisements and editorial material in national and local newspapers/publications, billboards and banner advertisements, YouTube videos explaining the registration process and how to use, and the agent network carried/distributed relevant educational material.

Distribution Structure: Telenor Pakistan Franchisees, the linchpin of GSM distribution, were established early on as the main players in Easypaisa's distribution strategy. Those 278 businesses were long-standing partners of Telenor Pakistan, responsible for the sales and distribution of scratch cards as well as electronic top-up. They had geographical exclusivity and were responsible for recruiting and serving retailers (agents), that is, selling and distributing airtime and SIMs to them. For Easypaisa, Franchisees were tasked with recruiting, training, and shuttling cash to mobile money agents. In addition to Franchisees and agents, Easypaisa is offered through Telenor Pakistan's 30 owned and operated sales and services centers and Tameer Bank's 40 bank branches. To support the Franchisees, Telenor Pakistan leveraged its GSM sales organization both centrally and regionally. The centralized unit set (key performance indicators) planned the roll out of products, launched below-the-line campaigns to drive usage, and set up training programs. The regional teams worked directly with the Franchisees and agents to set targets, follow up on performance, develop trade marketing activities to ensure local visibility, conduct trainings, and offer general support to the franchisees in developing their business.

13.3.7.4 Technical Solution

Telenor Pakistan selected Fundamo's[*] Platform to roll out its mobile money solution. Figure 13.7 shows a high level technical architecture of Easypaisa. The platform plays a central role in the

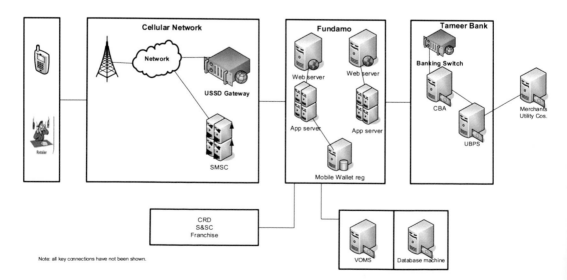

FIGURE 13.7   Easypaisa architecture (block level).

---

* Fundamo is now part of Visa, Inc.

whole scheme as it connects the Telenor's GSM network to Tameer Bank. The transaction takes place either using USSD or SMS. The user can access the bank's CBA (core business application) or UBPS (utility Bills Payment Server) through Fundamo's platform. Core banking is a banking service provided by a group of networked bank branches where a customer may access their bank account and perform basic transactions from any of the member branch offices. Using UBPS, customers can pay utility bills from their nearest Easypaisa shop, Telenor franchise, Telenor Sales and Service Center or Tameer Microfinance Bank branch.

The Easypaisa ATM cards provide 24×7 access to mobile accounts using TPS IRIS Enterprise Payment Switch. IRIS multi-channel switch provides connectivity with the 1LINK ATM network and its integration with Fundamo's m-wallet platform allows customers to avail e-banking services through ATMs and bank's POS machines. The platform is also connected to Telenor's Pakistan CRD (customer relation department), S&SC (sales and service centers), and Franchises. At the back end, it is also linked to the VOMS (voucher management system), database machine, CRM (customer relationship management), and billing systems.

13.3.7.5   Major Incident and Recovery

Engineering always brings challenges and heartburns and it is not surprising that Easypaisa gave the same to Telenor Pakistan. One fine evening in the year 2011, the system started to crumple. It was noted in application logs that the application component was unable to establish connection with the database machine. The database machine runs database workloads including online transaction processing and data warehousing. Although an important component, this database machine is a third-party box and not part of the Fundamo system.

All mobile banking services had to be brought down and connectivity with the database machine was discontinued. As an interim measure, a backup database was brought online in case recovery of the database machine failed or took longer than anticipated. An outage was provoked on the production in the process of switching over activity to point Easypaisa services back to original database machine from a backup database. All required parameters were reviewed before switching on the services. After resumption of services, throttling on USSD traffic was increased gradually and traffic was fully operational after 28 hours. The root cause was a bug in the database machine server software. Close to 78,000 customers were affected and more than 357,000 transactions, resulting in considerable revenue loss and customer dissatisfaction. The proposed way forward was the installation of disaster recovery mechanism which was approved and got implemented the next year (2012).

13.3.7.6   Result and Roadmap

Easypaisa resulted in higher ARPU and solidifying a second spot in terms of overall cellular market share for the operator. It also provided it with another means to get new customers and a source of revenue [24].

In a nutshell, to launch Easypaisa, U.S.D 7 (million) was invested in the technology platform, national marketing campaign, organizational structure, and agent training. An additional budget was ring fenced* for expected losses over the initial years of operations [23]. The investment paid off, gaining 85% of the market share on OTC money transfer with the help 25000 agents and 6 million unique users with 90 million transactions that provided revenue of PKR 5.7 billion by the end of 2013.

The next target for the operator is to generate up to 10% of the total revenue from mobile financial services, that is, Easypaisa. There are also plans to upgrade the basic platform to a second-generation Enterprise Edition to serve 10 million customers monthly and to become the largest retail bank in the country.

---

* The hypothecation of a tax is the dedication of the revenue from a specific tax for a particular expenditure purpose.

13.4 mHEALTH

Mobile Health, more commonly known as mhealth, is a term used for the practice of medicine and public health supported by mobile devices. Although the definitions vary, the World Health Organization's (WHO) Global Observatory defines mHealth as medical and public health practices supported by mobile devices, such as mobile phones, patient monitoring devices, personal digital assistants (PDAs), and other wireless devices. WHO considers mHealth to be a component of eHealth. WHO defines eHealth as the use of ICTs for health [27].

mHealth has been made possible due to the exponential increase in the use of mobile devices worldwide over the last ten years, particularly in the developing countries. The declining costs of mobile devices, growth in subscriptions, digitalization, and rapid advances in technology have driven an explosion of mHealth pilot projects and programs across the globe since 2005 [28]. mHealth has been incorporated into the field of healthcare in an attempt to address the wide variety of challenges, particularly in developing countries, such as skilled worker shortages, malpractice due to quacks, lack of timely reporting for surveillance and diagnostics, an influx of counterfeit drugs, poor treatment adherence, and poor inventory and supply chain management [29]. A vast majority of WHO member states have at least one mHealth initiative in their respective countries.

13.4.1 mHEALTH VALUE CHAIN

The mHealth value chain involves six key segments of the society, namely device/chip vendors, application developers, network operators, healthcare professionals, healthcare insurers (including government health systems), and patients. Each of these segments and subsegments has distinct, although overlapping, needs, which result in varying approaches and business models for extracting market value.

Figure 13.8 depicts an end-to-end mHealth value chain starting from the collection device all the way up to the clinician. The personal data from instruments such as (thermometer, insulin pump, etc.) enabled with any WPAN technology like Bluetooth, ZigBee, and so on can send the data to the mobile device. The mobile device, using the embedded health module and mHealth application, then sends it to the network operator. The mobile network having an mHealth application server passes the information down to the clinician via a healthcare administrative system. The clinician then routes back the appropriate response to the patient through the network [30,31].

More specifically, the chip vendors have the role of developing embedded modules for the devices to enable mHealth related applications. The infrastructure vendors are responsible for enabling mHealth communications through their systems. The application developers are responsible for creating mHealth applications that can work with devices/chipsets, systems, and networks.

The MNOs have the central role in the overall value chain. Their core contribution is their 2G cellular and 3G/4G mobile broadband network coverage. Additionally, they also have other valuable assets including M2M experience, strong consumer brands, large customer bases, and broad retail distribution footprints. MNOs are also best positioned to provide services such as authentication and security, as well as billing and customer care on behalf of healthcare organizations.

The health care professionals and administrations (government and civilian) are under constant pressure, perhaps more than anyone else in the value chain, to enhance the effectiveness of the health care

FIGURE 13.8   mHealth value chain.

and health care delivery systems. Hospital administrators are likely to be most concerned with maximizing the return on investment for mobile healthcare technology, while doctors and patients will place more importance on rapid diagnoses and healthy recovery. Insurers lie between healthcare professionals and patients and focus on providing the most cost-effective care with minimal follow up treatments. Nongovernmental organizations (NGOs) and health care foundations also have the responsibility of delivering healthcare services to remote villages of underdeveloped and developing countries.

13.4.2 mHEALTH ARCHITECTURE

GSMA has defined reference architecture for connected mHealth devices in [32]. The architecture consists of four building blocks, namely mHealth patient device (yellow block), mHealth clinician device (blue block), mobile network (gray cloud), and mHealth platform (light blue box). On the far right hand side of the mHealth platform are two key elements of the Healthcare IT system (HIT), that is, the Personal Health Record (PHR) and the Electronic Health Record (EHR). Figure 13.9 shows these main building blocks and interfaces of the representative reference architecture depicting all the key capabilities of a mobile network that can support a mobile health solution. This section provides some key capabilities of this architecture and for details readers can refer to [32].

mHealth Device: The mHealth device could be a sensor with embedded cellular connectivity, a device consisting of one or more mobile health sensors that connect to a mobile health Gateway running mHealth application (with cellular connectivity) through PAN interface, a device designed to connect via a short-range technology (e.g., Bluetooth) to a mobile phone, and so on. Additionally,

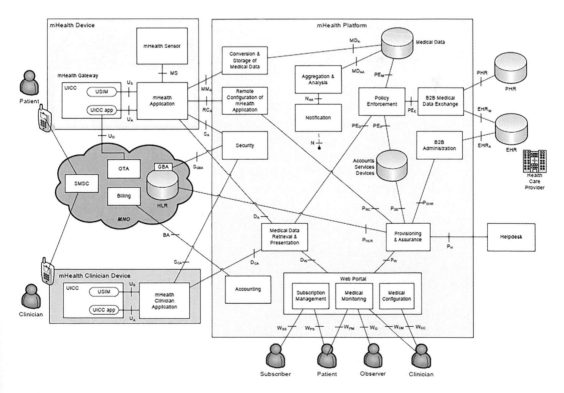

FIGURE 13.9 Main building blocks and interfaces of the mobile health reference architecture. (From GSMA 2011. Connected Mobile Health Devices: A Reference Architecture. Version 1.0 [32].)

there are a number of different interfaces that could be used by the device to communicate with the mHealth platform. The GSMA recommends that an interface shall conform to an internationally recognized healthcare messaging standard; for example, IEEE 11073, IHE PCD-01 or DICOM*.

Clinician Device: The mHealth clinician device (containing the mHealth clinician application) enables a clinician to view the data that is being sent by a patient's device and also allows the clinician to send the reply. It is likely to be a secured smartphone or a tablet.

Mobile Network: The mobile network manages communications between patients and clinicians. It manages subscriptions, billing, roaming, and so on, which assist in the delivery of mHealth solutions.

mHealth Platform: The platform consists of a number of blocks as shown in Figure 13.9. For instance:

- The *Conversion and Storage of Medical Data* capability receives data from the mobile health devices, converts this data when necessary, and stores the data in the medical data database for further analysis and processing while adhering to relevant healthcare privacy and security standards.
- The *B2B Medical Data Exchange* capability takes care of the automated information exchange between the mobile health platform and the databases of the healthcare providers.
- The *B2B Administration* provides administration functionality to the healthcare provider; for instance, accounting, audit control, and so on.

13.4.3 mHealth Applications

The most widely adopted application of mHealth is text messaging due to its ease of use, low cost, and availability in almost every phone on the earth. Tens of mHealth solutions are available; the key ones that are getting wider attention include family planning, newborn health, and tobacco control.

Family Planning: The mHealth programs for family planning provide a range of mechanisms that make use of all types of mobile phones (i.e., basic, feature, and smart). These programs at a high level can be categorized into two groups of mHealth practice, namely client-centered and provider and health system-focused. The client-centered programs are designed to provide health information and support directly to clients or the general public, whereas the latter one provides training, counseling/job aids, and performance support for health workers, systems, data, and program management [28].

Newborn mHealth: The day a baby is born is the most dangerous day of his/her life, as over 1 million children die each year on the day they are born [33]. The three main causes of newborn death are prematurity, complications during childbirth, and neonatal infections. This risk calls for dedicating resources and time and delivering quality care at the time of birth and weeks after birth [34]. These risks can be addressed to some extent by using mHealth solutions.

Tobacco Control: WHO's Tobacco Free Initiative (TFI) was incepted in 1988 and it has been making steady progress. The WHO FCTC (World Health Organization Framework Convention on Tobacco Control) is a global public health treaty adopted in 2003 by countries across the globe as an

---

* CEN ISO/IEEE 11073 Health informatics - Medical/health device communication standards enable communication between medical, health care, and wellness devices and with external computer systems. Integrating the Healthcare Enterprise (IHE) is a nonprofit organization based in the U.S. state of Illinois. Digital Imaging and Communications in Medicine (DICOM) standards specify many different formats for image data [5].

agreement to implement policies that work toward tobacco cessation [5,35]. The Tobacco Control and Mobile Health is a recent initiative adding the capability of mobile devices into the overall picture of TFI [36]. Again, the best way to create awareness is via SMS, which has a far greater reach than through TV, print or other media campaigns. WHO is developing geo-tagging/geo-fencing projects to create images of smoke free cities and locations over the Internet. This process will use both mobile phones and Internet to promote awareness.

13.4.4 CHALLENGES

mHealth brings challenges as it tends to mingle two very different industries. Mobile communications is a consumer-focused industry driven by the demands of the users while healthcare pursues a rigid policy where the medical practitioners are in charge and attempt to take a paternalistic approach [37,38].

The healthcare industry has been working toward a patient-centric culture by making them an active participant in their own care. There has been an increased focus on prevention and promotion of good health which can be acquired without going to a hospital. ICT (eHealth/mHealth) can play an important role, particularly in the well-being of poor patients [38].

The key challenges are as follows:

Privacy and Security: mHealth solutions and devices collect large quantities of information (e.g., data stored by the user on the device and data from different sensors, including location) and process it. Data and privacy protection is a challenging task when it traverses through mobile networks and across continents. In many developing countries, the law exists on these items but the implementation is far from over. Loss or theft of devices storing sensitive information can be a serious security issue. Thus, mHealth solutions may contain specific and suitable security safeguards such as the encryption of patient data and appropriate patient/clinician authentication mechanisms to mitigate security risks [37].

Regulatory: Such issues arise in mHealth largely as a result of different regulatory motivations for health and ICT sectors. The regulation of the consumer-centric ICT industry places emphasis on fostering competition and innovation and applying just enough regulation. The clinician-centric health sector on the other hand applies safety first and at least do not harm principles.

Reimbursement Models: The current reimbursement payment models incentivize more on work done rather than outcomes achieved and that perhaps needs to be changed. Some national legislation still provides that a medical act can only be performed with the physical presence of both the patient and the doctor; preventing the enablement of mHealth solutions.

Standards and Interoperability: The absence of standards for interoperability* between mHealth applications impedes innovation, economies of scale, and good utilization of funds. Interoperable technologies are supportive of innovations and lead to open and patient-centered systems to support high-quality care as people move across organizational and national boundaries.

Lack of Education and Information: The lack of education, information, and information sharing is a harmful factor for the promotion of mHealth, particularly in developing countries. People who live in villages and remote areas in many parts of the world, where mHealth is needed the most, usually have the highest illiteracy rate, which further aggravates the

---

* Semantic Health study definition "Interoperability is where two or more e-Health applications (e.g., EHRs) can exchange, understand, and act on citizen/patient and other health related information and knowledge among linguistically and culturally disparate clinicians, patients, and other actors or organizations within and across health system jurisdictions, in a collaborative manner."

challenge. Governments/Regulators may establish programs for the promotion of mHealth to build trust and provide training to healthcare professionals and society at large.

13.4.5 mHEALTH MARKET POTENTIAL

In recent years, mHealth has emerged as a complementary way of delivering healthcare building upon the ubiquitous connectivity of mobile networks along with proliferation of mobile devices across the globe. The coexistence between wireless communication technologies and healthcare systems on the one hand and health and social care on the other hand, is creating new business opportunities.

mHealth in the high-income countries is driven by the notion to cut healthcare costs while in developing countries it is mainly boosted by the need for access to primary care. A joint analysis by GSMA and PwC (PricewaterhouseCoopers) projects that the global mHealth market will reach U.S. $23 billion in 2017–2018. Most of the market share will be in Europe and the Asia-Pacific regions with a 30% share each, followed by North America with 28%. Latin America and Africa are expected to have smaller markets with estimated shares of 7% and 5%, respectively. The report also predicts that mobile operators are expected to be the key beneficiaries of the expected growth in the mHealth market getting about 50% share of the overall market, corresponding to U.S. $11.5 billion, in 2017. This is followed by device vendors (29%), content/application players (11%), and healthcare providers (10%) [37,39,40]. According to Statista, mHealth is forecast to be valued at around 25.39 billion U.S. dollars in 2020 [41].

13.5 MASSIVE INTERNET OF THINGS

After connecting people anytime and everywhere, one of the next steps in the evolution of Telecommunications is to connect things/objects. The IoT is broadly described as an interconnected network of physical objects/things that will deliver new services within and across industries. A thing in IoT is an object that can be assigned an IP address with the ability to transfer data over a network. The thing could be person with a heart monitor implant, an automobile with a built-in sensor to alert the driver about the oil and filter changes, a home/building with smart electric meter, and so on.

There are many loose definitions of the IoT leading to some fragmentation within the industry. Here are some examples:

- According to ITU-T, IoT is a global infrastructure for the information society, enabling advanced services by interconnecting (physical and virtual) things based on existing and evolving interoperable information and communication technologies [42].
- According to EU FP7 CASAGRAS*, it is a global network infrastructure, linking physical and virtual objects through the exploitation of data capture and communication capabilities. This infrastructure includes existing and evolving Internet and network developments. It will offer specific object identification and sensor and connection capability as the basis for the development of independent cooperative services and applications. These will be characterized by a high degree of autonomous data capture, event transfer, network connectivity, and interoperability [43,44].
- The term massive IoT is also getting attention and in the context of 5G refers to the interconnection of 10's of billions of things. ITU called it mMTC, which refers to having at least 1 million devices per km^2.

---

* CASAGRAS (Coordination and Support Action for Global RFID-related Activities and Standardization) is a European Framework 7 project.

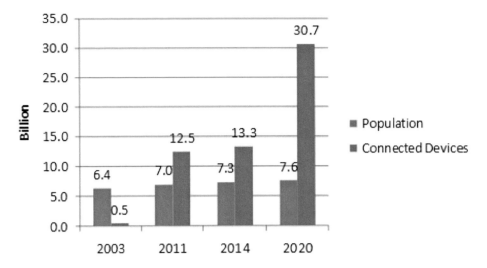

FIGURE 13.10 IoT to massive IoT.

The three key users of IoT include businesses, consumers, and governments forecast to create a U.S.D 13 trillion return on investment by 2025 [45]. There were about 6.36 billion humans and about 500 million devices connected to the Internet (mostly PCs and a few smart phones) in 2003 [46,47]. In 2011, there were 7 billion human beings and 12.5 billion connected devices [48]. In 2014, there were 7.3 billion human beings and 13.3 billion connected devices [49]. The world population is expected to reach 7.6 billion and there will be approximately 30.7 billion connected devices by 2020 [50,51]. This trend (Figure 13.10) shows that there are many more connected devices than human beings and it is expected to significantly increase in the future making today's IoT tomorrow's massive IoT.

13.5.1 M2M versus IoT

Quite often, the terms IoT and M2M are used interchangeably. However, it is certain that these two concepts do indeed have slightly different meanings and it is important to identify the differences before diving into the details.

M2M is defined as the data exchange between devices using a communication channel and processing the data without (or minimal) assistance of humans, for example, enabling a sensor or meter to communicate the data it records (such as temperature, inventory level, etc.) to a remote computer for processing [5].

M2M with Internet protocols could be considered a subset of the IoT [52]. M2M is about connecting and communicating with a thing, where a thing can be machine, device or a sensor with little human interaction. IoT on the other hand, represents things connecting with systems, people, and other things [53].

The concepts are similar, but generally, the industry reserves the M2M term for more industrial type applications where there is little human involvement and includes most consumer applications under the broader umbrella term of IoT. IoT is a concept that will create opportunities and threats in just about every industry prevailing today.

13.5.2 Ecosystem

IoT, an under-development ecosystem, involves a number of vertical markets where one company of the value chain may be providing more than one part. The key reason for this is a general lack of standardization within many areas of connected devices and harmony between application software

TABLE 13.4
IoT Ecosystem Products

Product/Player	Description
Chip	For providing connectivity based on radio protocols and for measuring environmental/electrical variables (i.e., sensors)
Microcontroller	For processing and storage of data on a chip
Module	Single package containing chip, sensor, and microcontroller
Device	A module with applicable software in a usable form factor
Short range technology	Use of WPAN (Wireless Personal Area Network) and WLAN (Wireless Local Area Network) technologies WiFi, Zigbee, NFC, and so on for connectivity
Long range technology	Use of cellular (2G/3G/4G/5G) networks (including radio, transmission, core, and billing) for connectivity/billing
Spectrum	Use of licensed or unlicensed spectrum for communications
Platform	For processing IoT data within the cellular network
IoT application	IoT application for the specific task
Application software	For presenting to the requested party in a suitable form

and solutions and between different verticals or end markets. This has forced many of the successful early players of IoT to vertically integrate in order to develop a full solution of hardware, software, and services designed for a specific vertical market [48].

IoT involves a number of distinct product and service players starting from chip manufacturers all the way up to IoT application developers as shown in Table 13.4 and Figure 13.11. The radio chips enable underlying connectivity, sensors support data gathering, while microcontrollers provide the processing of the collected data. The IoT modules combine the radio chip, sensor, and microcontroller along with storage capability and make it insertable into a device. The device with the IoT application (software), which may be considered as part of the wireless sensor network (WSN), connects using a wireless medium (WiFi, Zigbee, NFC, etc.) to a gateway. The gateway in turn is connected to the wireless cellular/broadband network, preferably with optical fiber. If an IoT device is embedded with an SIM that can connect to the cellular network, then a WSN will not be needed.

The network provider using its wireless access spectrum, transmission medium, and billing/provisioning elements authorizes, bills, and sends the data to the requested party (could be a system, person with an appropriate device or a thing) via Internet. The network may also have a platform (server) embedded with required module and software for processing and dispatching this data to the requested party via Internet. Finally, the application software would present all the information gathered in a usable and analyzable format to the requested party [48,54].

13.5.3 STANDARDIZATION

The current IoT related standards' landscape is rather fragmented and its success in the future is dependent on the development of interoperable global standards. Tens of organizations, alliances, and SDOs are involved in developing bits and pieces of the IoT standards, but these lack cohesiveness.

IoT standardization is rather complex as it involves a number of players within and outside the telecom industry. The requirements from these different industries often come from legislation or regulatory activities. Thus, any IoT related standardization may need to pay attention not only to technologies and protocols, but also to legislative and regulatory measures of the particular applied sector [55].

The standardization activity is taking place primarily at five levels/layers. At the radio access layer, 3GPP is developing standards for connectivity and radio network optimization. At the link layer, IEEE, and so on have developed many standards, while the network/Internet layer has been under the radar of IETF and so on. For the service layer, M2M, OCF, AllSeen, and so on are developing

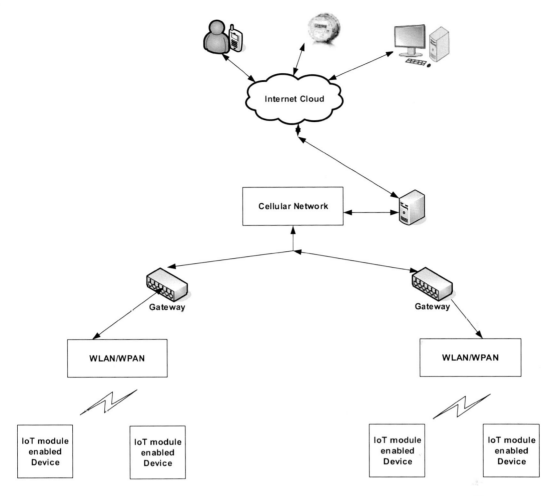

FIGURE 13.11   IoT value chain.

frameworks for enabling IoT services, and finally, for developing protocols for IoT applications (at the application layer) OASIS, OMA, W3C, and so on are actively engaged*.

A very few standardization activities are as follows, details can be found at [56,57].

13.5.3.1   ITU

One relevant standard is ITU-T Y.2060 that was published in 2012 [42]. It clarifies the concept of IoT by providing a reference model, ecosystem, and business models. Figure 13.12 shows the IoT reference model that consists of four layers as well as management capabilities and security capabilities associated with these four layers.

- *Application layer* contains IoT applications.
- *Service support and application support layer* for providing generic (data processing or data storage) or specific capabilities.
- *Network layer* for networking capabilities such as access and transport resource control functions, mobility management or authentication, authorization, and accounting. It also provides transport capabilities for the connectivity of control and application specific data.

---

* OCF: Open Connectivity Foundation; OASIS: Organization for the Advancement of Structured Information Standards; OMA: Open Mobile Alliance; W3C: World Wide Web Consortium.

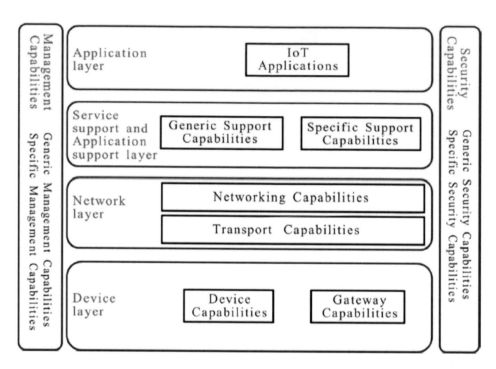

FIGURE 13.12   IoT reference model. (From ITU-T 2012. Recommendation ITU-T Y.2060 Overview of the Internet of Things [42].)

- *Device layer* include device capabilities such as interaction with the communications network and ad-hoc networking, and gateway capabilities such as support for multiple interfaces for connectivity with wired and wireless networks and protocol conversion. An example of protocol conversion is ZigBee technology protocol at the device layer and a 3G technology protocol at the network layer.

In a similar way to traditional communication networks, IoT management capabilities support the traditional FCAPS classes, that is, fault management, configuration management, accounting management, performance management, and security management. The standard specifies essential generic management capabilities such as device management, local network topology management, and traffic and congestion management. It further defines specific management capabilities that are closely coupled with application-specific requirements, for example, smart grid power transmission line monitoring requirements.

The standard also highlighted generic security capabilities that are independent of applications and are applicable at application, network, and device layers. The specific security capabilities are closely coupled with application-specific requirements, for example, mobile payment.

13.5.3.2   3GPP

3GPP in Rel-13 has introduced eMTC and NB-IoT (narrowband IoT) which are collectively referred to as LTE IoT. These technologies are optimized for lower complexity/power, and at the same time are effective in deeper coverage and higher device density areas. These are designed to seamlessly coexist with other LTE services as well [58].

The same release brought two scaled down versions of UE categories, that is, LTE IoT Cat-M1 and LTE IoT Cat-NB1. The former delivers data rates up to 1 Mbps in 1.4 MHz bandwidth in the existing LTE FDD/TDD spectrum. The latter category delivers data in 10's of kbps in the 200 kHz bandwidth in the LTE FDD spectrum.

It is expected that a 5G NR based massive IoT solution may be available to address the IMT-2020 requirements of ITU. Rel-14 and beyond are working for the enablement of massive IoT.

The world around us is full of smart innovations and IoT/M2M devices and these are not waiting for any standards. Focusing on one standard or a set of agreed upon standards for IoT is required, but the task is far from over. At the same time, the agreed upon standards do not necessarily mean that the objective of interoperability is achieved. The mobile communications industry has been successful not only because of its global standards, but also it mandates interoperability through certification of mobile devices and infrastructure equipment and the same needs to happen for IoT as well.

13.5.4 APPLICATIONS

IoT, potentially, will find its place in almost all the industries and create value. Starting from smart meters, smart homes, health care, and to the extent of fleet tracking, IoT applications can be used. IoT relates to the use of the Internet for smart living. The main IoT applications that span across a number of industries are referred to as vertical domains [59]. These include but are not limited to health, transport, energy, security, communications, and infotainment as shown in Figure 13.13. The vision of a pervasive IoT requires the integration of the various vertical domains into a single, unified, horizontal domain which is often referred to as smart life. This vision creates significant opportunities in the vertical markets of consumer electronics, automotive electronics, medical applications, communication, and so on. The applications in these areas directly benefit from the More-Moore and More-than-Moore semiconductor technologies, communications, networks, and software developments.

Some examples of IoT applications are as follows [48,54,55]:

IoT applications for smart cities

- Smart parking allows the monitoring of parking spaces' availability in the city.
- Smart lighting enables intelligent and weather adaptive lighting in street lights.
- Waste management provides detection of rubbish levels in containers to optimize the trash collection routes.

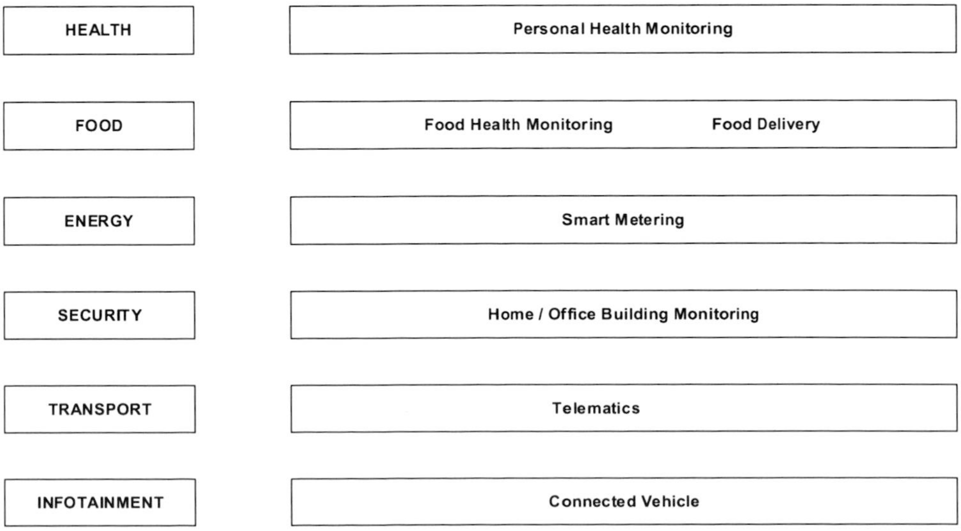

HEALTH	Personal Health Monitoring
FOOD	Food Health Monitoring Food Delivery
ENERGY	Smart Metering
SECURITY	Home / Office Building Monitoring
TRANSPORT	Telematics
INFOTAINMENT	Connected Vehicle

FIGURE 13.13 Application matrix: Societal needs versus market segments (vertical domains).

IoT Application for Protecting the Environment

- Forest fire detection allows monitoring of combustion gases and pre-emptive fire conditions to define alert zones.
- Air pollution helps in controlling CO_2 emissions from factories, pollution emitted by cars, and toxic gases generated at farms.

IoT Applications for Smart Grid

- Smart grid helps in monitoring energy consumption and leakage.
- Tank level helps in monitoring of water, oil, and gas levels in storage tanks.

13.5.5 CHALLENGES/SOLUTIONS

IoT has been criticized for being developed rapidly without giving appropriate due consideration to security, legal, and regulatory challenges. Concerns have also been raised regarding the ability of IoT to erode people's control over their own lives [60]. These are in addition to the technical standardization challenges that may provide a platform for harmonization across industries.

Cyber attacks can become an increasingly physical (rather than simply virtual) threat. Thus, integrity and confidentiality of data is a must which can be addressed to some extent by user and device authentication. These measures may also be helpful in avoiding financial loss and life-threatening situations [61].

An IoT system is composed of independent systems supported by more than one industry, which are combined together in order to interact and provide a given service which cannot be provided by the individual systems when not cooperating. The underlying challenge is that there are no clear and agreed upon architectures for building a connected system. For example, the light switch may have one level of security while the TV remote control has another. Thus, strong cooperation among all players/industries along with clear standards (architectures) are needed to maximize the ROI on IoT [55,62].

13.6 CONCLUSION

The success of MFS and mHealth has been extraordinary, particularly in developing nations primarily due to their lack of facilities, manpower, and poverty. Turkey has been a model of success in tobacco control as declared on 2013 Tobacco Day by WHO. Turkey is the only country in the world to have received three WHO awards for achievements in tobacco control, which has been made possible via mobile technology. Similarly, Telenor Pakistan's Easypaisa is an award-receiving mobile money service serving millions of users.

The IoT ecosystem has made a lot progress; however, work needs to be done for developing an agreed upon architecture for building connected systems. IoT can turn out to be a killer application if not at least a major revenue earner for 5G players.

PROBLEMS

1. What is a service creation environment in the ITU-T defined intelligent network?
2. Define SDP?
3. Define IMS and its difference from SDP?
4. What are the key components of SDP?
5. Discuss in a group of two whether or not SDP and/or IMS will be required for 5G networks?
6. Why is mobile money is needed for both developed and developing nations?
7. Define mobile payment types?
8. Describe the three mobile money transfer models. Give at least 2 examples of each model?

9. Describe mobile wallet with examples?
10. Discuss in a group of two the key reasons for the success of Easypaisa.
11. Differentiate between mHealth and eHealth?
12. What are the key segments that are involved in the value chain of mHealth?
13. What are the basic building blocks of mHealth as per GSMA reference architecture?
14. Define client-centered and provider and health system-focused family planning programs?
15. Define Internet of Things as per ITU-T?
16. What is meant by massive IoT?
17. Describe the differences between M2M and IoT?
18. Which IoT technologies have been introduced by 3GPP in Rel-13?
19. Discuss in a group whether IoT requires standardization or not?
20. How can IoT help in protecting the environment?

REFERENCES

1. Asif, S. 2011. *Next Generation Mobile Communications Ecosystem: Technology Management for Mobile Communications.* Wiley Inc., UK.
2. Alan Quayle 2008. Defining SOA, SDP and IMS; and How They Fit Together. http://alanquayle.com/2008/05/defining-soa-sdp-and-ims-and-h/
3. Heavy Reading 2005. SDP and IMS: Perfect Together? http://www.heavyreading.com/servsoftware/details.asp?sku_id=924&skuitem_itemid=851
4. Metaswitch Network 2011. Service Delivery Platform.
5. Wikipedia. http://en.wikipedia.org/wiki/Main_Page.
6. Nokia Siemens Networks 2008. Service Delivery Framework.
7. Koukal, M. and Bestak, R. 2006. Architecture of IP Multimedia Subsystem. Multimedia Signal Processing and Communications. *48th International Symposium EL-MAR-2006*, Zadar, Croatia, June 07–09, 2006, pp. 323–326.
8. Khartabil, H. et al. 2006. *The IMS: IP Multimedia Concepts and Services.* John Wiley & Sons, New York.
9. Willie, W.L. 2002. *Broadband Wireless Mobile.* John Wiley & Sons, New York.
10. Service Requirements for the Internet Protocol (IP) Multimedia Core Network Subsystem; Stage 1. Technical Specification (Release 8), 3GPP TT 22.228 (V8.1.0), Technical Specification Group Services and System Aspects, 3GPP, March 2006.
11. IP Multimedia Subsystem (IMS); Stage 2. Technical Specification (Release 6), 3GPP TS 23.228 (V6.9.0), Technical Specification Group Services and System Aspects, 3GPP, March 2005.
12. Light Reading Services Software Insider 2005. SDP and IMS: Perfect Together?
13. Maes, S.H. 2010. Understanding the Relationship between SDP and the Cloud. *IARIA Cloud Computing 2010: The First International Conference on Cloud Computing, GRIDs, and Virtualization*, Lisbon, Portugal, Nov. 21–26, 2010, pp. 159–163.
14. Zhang, W. et al. 2013. Architecture and Key Issues of IMS-based Cloud Computing. *IEEE 6th International Conference on Cloud Computing*, Santa Clara, CA, USA, 28 Jun–03 Jul 2013, pp. 629–635.
15. Wang, P. et al. 2013. A Survey of Mobile Cloud Computing: Architecture, Applications, and Approaches. *Wiley Wireless Communications and Mobile Computing*, 18(13):1587–1611. First published online: 11 Oct 2011 DOI: 10.1002/wcm.1203.
16. ITU-T Technology Watch Report 2013. The Mobile Money Revolution Part 1: NFC Mobile Payments.
17. World Bank 2017. Financial Inclusion. http://www.worldbank.org/en/topic/financialinclusion/overview
18. GSMA 2017. State of the Industry Report on Mobile Money—Decade Edition: 2006–2016.
19. Smart Card Alliance Payments Council 2011. Mobile Payments and NFC Landscape: A US Perspective. Publication Number: PC-11002.
20. ITU-T Technology Watch Report 2013. The Mobile Money Revolution Part 2: Financial Inclusion Enabler.
21. World Bank 2012. Maximizing Mobile. Report on Information and Communications for 2012. http://www.worldbank.org/ict/ic4d2012
22. Fokus, F. and Stamatis K. 2004. Mobile Payment: A Journey Through Existing Procedures and Standardization Initiatives. *IEEE Communications Surveys & Tutorials*, 6(04):44–46.
23. MaCarty, M.Y. and Bjaerum, R. 2014. Mobile Money for the Unbanked, Easypaisa: Mobile Money Innovation in Pakistan. GSMA.

24. Fundamo (now part of Visa) 2012. Case Study: Telenor Easypaisa—The Enterprise Mobile Financial Services Platform.
25. Intermedia 2013. Mobile Money in Pakistan—Use, Barriers and Opportunities. The Financial Inclusion Tracker Surveys Project.
26. Gallup Pakistan, Gallup International Association 2013. Use of Mobile Money in Pakistan—Findings from FITS (Financial Inclusion Tracker Surveys Project) study.
27. World Health Organization 2011. mHealth—New Horizons for Health Through Mobile Technologies—Global Observatory for eHealth series—Volume 3.
28. High-Impact Practices in Family Planning 2013. *mHealth: Mobile Technology to Strengthen Family Planning Programs*. USAID, Washington, DC.
29. Lemaire, J. 2011. Scaling up Mobile Health—Elements Necessary for the Successful Scale of mHealth in Developing Countries. Advanced Development for Africa (Actevis Consulting Group).
30. Babu, S. 2012. MHealth Value Chain. http://www.selfgrowth.com/articles/mhealth-value-chain
31. GSMA, TMNG Global 2010. Mobile Technology's Promise for Healthcare.
32. GSMA 2011. Connected Mobile Health Devices: A Reference Architecture. Version 1.0.
33. Save the Children 2013. Surviving the First Day—State of the World's Mothers. 2013.
34. Keisling, K. 2014. *mHealth Field Guide for Newborn Health*. CORE Group, Washington, DC.
35. WFC Policy Statement. Support for WHO Tobacco Frees Initiative. Approved by the Assembly of the World Federation of Chiropractic.
36. Pujari, S. 2011. *Tobacco Control and Mobile Health*. World Health Organization, Geneva, Switzerland.
37. European Commission 2014. Green Paper on Mobile Health ("mHealth"). COM(2014) 219 final.
38. GSMA and PA Consulting Group 2011. Policy and Regulation for Innovation in Mobile Health.
39. PwC 2013. mHealth Insights—The Global mHealth Market Opportunity and Sustainable Reimbursement Models.
40. PwC and GSMA 2012. Touching Lives through Mobile Health: Assessment of the Global Market Opportunity.
41. Statista 2017. mHealth (Mobile Health) Industry Market Size Projection from 2012 to 2020 (in Billion U.S. Dollars). https://www.statista.com/statistics/295771/mhealth-global-market-size/
42. ITU-T 2012. Recommendation ITU-T Y.2060 Overview of the Internet of things.
43. rfidglobal.eu.
44. EU Project Number 216803—Final Report: RFID and the Inclusive Model for the Internet of Things. CASAGRAS (Coordination and Support Action for Global RFID-Related Activities and Standardization).
45. Business Insider 2016. Here's How the Internet of Things will Explode by 2020. http://www.businessinsider.com/iot-ecosystem-internet-of-things-forecasts-and-business-opportunities-2016-2
46. Evans, D. 2011. *The Internet of Things—How the Next Evolution of the Internet Is Changing Everything*. Cisco, California, USA.
47. Colony, G. 2003. Forrester CEO: Web Services Next IT storm. http://www.infoworld.com/article/2681101/operating-systems/forrester-ceo--web-services-next-it-storm.html
48. Leopold, S. et al. 2014. *The Internet of Things – A Study in Hype, Reality Disruption, Growth*. Raymond James & Associates, Inc, Florida, USA.
49. Machina Research 2015. Service Provider Opportunities & Strategies in the Internet of Things. Research Report Sponsored by Cisco.
50. Lucero, S. 2016. *IoT Platforms: Enabling the Internet of Things*. IHS, London, UK.
51. Worldometers 2017. World Population by Year. http://www.worldometers.info/world-population/world-population-by-year
52. Telefonica 2013. What Is the Difference between M2M and IoT? https://m2m.telefonica.com/m2m-media/m2m-blog/item/514-difference-m2m-internet-of-things
53. Axeda 2014. The Connected Effect—IoT vs. M2M. There's a Difference. http://blog.axeda.com/archived-axeda-blog-/tabid/90718/bid/104683/IoT-vs-M2M-There-s-a-Difference.aspx
54. Warma, H. et al. 2013. Internet of Things Market, Value Networks, and Business Models: State of the Art Report. Technical Reports TR-39, Department of Computer Science and Information Systems, University of Jyvaskyla.
55. Vermesan, O., and Friess, P. 2013. *Internet of Things: Converging Technologies for Smart Environments and Integrated Ecosystems*. River Publishers, Aalborg, Denmark.
56. Postscapes—IoT Standards and Protocols. https://www.postscapes.com/internet-of-things-protocols/
57. TechBeacon 2015. The State of IoT Standards: Stand by for the Big Shakeout. https://techbeacon.com/state-iot-standards-stand-big-shakeout

58. Qualcomm Technologies, Inc. 2017. Leading the LTE IoT Evolution to Connect the Massive Internet of Things.

59. Vermesan, O., Friess, P., Guillemin, P. and Gusmeroli, S. et al. "Internet of Things Strategic Research Agenda", Chapter 2 in Internet of Things—Global Technological and Societal Trends, River Publishers, 2011, ISBN 978-87-92329-67-7.

60. Crump, C., and Harwood, M. 2014. Tomgram: Crump and Harwood, The Net Closes Around Us. TomDispatch. http://www.tomdispatch.com/post/175822/tomgram%3A_crump_and_harwood%2C_ the_net_closes_around_us/

61. GSMA 2011. Connected Life—GSMA Position Paper, GSMA White Paper. http://www.gsma.com/ documents/gsma-connected-life-position-paper/20440

62. Sarma, S. 2016. The Internet of Things: Roadmap to a Connected World. MIT Technology Review, Massachusetts, USA. https://www.technologyreview.com/s/601013/the-internet-of-things-roadmap -to-a-connected-world/

14 Burning Challenges

Technological advancements on one hand have brought benefits and on the other a number of challenges. These advancements have enabled the migration from voice to data centric networks that have been both rewarding and challenging. The rewards as we all know are in the form of faster speeds, always-on connectivity, better user experience, hundreds of useful applications, socio-economic benefits and so on. The challenge on the other hand produces innovations and may lead to the inception of next generation technology.

This chapter focuses on some key challenges of current and future networks and also provides some solutions to address the same. The discussion revolves around signaling storms that were witnessed in multiple 3G networks during the beginning of the decade due to the enormous use of smart phones, and their frequency could increase in future networks if they are not addressed in an effective manner. In the next section, HetNets will be discussed, which is quite interesting as the industry moves from 3G/4G to 5G networks. Another daunting challenge that the industry will face in 5G networks revolves around D2D communications. Various 5G related forums are looking into challenges related to D2D and a summary of such is provided in the chapter. Last but not least is the big data challenge that has risen due to the enormous growth of data over the past decade and this will increase as the world heads toward 5G.

14.1 SIGNALING STORM

During the inception years of broadband, operators easily handled the data traffic through readily available network optimization techniques. Additionally, at that time, the main generating sources of data were battery efficient desktops and laptops. However, the exponential growth of smart phones along with bandwidth hungry applications has shifted the optimization focus equally toward connected devices.

Smart phones have created a seismic shift in the way people interact with information and entertainment. Along with the positive side, these are also responsible for creating signaling surges in mobile networks. Many popular applications such as messaging, email, Facebook, and Twitter are designed to touch base with the network for updates at regular intervals. These applications work independently and are not aware of the network conditions, resulting in generating uncoordinated connection requests and a huge amount of signaling (which has been referred to by some as signaling storms) [1].

Figure 14.1 shows the signaling generated by playing a multiplayer poker game online for 30 minutes via laptop versus smart phone (iPhone version 4.1). The baseline activity was also measured on the iPhone version 4.1 with weather and email applications running in the background for 30 minutes. It can be seen that the laptop generated 188 signaling messages during the half hour while the smart phone generated more than 10 times that amount at a staggering 1996 signals during that half hour. This is 66 signals per minute or an average of one signal per second [2].

Figure 14.2 shows the true story of the signaling storm that took place in the Telenor Norway network in June 2011 [3]. An extreme high load on signaling processors located in RNC was observed with 2 million BHCA (busy hour call attempts). The load on the signaling processor unit was, in long periods, observed to be between 70% and 100%, while the dimensioning criterion requires that the signaling shall be below 70% at peaks.* The cause was very high penetration (more than 50%) of smart phones and the result was network outage for more than 18 hours over the Pentecost holiday weekend. Telenor then reactively implemented a number of measures, primarily capacity augmentation and protective mechanisms to protect against similar events in the future [4].

---

* The BHCA figures are hypothetical, but BHCA number per subscriber was the main culprit.

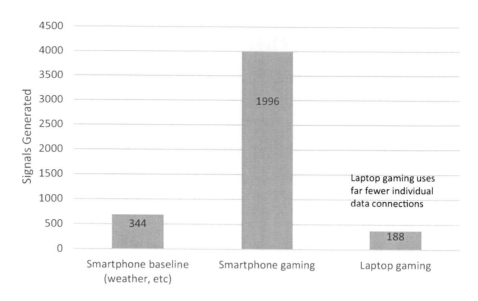

FIGURE 14.1   Signaling generated via online multiplayer poker game. (From Nokia Siemens Networks Smart Labs 2011. Understanding Smartphone Behavior in the Network [2].)

14.1.1   3G Networks—Signaling Technicalities

Signaling manages data sessions, on/off connectivity exercises, and mobility, thus places heavy demand on the control plane both at the radio interface and in the core network. The 3G devices switch from a connected state to a disconnected state to save the battery. This results in inefficiency as only a small amount of data is sent at every opening/closing of the time slots and CPUs in RNCs can become overloaded [1].

In 3G UMTS networks, the RRC layer handles connection establishment, maintenance, and release functions between the UEs and the UTRAN (UMTS Terrestrial Radio Access Network). The operation of RRC relies on states where each state is differentiated with the quantity of radio resources it requires to function as shown in Figure 14.3 and described in 3GPP TS 25.313. During the idle mode, the lowest energy is consumed and both network and UE can see each other but no communication takes place between them, while the data transfer can only take place during the connected mode. A high-level description of the four RRC states is as follows [1,2,6–10]:

- Cell_PCH (Paging Channel): During this state, the UE monitors the network for messages and informs the network whenever it camps in a new cell. The UE does not transfer data and saves energy using DRX.* However, whenever UE needs to send the "cell update" message, it changes temporarily to the CELL_FACH state. The control connection in this state is not lost, that is, the UE still has an RRC connection but seldom uses it.
- URA_PCH (UTRAN Registration Area Paging Channel): This state is identical to Cell_PCH, except that in URA-PCH, the updates are only sent when UE changes URA instead of a cell change. During this state, UE transmits even less frequently that in the CELL_PCH state.
- Cell_FACH (Forward Access Channel): In this state, the UE is in connected mode using common or shared channels. It is used to send/receive smaller amounts of data in a shared channel.

---

* Discontinuous Reception: DRX is used by UE in order to reduce power consumption in UMTS.

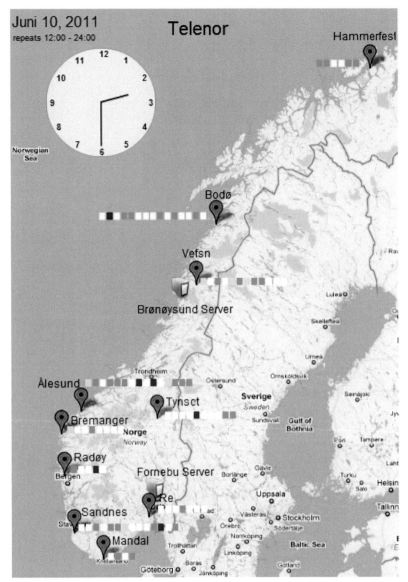

Note: Each color box denotes a machine node in a municipality. The animation splits all nodes into three categories: Green: packet_loss_rate <= 2%, Yellow: 2% < packet_loss_rate <= 80%, Red: packet_loss_rate > 80%; White: no measurements from that node for the specified network interface.

FIGURE 14.2 Signaling storm in Telenor Norway network in June 2011 (a simulation). (From Simula Research Laboratory 2011. Watch Telenor's Network Go Down in Real Time. http://nevada.simula.no/animations/20110610/aServerWide.php?server=brsund [5].)

- Cell_DCH (Dedicated Channel): In this state, the UE is in connected mode using a dedicated channel for transmission and reception of large data volumes. It consumes the most battery power and it is full power state.

The network keeps the UE for a certain duration in the Cell_DCH state in case there is some more information that needs to be transmitted. After some time of inactivity, the network places the UE

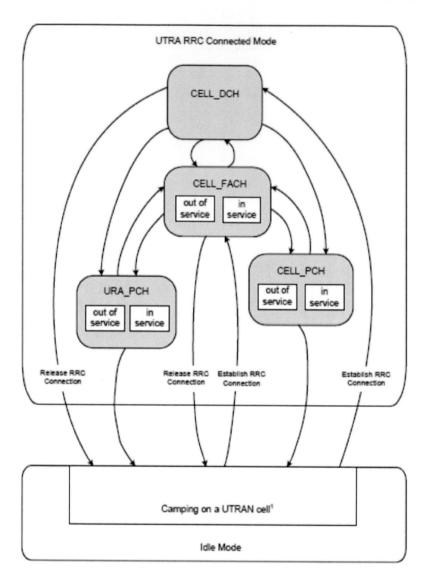

FIGURE 14.3 UTRA RRC connected mode. (From 3GPP TS 25.331 (V14.3.0) 2010. Radio Resource Control (RRC); Protocol Specification (Release 14). Technical Specification Group Radio Access Network, 3GPP, June. [9])

in the Cell_FACH state. Similarly, after some period of inactivity, the network transitions the UE to a lower energy state (Cell_PCH or URA_PCH) and from there to idle. Cell_PCH or URA_PCH may not be present in every network [6].

Each time a UE switches from one state to another, it generates signaling traffic. The round trip from idle to active and back to idle roughly takes 30 messages and 2 seconds without the presence of PCH states. If the PCH state in present, then it could downsize messages and duration to half of that size. The transitions between states are based on certain timing thresholds which are settable. A key goal is to optimize timers to improve user experience and network performance [2].

Several solutions have been designed and implemented to optimize signaling in 3G networks. Some key ones are as follows.

Continuous Packet Connectivity: The main purpose of this feature is to reduce control channel overhead for low bit rate "always on" types of applications. It increases the number of HSPA users that can be kept efficiently in active mode (i.e., CELL_DCH state).

Enhanced Cell_FACH State: This allows UE to send signaling on the HS-DSCH (High Speed Downlink Shared Channel) which is required to move users from Cell_FACH to Cell_DCH. This reduces setup time and improves user experience.

Enhanced UE DRX: The UE DRX was introduced for the Cell_FACH state in Rel-8. It enables UE to restrict the downlink reception times and thus reduce battery consumption by allowing the UEs to go into "sleep mode" during periods of time when downlink reception has been restricted.

Fast Dormancy: This is a 3GPP Rel-8 feature where UE sends a "Signaling Connection Release Indication" message to the network with the information element "Signaling Connection Release Indication Cause" present and set to "UE Requested PS Data session end." At this moment, the network can then decide to do nothing, to release the mobile to idle or to put the connection into Cell_/URA_PCH state. The evolution has been toward network controlled fast dormancy which is enabled in both networks and mobile phones. The key motivations - to reduce network overhead for exchanging signaling message with UE and UE battery consumption [9,11].

14.1.2 LTE Networks—Signal Technicalities in LT

LTE offers improvements over 3G to handle signaling traffic in various ways. However, a higher use of dedicated bearers and a richer mix of applications have generated more signaling related challenges [1].

To improve upon 3G signaling practices, LTE was designed with only two RRC states with fewer state transitions between them. Figure 14.4 show the simplified state model of LTE in which the radio is either active or idle in comparison to the complex 3G UMTS model. The active state is known as ECM-Connected (in MME) and RRC_Connected (in the eNodeB) and idles are known as ECM-IDLE and RRC-IDLE, respectively. The ECM states describe the signaling connectivity between the UE and the EPC. Overall, this requires much less messaging than 3G resulting in shorter setup times (lower latency) and a better user experience, however, there are still challenges.

The challenge is that each active-idle transition in LTE still generates signaling events between the device, the eNodeB, and the EPC. This is because an S1 (interface between eNodeB and MME) connection establish/release procedure is required during each transition. LTE requires ~10 messages as compared to ~30 messages of 3G which are needed to establish a connection. The situation gets more problematic for a network triggered service which may need 15–23 messages as the device has to be paged first and located before the connection can be established. The timer for S1, which is configurable, becomes critical and requires careful planning and optimization [1].

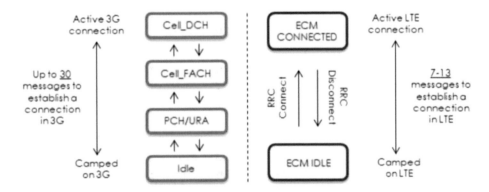

FIGURE 14.4 3G UMTS and LTE state models. (From Gabriel, B. 2012. The Evolution of the Signaling Challenge in 3G & 4G Networks. Heavy Reading, June. [1])

Another key reason for less signaling in LTE is that access control in 3G is handled by RNC that manages many UEs while access control in LTE is in a eNodeB that handles fewer terminals. Thus, signaling traffic can be thought of as less problematic in LTE, however, battery power is still a concern. Last but not least is the presence of nine different bearers (equivalent to nine PDP contexts in 3G), that may all be active at the same time on the same device. Thus, that will multiply signaling in proportion to the number of bearers and the frequency with which they are set up along with the default EPS bearer which is a must. For example, VoLTE has its own bearer type and will drive differing signaling demands. However, in practice, 3–5 bearers are usually active in a network [1,12–14].

The key techniques to reduce signaling and power consumption include single MME/SGSN node, DRX, and DRA (Dynamic Routing Agent).

Single MME/SGSN Node: The critical elements involved in the UMTS/LTE handovers are SGSN and MME that handle communications for redirecting a session as the user moves between LTE and 3G. This involves a reasonable amount of signaling which takes place over the S3 (interface between SGSN and MME). To optimize the process, combined SSGN-MME products (described in Chapter 7) have been offered that internalize the S3 interface and reduce signaling load [1].

Paging and Tracking Area Management: Paging is required to locate a UE in the network when it is in an idle state as its exact location in the network is unknown. Paging procedures are signaling messages that take place among UE, eNodeB, and MME. It may be noted that in LTE, the signaling traffic generated by MME paging the UE is significant (25–35% of the total MME signaling). Though not as significant as the MME paging process, the Tracking Area Update (TAU) procedure is another source of signal generation in LTE. This type of signal generation occurs when there is frequent movement of UE from one TA* to another or when the periodic TAU timer expires.

Techniques have been implemented to address the signaling challenges associated with these processes. Paging associated signaling can be reduced by tailoring paging policies/algorithms for each type of service. MME signaling load can also be reduced by using dynamic TA list management techniques. These techniques can optimize the number of tracking areas in the UE TA list by monitoring its cyclic patterns. The regular updates and optimization of the UE TA list reduces the number of TAU requests that needs to be generated, lessens the toggling effect at TA boundaries, and lowers battery consumption [8].

DRX: LTE employs a discontinuous reception mechanism, making UE wake up only periodically to receive or transmit data to save battery power. 3GPP has defined short DRX cycle and long DRX cycle parameters enabling LTE devices to enter the idle state quickly after data transmission is complete. DRX configuration (short and long) can be differentiated based on service and UE types to minimize consumption of power.

Diameter versus DRA: Diameter is a vital protocol for LTE and it is used on multiple interfaces. However, if it is implemented inefficiently, it may result in overwhelming diameter messages on concerned nodes. To manage this, operators are adding DRAs as part of their next-generation signaling core. It may be noted as well that diameter enabled architecture becomes increasingly complex as

---

* TA (tracking area) represents a group of contiguous cells within E-UTRAN used by the MME to track and locate the UE when in idle mode as it moves through the network. A TA list is a group of adjacent TAs that is managed by the MME and periodically sent to the UE. The MME sends paging messages to the cells that are included in either the TA or TA list in which the UE is registered. TAU procedures can generate a lot of signalling if a UE is moving along the border between TAs that are not all part of its TA list, especially when the TA size is large (e.g., 50-100 eNodeBs in one TA). This is known as a "toggling" effect because of the multiple registrations with the MME that occur as the UE moves in and out of TA boundaries which generates additional TAU signaling [8].

new nodes get added. DRA on the other hand acts a center point where all the nodes get connected allowing load balancing and optimized routing. DRA makes more sense for operators which have rapid and extensive LTE/LTE-Advanced plans. For operators who are slowly transitioning to LTE, investment in DRA can be postponed and diameter capabilities can be used in the transition phase.

Signaling in the Future: There will be approximately 3 billion LTE subscribers by the middle of 2018 worldwide, and this figure is expected to reach 5 billion by 2021. Considering single data connection setups for 50% subscriptions simultaneously may result in around 25 billion messages.

We assume 5G will also have the same two states (idle and connected) as LTE, but half the number of required messages (i.e. $\sim$ 5 messages) for setting up a connection. By 2022, there are expected to be 500 million 5G devices. Assuming 50% of these devices are setting up connections simultaneously, it means that 1.25 billion messages may be generated.

In addition, there could be 15.5 billion IoT devices which operate using unlicensed noncellular radio technologies* such as WiFi, Bluetooth, and ZigBee with a typical range of up to 100 m.

In a nutshell, this set ups a stage for a near perfect signaling storm for the future.

14.1.3 CONCLUDING REMARKS

In the coming years, the IoT and social media will become more influential resulting in more short-lived connections and, therefore, significant signaling load on 4G LTE networks. LTE has inherent advantages over 3G, notably in the radio design, but this alone will not effectively address the problem. To reduce the signaling load, all players (operators, network vendors, device and chipset designers, and application developers) have to play their respective roles.

An abundance of small cells and IoT are all set to create more severe and perfect signaling storms in 5G networks. Robust techniques need to be developed to avoid these potential debacles. Finally, the industry has to work together to stop the next storm of signaling from happening in 4G/5G networks.

14.2 HYPERDENSE HETNETS

A HetNet is a network comprised of various wireless access technologies and cell types each having different capabilities, constraints, and operating functionalities. It consists of a mix of macrocells, microcells, and low power nodes such as picocells and femtocells with the goal to bring the network closer to the end users [15–25]. A sample HetNet is shown in Figure 14.5.

The concept of HetNet was introduced during the standardization of 4G (LTE-Advanced) technology and discussions are ongoing for incorporating HetNets into 5G. HetNets have been deployed to address coverage and capacity demands in a cost effective manner. The key driver behind the uptake of HetNets is the belief that adding small cells to the grid of macrocells is one of the best solutions to address the demands of wireless data and indeed it has been to some extent.

Today's HetNets are evolving what has been termed as a hyperdense network. A hyperdense network is one in which large numbers of small cells are deployed per square kilometer, usually in a restricted geographical area, and typically with large numbers of carriers. There is as such no consensus on the number of small cells that will make a network a hyperdense network, but it will be in the 100s. In such a scenario, a mobile device can connect to multiple base stations or at least frequently have a base station to itself. This requires a paradigm shift in the design of future 5G networks where small cells will be found in abundance.

A 5G hyperdense network will further aggravate existing HetNet challenges and may bring new ones as well. This section primarily looks into key backhaul, mobility management, site placement, and security challenges.

---

* May include certain fixed line LANs and powerline technologies.

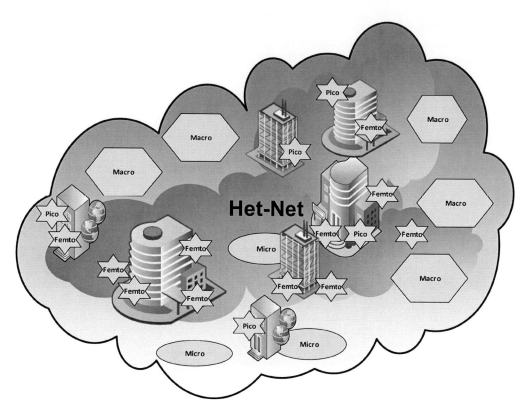

FIGURE 14.5   A typical HetNet.

Backhaul: Backhaul requires the availability of suitable media and power at the right place, time, and cost. For example, running a fiber optic cable to a lamp post (hosting a picocell) can be expensive even if the distance is only a few hundred feet.

Today's cellular backhaul is usually provided via line of sight microwave radios, Ethernet, fiber, copper, and third party leased lines. However, for small cells, access to wired networks like DSL and cable will also be required (Figure 14.6). The problem gets worse in countries which have weak DSL infrastructure, thus delaying the deployment of small cells.

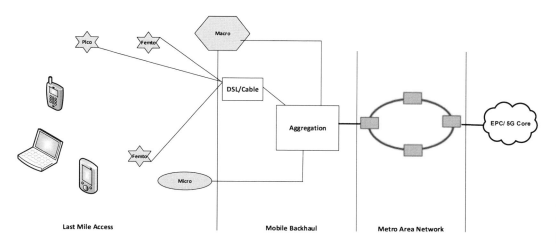

FIGURE 14.6   HetNet backhaul.

More and more small cells are getting located close to each other making backhaul more challenging. Backhaul requirements will become more complicated with hyperdense HetNets, thus the following may be considered:

- Use of standardized agreements with fiber/copper providers
- Use of partnership model with fiber providers
- Use of better traffic compression models on existing copper infrastructure when fiber is unavailable or cost prohibitive

In a nutshell, one size fits all will not work and a combination of wireless and wired technologies will be used to address the problem.

Mobility Management: Mobility related challenges are similar as in macro networks such as intercell handovers and idle mode mobility. However, with small cells, the frequency of these handovers and state changes (as described in the signaling storm section) gets quite high. This increases the probability of handover failures along with user session disconnections. This can be mitigated by keeping higher-speed UE's sequestered on the macrocell layer, but achieving this requires speed-based traffic steering which is a challenge.

3GPP TR 36.839 [19] shows that in a HetNet, the outbound mobility from a small cell has a relatively higher failure rate than other types of HetNet mobility. In particular, the failure rate is high for fast moving (60 mph or more) devices. Studies have also shown that besides a higher handover failure rate as compared to macrocell based network at high speeds, the ping-pong effect (leaving and then re-entering the same cell) is actually worse with low mobility. Thus, in HetNets there is a need to have speed-based mechanisms to improve mobility robustness.

Site Placement: An obvious difference in a HetNet environment is the placement of BTS and their corresponding coverage. The macrocells are normally modeled on a hexagonal grid and are placed on the ground. Small cells are required to be placed in more precise locations (such as on light poles, rooftops, parking lots, etc.) than macrocells, resulting in less flexibility in choosing sites and gaining access to them. This requires better cost effective installation techniques on existing structures while maintaining an aesthetic sense, a simplified process for access to sites, and a fast track site acquisition process allowing approval for groups of small cells.

Good and accurate location planning for small cells is required to maximize capacity. The best traffic location results are currently provided by network-based solutions which can determine the UE position within a few meters. For a HetNet environment, beside traditional methods, it is critical to identify wired network reliability and methods to reduce drive testing and optimize capacity.

Capacity management in HetNets is similar to multi-RAT macro only networks. With multi-RAT HetNet, UE during its camping decision needs to choose the RAT and also choose between macro and small cell layers. This process needs to be optimally controlled to achieve optimal load distribution and increase capacity.

In HetNets, the downlink and uplink may be considered as separate networks because the asymmetry between the two is potentially much higher than in the macro network. For instance, a femto cell coverage area might be limited to a single house or even to some part of a single floor of a single building. This will require different models for interference, cell association, and throughput. Interestingly, a very simple sub optimal approach known as biasing is preferred by the industry. In biasing, small base stations are preferred by some amount known as the bias value, to account for the fact that they are lightly loaded; then the usual max-SINR association is used with the biased SINRs. More work is needed to better understand how to optimize (and adapt) biasing for HetNets, particularly under realistic loading models and diverse types of traffic (e.g., balancing QoS for data, VoIP, and video streaming).

Security Challenges: HetNet brings a number of additional security challenges which are physical in nature as well as related to information. Small cells can easily be tampered with and potentially compromised by unauthorized people since these are primarily deployed in public areas. Thus strong access measurements need to be placed to prevent misuse of units.

14.2.1 CONCLUDING REMARKS

HetNets are/will be instrumental in meeting the needs of coverage and capacity of increasingly data centric networks in a cost and operationally efficient fashion. HetNets may also make the overall customer experience smoother and more predictable. HetNets will continue to evolve in 3GPP for both LTE and 5G technologies to meet the deployment and operational challenges of these networks. However, HetNets bring new challenges as identified in this section which require cost effective solutions.

The advent of 5G will take network heterogeneity to a much higher level. This is due to the greater presence of small cells, frequency band combinations, and radio/transmission technologies. Thus, devices will have a larger selection of connectivity options at their disposal which are of different characteristics in terms of the quality of service, power consumption, billing, and so on. Therefore, HetNets with 5G will present a number of challenges and at the same time opportunities for users and service providers.

14.3 D2D COMMUNICATIONS

D2D communications is a method of two-way information sharing between devices either by sharing the licensed cellular spectrum with cellular communications or by using the unlicensed spectrum as stated in chapter 12. D2D communication promotes a low power, high data rate, and low latency services between end users via peer to peer channel. This channel has different characteristics compared to the conventional cellular channels of existing propagation models. One difference is the height of the terminal, that is, in current models one end of the link is situated higher than the other end of the link. In D2D, such asymmetry does not exist. This has an impact on at least path loss and shadowing. The other key difference between cellular and D2D models is that in D2D both ends of the link can be moving, whereas in cellular models the base station is static. This has an impact on the temporal evolution of the channel and overall D2D architecture [26].

Some key challenges and limited solutions are as follows.

Power Control, Resource Allocation, and Interference Management: Power control and resource allocation of D2D connections can be either distributively determined by the UEs themselves or centrally performed by the base station. The latter is a well-known case performed in 3G and LTE systems (not in terms of D2D communications) where either dedicated or shared resources are allocated to minimize interference. In the former case, dedicated resources have to be allocated to all the D2D connections statically or semi-statically so that no interference occurs in the path of cellular connections.

Under inband D2D communication, UEs can reuse the same uplink/downlink resources that are used for regular cellular communications in the same cell. Therefore, it is important to design the D2D mechanism in such a manner that D2D users do not disrupt the cellular services. Interference management is usually addressed by power and resource allocation schemes, although the characteristics of D2D interference are not yet well understood [27]. Significant research has been performed on the centralized resource allocation and power control algorithms considering mutual interference between D2D and cellular connections, where D2D communication is considered as an underlay to LTE-Advanced networks [28–31].

In inband D2D, the transmission power should be properly regulated so that the D2D transmitter does not interfere with the cellular UE communication while maintaining the minimum SINR requirement of the D2D receiver. In outband D2D, the interference between D2D and cellular user is not of concern. Therefore, power allocation may seem irrelevant in outband D2D. However, with increased occupancy of ISM bands, efficient power allocation becomes crucial for avoiding congestion, collision issues, and intersystem interference.

The interference can be efficiently managed if the D2D users communicate over resource blocks that are not used by nearby interfering cellular UEs. This can be easily being done in overlaid inband D2D where part of the cellular resources is dedicated for D2D communications. However, it will be challenging in underlaid inband D2D where both share the same resources. Resource allocation for outband D2D simply consists of avoiding ISM bands which are currently used by other D2D users, WiFi hotspots, and so on [27].

Energy Management: D2D communication can potentially degrade the energy efficiency of the UE, since the D2D UE has to listen not only to the BS but also to its peers. If the device discovery protocol forces the UE to wake up very often to listen for pairing requests or to transmit the discovery messages frequently, the battery life of the UE may be significantly reduced. Thus, trade-off between UE's power consumption and discovery speed of the UEs need to be further studied [27].

Architecture and Mobility Management: It is necessary to investigate how future 5G communications architecture will look in order to address this challenge. Nevertheless, 3GPP has provided a technical requirement document but not a formal standard (technical specification) on the architecture of D2D.

Many research papers have been produced on two mobility management aspects of D2D. For example, in [32], two mobility management mechanisms have been proposed to reduce latency and signaling overhead. In the *D2D-Aware Handover* proposal, it is recommended that a D2D pair be controlled by the same BS. If one of the UEs starts to move and gets better SINR from the close-by BS_2, the authors suggested that it should not handover to it and instead a joint handover of both UEs takes place when the conditions improve. In the second case, referred to as D2D-Triggered *Handover* proposal, the authors suggested clustering the members of a D2D group within a minimum number of cells or BSs. This will reduce network signaling overhead caused by the inter-BS information exchange, such as that related to D2D radio resource usage. The solution targets the scenarios where D2D groups are dynamically formed by more than two D2D UEs. The solution can be applied when the D2D UEs taking part in a D2D group are varying in time, for instance, due to the mobility.

Both proposals have drawbacks. In the former case, the forceful stickiness to BS_1 will increase inter-BS information exchange, and in the latter one, it will put pressure on backhaul requirements.

Theoretical, Simulation, and Test Bed: There is a lack of experimental (lab) test beds that support D2D and current popular simulators such as OPNET (Optimized Network Engineering Tools), Omnet++ (Objective Modular Network Testbed in C++), NS3 (network simulator) do not as such support D2D communications.

14.3.1 Concluding Remarks

The operator controlled D2D communications should enable the operators to control their networks in order to provide better user experience and make profit accordingly. At the same time, they should be flexible and low cost to compete with free (outband) D2D communications. The operators still face several challenges in providing such a D2D solution that can address the above two contradicting objectives simultaneously [33,34].

Bottom line, a lot of work needs to be done to make D2D a reality.

14.4 BIG DATA

Big data is one of the hottest topics in the industry and academia with an enormous wealth of reports and papers. It cuts across many industries including IT, Telecom, content providers, social media providers, governments, and so on. The focus in this section is primarily on the big challenges that the world at large is facing on this discovery [35–42].

Big data is a term applied to data sets where size is beyond the capability of commonly used software tools to capture, manage, and process. The term big data tends to be used in multiple ways, often referring to both the type of data being managed as well as the technology used to store and process it.

The data comes from everywhere, sensors used to gather climate information, posts on social media sites, digital pictures and videos posted online, transaction records of online purchases, and from cell phone GPS signals to name a few. This and a similar amount of available digital data at the global level grew from 150 Exabytes (1000^6) in 2005 to 1200 Exabytes in 2010. It is expected to be more than 35 Zettabytes (1000^7) by 2020, from less than 1 Zettabyte* in 2009. There are variations in big data forecasts, however, these all predict a very high growth in the coming years. The world of big data is increasingly defined by the 4 'Vs,' that is, these 'Vs' become a reasonable test as to whether a big data approach is the right one to adopt for a particular set of data. The Vs include volume (the amount or magnitude of the data), velocity (speed at which data is input or output), variety (the range of data types, structures, and sources) and veracity (the accuracy and contextual usefulness (value) of data). The dominant big data technologies that are commercially used for processing are Apache's Hadoop and No-SQL† databases. The details of these two technologies can be found in [35–42].

The field of big data faces several challenges; some relate to the data acquisition, privacy, and security and others to its analysis.

Privacy and Cyber Security: Privacy is the most sensitive issue, with conceptual, legal, and technological implications. How data can be acquired and presented, what can be shared, and with whom it can be shared, are some critical questions that do not have an easy answer. Privacy is an overarching concern that has a wide range of implications for anyone wishing to explore the use of big data for development—vis-à-vis data acquisition, storage, retention, use, and presentation.

Cyber security is another challenge which requires constant attention due to ever-changing data management technologies and diversity of data analytics. Big data to some extent is a double-edged sword for a system's cyber security. On one hand, its unique analytic features can help organizations enhance and expand the functions of their cyber security systems; on the other hand, however, the availability and accessibility of a large amount of real-time data from different sources creates new risks of system intrusions.

Data Policies: As more and more digitized data travels across organizations, it becomes important to devise a set of policies on topics like privacy, security, intellectual property, and liability. Big data raises a number of legal issues, in particular when it can be copied and combined with other data and perhaps involving two or more countries. The same piece of data can be used simultaneously by more than one person. All of these are unique characteristics of data compared with physical assets, which raises a number of questions in terms of intellectual property rights and fair use. In many countries, there are laws which prohibit the transfer of data outside its geographical boundaries such as in Pakistan.

Data Acquisition: Gaining access to data and to correct data is vital for companies to succeed. In some cases, organizations will be able to purchase access to the data. In other cases, however, gaining access to third-party data is often not straightforward as economic incentives may be forbidding stakeholders to share data. A stakeholder that holds a certain dataset might consider it to be the source of a key competitive advantage and thus would be reluctant to share it with other stakeholders. Organizations looking for data may have to find ways to offer compelling value propositions to holders of valuable data.

Industry Structure: The public sector faces more difficult barriers than other sectors in the way of capturing the potential value from using big data. The key reasons are bureaucracy, red tape, lack of transparency, and lack of competitive pressure. The profit greedy private sector is seeking new ways

---

* 1 Zettabyte is 1 trillion gigabytes or the equivalent of 250 billion DVDs (Digital Video Disc).
† NoSQL or Not Only SQL databases are heavily used in big data applications.

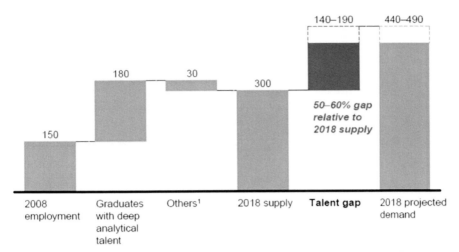

FIGURE 14.7 Demand-supply gaps for data scientists and data managers in the U.S. by 2018. (From James, M. et al. 2011. *Big Data: The Next Frontier for Innovation, Competition, and Productivity.* McKinsey & Company—McKinsey Global Institute [42].)

to handle big data at all costs. For example, Telenor Pakistan created the first Business Intelligent Unit in the local telecom industry analyzing more than 105TB of data [43]. This has become a source of competitive differentiation in one of the lowest ARPU and highly competitive markets.

Organizational Challenges: Big data brings a number of organizational challenges including availability of talent, leadership mindset, budget, and others.

- *Talent*: Talent in this arena is scarce and it is not easy to find many qualified data scientists, for example, as compared to finding RF engineers. A data scientist is the buzz word with no concrete definition. According to a McKinsey Global Institute 2011 report, the U.S. alone may face a 50% to 60% gap between supply and the requisite demand of deep analytic talent and data managers, respectively, by 2018 as shown in Figure 14.7. To fill this gap, a number of universities have started to offer formal training or advanced analytics degrees. On the other hand, companies also look for crowdsourcing options, where the needed services, ideas or content are obtained by soliciting contributions from a large pool of experts, especially from an online community, rather than from traditional employees or suppliers. Crowdsourcing helps organizations that cannot afford to hire or have no need to hire full time data scientists to solve their problems; it takes advantage of wisdom from a very large pool of data scientists to find the best solution to their problems at a much lower cost, resulting in a win-win solution.
- *Disoriented Leadership*: Organizational leaders often lack the understanding of the value in big data as well as how to unlock this value. Who will lead the big data initiative in the organization is a question mark. Whether it is CTO (Chief Technical/Technology Officer), CIO (Chief Information Officer) or new CxO (Chief x Officer) in the form of CDO (Chief Data Officer) is debatable. In addition, many organizations do not structure workflows and incentives in ways that optimize the use of big data to make better decisions and take more informed action.
- *Budget*: Big data storage and analysis require more computational power than the traditional data processing. It requires analytics servers, high power computing servers, and applications that could put a dent in a company's budget.

14.4.1 Concluding Remarks

Big data will become more challenging with IoT which is expected to be a norm in 5G networks. The right technologies with organizational matureness are a necessity for tackling the challenge. In conclusion, analytics is the key to extracting the right value from the enormous pile of data and with the help of God-gifted intuition.

PROBLEMS

1. Define signalling storms as related to cellular networks?
2. What was the main cause that resulted in the signalling surge in the Telenor Network and what was the main remedy the company took to resolve it?
3. Describe UTRA RRC connected states?
4. Briefly describe the role of fast dormancy in signalling storms?
5. Why does LTE technology have less signalling than 3G UMTS?
6. Could combining MME and SGSN in one platform result in less signalling? If so, explain the reasons?
7. Discuss the effect of paging and TA on signalling?
8. Explain the difference between diameter protocol and dynamic routing agent?
9. Define HetNets?
10. Could mobile backhaul become a bottleneck for HetNets?
11. Why is site planning more complex in HetNets/small cells?
12. What is D2D communications?
13. What are the key challenges associated with D2D communications?
14. What is big data?
15. What is crowdsourcing and how could it help organizations?

REFERENCES

1. Gabriel, B. 2012. The Evolution of the Signaling Challenge in 3G & 4G Networks. Heavy Reading, June.
2. Nokia Siemens Networks Smart Labs 2011. Understanding Smartphone Behavior in the Network.
3. Fisher, A.D. 2011. Signal storm' Caused Telenor Outages. Cloudberry Media. http://www.newsinenglish. no/2011/06/16/signal-storm-caused-telenor-outages/
4. Cellular News 2011. Internal Signaling Traffic Blamed for Telenor's Nationwide Outage Last Week. http://www.cellular-news.com/story/49606.php
5. Simula Research Laboratory 2011. Watch Telenor's Network Go Down in Real Time. http://nevada. simula.no/animations/20110610/aServerWide.php?server=brsund
6. GSM Association Official Document TS.18 2011. Fast Dormancy Best Practises, Version 1.0, GSMA.
7. Huawei 2012. Smartphone Solutions, Issue 2.0.
8. Nowoswiat, D. and Millikne, G. 2013. Managing LTE Core Network Signaling Traffic, Alcatel-Lucent. http://www2.alcatel-lucent.com/techzine/managing-lte-core-network-signaling-traffic/
9. 3GPP TS 25.331 (V14.3.0) 2010. Radio Resource Control (RRC); Protocol Specification (Release 14). Technical Specification Group Radio Access Network, 3GPP, June.
10. telecomHall 2016. What Are Modes, States and Transitions in GSM, UMTS and LTE? http://www. telecomhall.com/what-are-modes-states-and-transitions-in-gsm-umts-and-lte.aspx
11. Share Technote UMTS Quick Reference. http://www.sharetechnote.com/html/Handbook_UMTS_ FastDormancy.html
12. RF Wireless World 2012. ECM vs EMM in LTE | Functions of ECM and EMM in LTE. http://www. rfwireless-world.com/Terminology/LTE-ECM-and-EMM.html
13. Barton, B. 2012. LTE and Beyond: EPS Mobility Management (EMM) and Connection Management (ECM) States. http://www.lteandbeyond.com/2012/12/EMM-ECM-eps-mobility-management-and-connection-management.html
14. Junxian, H. et al. 2012. A Close Examination of Performance and Power Characteristics of 4G LTE Networks. *The 10th International Conference on Mobile Systems, Applications, and Services*, Low Wood Bay, Lake District, UK, June 25–29, 2012. Association for Computing Machinery, Inc. (ACM).

15. Marios, K. et al. 2011. Enhanced Inter-Cell Interference Coordination Challenges in Heterogeneous Networks. Cornell University, arXiv:1112.1597v1 [cs.NI].
16. Telecom Intelligence 2013. Smart Capacity Management in a HetNet World.
17. Durga, P.M. 2012. Heterogeneous Networks in 3G and 4G. *IEEE Communications Theory Workshop*, Maui, Hawaii, USA, May 14–16, 2012.
18. 4G Americas 2012. Developing and Integrating a High Performance Het-Net.
19. 3GPP TR 36.839 (V11.0.0) 2012. Mobility Enhancements in Heterogeneous Networks (Release 11). Technical Specification Group Radio Access Network, 3GPP, September.
20. Jeffrey, G.A. 2013. Seven Ways that HetNets Are a Cellular Paradigm Shift. *IEEE Communications Magazine*, 51(3):136–144.
21. Zheng, Z., Dowhuszko, A.A. and Hämäläi, J. 2013. Interference Management for LTE-Advanced Het-Nets: Stochastic Scheduling Approach in Frequency Domain. *Transactions on Emerging Telecommunications Technologies*, 24:4–17. DOI: 10.1002/ett.2570.
22. Thomas, D.N. et al. 2012. Analytical Evaluation of Fractional Frequency Reuse for Heterogeneous Cellular Networks. *IEEE Transactions Communications*, 60(7):2029–2039.
23. Wikipedia. http://en.wikipedia.org/wiki/Main_Page.
24. 3GPP TR 36.814 (V9.0.0) 2010. Evolved Universal Terrestrial Radio Access (E-UTRA); Further Advancements for E-UTRA Physical Layer Aspects (Release 9). Technical Specification Group Radio Access Network, 3GPP, March.
25. Arsah, A., Qing, W. and Vincenzo, M. 2013. A Survey on Device-to-Device Communication in Cellular Networks. Cornell University, arXiv:1310.0720v2 [cs.GT].
26. Tero, I. ct al. 2013. Flexible Scalable Solutions for Dense Small Cell Networks. FP7 Project ICT-317669 METIS.
27. Sami, H. et al. 2010. Device-to-Device (D2D) Communication in Cellular Network—Performance Analysis of Optimum and Practical Communication Mode Selection. *IEEE Wireless Communications and Networking Conference*, Sydney, Australia, April, 18–21, 2010, pp. 1–6.
28. Asadi, A., Wang, Q. and Mancuso, V. 2014. A Survey on Device-to-Device communication in Cellular Networks. Cornell University, arXiv: 1310.0720v6 [cs.GT].
29. Yu, C.-H. et al. 2011. Resource Sharing Optimization for Device-to-Device Communication Underlaying Cellular Networks. *IEEE Transactions Wireless Communications*, 10(8):2752–2763.
30. Hakola, S. et al. 2010. Device-to-Device (D2D) Communication in Cellular Network—Performance Analysis of Optimum and Practical Communication Mode Selection. *Proeedings IEEE Wireless Communications and Networking Conference*.
31. Zulhasnine, M., Huang, C. and Srinivasan, A. 2010. Efficient Resource Allocation for Device-to-Device Communication Underlaying LTE Network. *Proceedings IEEE Wireless Communications and Networking Conference*.
32. Hanis, P. et al. 2009. Interference-Aware Resource Allocation for Device-to-Device Radio Underlaying Cellular Networks. *IEEE 69th Vehicular Technology Conference*.
33. Osman, N.C.Y. et al. Smart Mobility Management for D2D Communications in 5G Networks, FP7 project ICT-317669 METIS.
34. Lei, L. et al. 2012. Operator Controlled Device-to-Device Communications in LTE-Advanced Networks. *IEEE Wireless Communications*, 19(3):96–104.
35. Oracle 2013. Information Management and Big Data A Reference Architecture.
36. UN Global Pulse 2012. Big Data for Development: Challenges & Opportunities.
37. Dirk, H. and Stefano, B. 2011. From Social Data Mining to Forecasting Socio-Economic Crises. Cornell University, arxiv: 1-66. New York, USA.
38. The Economist 2011. The Leaky Corporation. http://www.economist.com/node/18226961#sthash.8FzFF4Ea.dpbs
39. EMC^2 (IDC Digital University). http://www.emc.com/leadership/programs/digital-universe.htm.
40. TM Forum 2013. Managing and Mining Big Data. tmforum Insights Research.
41. Jean, Y. 2013. Big Data, Bigger Opportunities—Data.gov's Roles: Promote, Lead, Contribute, and Collaborate in the Era of Big Data. Data.gov, U.S. General Services Administration.
42. James, M. et al. 2011. *Big Data: The Next Frontier for Innovation, Competition, and Productivity*. McKinsey & Company—McKinsey Global Institute, New York, USA.
43. James, T. 2012. Beyond the Data Warehouse—Telenor Pakistan's Journey into Advanced Analytics. Decision Management Solutions.

15 Weak, Good, and Best Industry Practices

Certain challenges and practices help in creating the demands for the future work of the telecom sector. The solutions to these technical and business hurdles come later in the form of standards and technological and business-savvy advancements.

The focus of this chapter is on three such practices, namely spectrum management (as it is the most important piece of the puzzle), energy management (as it is and will be extremely important for both developing and developed nations), and patent portfolio management (to lead the innovation race, to make money for the business. and importantly. to reward the inventors).

These three and other such practices can be considered under the domain of technology management at a broader level. The technology management for companies is about sustaining and improving a company's competitiveness in the long term. The details on many similar practices can be found in [1].

15.1 SPECTRUM MANAGEMENT

The two most common challenges with wireless communications are weak signals and interference. Weak signals are more associated with transmit/receive power levels whereas interference may happen due to refraction/defraction of a signal via objects and obstacles. Interference can also be encountered if the radio spectrum is not properly regulated which leads to the concept of spectrum management.

Spectrum management is the overall process of regulating and administering access to and use of the electromagnetic spectrum (or radio frequency in the case of mobile communications). A primary goal of spectrum management is to ensure optimal use of the radio spectrum in social, economic, and technical terms [2]. Overall, its purpose is to mitigate radio spectrum pollution, rationalize and optimize the use of the spectrum, avoid and solve interference, allocate and assign in accordance with ITU-R regulations, assist the introduction of new wireless technologies, and coordinate wireless communications with neighbors and other administrations [3].

In the early days of wireless communications, interference was not a problem since there were plenty of frequency bands available and demand was less. If the interference problem arose, it was solved by allocating separate bands (which may also well apart) to each operator operating in the same geographic region. This way of spectrum management was reasonably successful for the last 70 or so years [3]. However, as we all know, this approach is no longer viable.

Frequency spectrum is a natural resource, similar to water, land, minerals, and so on. However, it is scarce and costly when it comes to mobile telecommunications. Spectrum is regulated at the country level, which follows the recommendations of ITU-R's WRC. The WRC, which is held every 3–4 years, is mandated to revise the radio regulations and allot radio frequencies for the world. Unlike some natural resources, it is scarce but reusable, thus it can be shared. However, before sharing can start, two steps need to take place, that is, allocation and then assignment*. For example, the 1885–2025 and 2110–2200 MHz frequency bands were allocated to IMT-2000 or 3G on a worldwide basis by WRC-1997 [4]. Following the recommendations of WRC, the 3GPP set 1920–1980/2110–2170 MHz for 3G and, for example, a 2×10 MHz (1920–1930/2110–2120 MHz) band

---

* Spectrum allocation refers to the allotment of a certain set of frequencies for a specific purpose at the regional or global level. Spectrum assignment refers to the assignment of a specific frequency band within a specific spectrum allocation to a particular user for certain period of time.

can be assigned to an operator within a country for a specific period of time to provide IMT2000 (3G). Once assigned, the operators can share it between themselves if the telecom policy of that specific country allows such spectrum sharing.

Furthermore, in addition to these two steps, identification for IMT is also becoming a necessary step before assignments can be made for mobile broadband within countries. The term IMT has been evolving. It started with IMT2000 (3G), moved to IMT-Advanced (4G), and currently refers to IMT2020 for future 5G systems/networks. Thus, the above-mentioned allocation (1885–2025 and 2110–2200 MHz) may not allow deployment of 5G as it was only allocated for 3G. Therefore, the 1427–1525 MHz band was specifically identified for IMT by WRC-2015, though it was already allocated to mobile service in one of the earlier WRCs.

As stated in Chapter 3, WRC-15 was not successful in getting enough frequency spectrum for IMT 2020 systems, that is, 5G. At the global level, WRC-2015 was only able to identify 51 MHz for IMT [5]. The key factors that inhibited reasonable allocations for IMT 2020 were (a) current substantial allocations in <6 GHz bands to various sectors including mobile, (b) lack of conclusive studies for the existence of mobile with other technologies such as broadcasting, and (c) geo-political reasons in some cases.

The following is a presentation of a solution to a geo-political impediment in spectrum allocation that was presented at the WRC-15.

15.1.1   SPECTRUM ALLOCATION—GEO-POLITICAL PERSPECTIVE CASE STUDY

It is imperative to understand geo-political challenges in assigning spectrum demands. The unsuccessful attempt at the WRC-2015 event to allocate the UHF band (Ultra High Frequency: 300 MHz to 3 GHz) for IMT is a prime example of geo-political issues impacting such allocations. It failed primarily due to unaddressed geo-political issues between nations and partially due to the presence of legacy services in this particular frequency band. The key ITU-R radio regulation/provision No. 9.21 [6] largely failed to address the challenge which requires an administration to secure agreements (from neighboring administration(s)) for allocation of a frequency band within its boundaries to a service, stating that it will not cause interference to the neighbors.

15.1.1.1   UHF Case at WRC-2015

The UHF band is heavily used for broadcasting services worldwide. Countries are slowly transitioning to digital broadcasting which requires less frequency spectrum. The newly unallocated spectrum can be used for mobile or other services. However, in many emerging and developing economies, analog TV is still the king with billions of viewers.

15.1.1.1.1   The Story

As stated above, one of the most interesting chapters of the WRC-2015, which was held from the 2nd to the 27th of November 2015, was the case of the UHF band (470–698 MHz). For the most part, Region 2, comprised of the Americas, was in sync while the other two were in disarray. There was a lot of behind-the- scenes discussion that took place to come to a consensus. However, many administrations had taken strict positions and were not willing to change their stance [5]. The positions of the various countries are as follows, while the details of this unresolved case can be found in [7].

Region 3:

- Bangladesh: Was in favor of allocating only 604–698 MHz for IMT.
- India: Initial position was in favor of allocating the entire band 470 to 698 MHz for IMT.
- New Zealand: Was in favor of allocating the entire band 470–698 MHz for IMT.
- Pakistan: Initial position was only in favor of allocating 604–698 MHz for IMT.
- Papua New Guinea (PNG): Was in favor of allocating the entire band 470–698 MHz for IMT.
- Australia, Iran, Indonesia, Malaysia, Thailand, and many others were against the allocation.

It was clearly understood by both camps that introduction of IMT services like 3G/4G/5G would harm the existing broadcasting services within the country and the neighboring administrations. Current studies have been inconclusive and point to interference from mobile into television services.

15.1.1.2 The Result

During the last week of plenary sessions, Iran opposed Pakistan, Pakistan opposed India, and Indonesia opposed PNG for this allocation. New Zealand, being a standalone island nation, was in good shape, and Bangladesh got lucky partly because of the absence of some of its neighbors from the forum and partly because it has good relations with India.

After Pakistan's opposition, India changed its stance and requested 604–698 MHz for IMT and approached Pakistan. As the world knows, Pakistan and India have fought three wars since their independence in 1947 from Britain and the issue of Kashmir is still unresolved, thus securing favors from the other nation is just like finding a needle in a haystack. This identification would have also had severe harmful implications for Pakistan's existing services. Keeping the repercussions in mind, Pakistan again blocked a second attempt by India and India in return recorded a statement against Pakistan which is part of the conference plenary meeting minutes.

15.1.1.3 Possible Remedy

The following two suggestions can be executed in parallel to find a common ground.

a. A step by step migration plan of UHF band from broadcasting to IMT applicable to concerned neighboring administrations is perhaps what is required. This roadmap may be shared and agreed among all neighboring administrations before WRC-2019. For example, while keeping 470–550 MHz for broadcasting, the rest can be gradually transitioned for IMT as shown in Table 15.1.

b. The corresponding ITU-R provision 9.21 may be amended by mandating the concerned administrations to submit an agreed implementation plan within six months when such a challenge arises.

15.1.2 Spectrum Assignment

After allocations, each country can assign the spectrum to respective operators/users as required. When a spectrum is licensed, a fee is charged using one of the following methods [8]:

Auction is the preferred method of assigning blocks of spectrum for dedicated use. In other words, it is a process where governments sell the rights (licenses) for transmitting/receiving signals over specific bands of the electromagnetic spectrum. The license to use a spectrum can be made technologically neutral (3G today, 4G tomorrow) or service neutral.

Administrative Incentive Pricing (AIP) manifests the opportunity cost of a spectrum for its efficient use. It can be introduced for a congested spectrum which is not suitable for auction such as the microwave radio frequency.

TABLE 15.1
UHF Band Migration from Broadcasting to IMT

Frequency (MHz)	Timelines
551–580	First and second years
581–610	Third and fourth years
611–640	Fifth and sixth years
641–670	Seventh and eight years
671–694/698	Ninth and tenth years

Administrative Cost Recovery (ACR) is appropriate where auctions and AIP are not suitable, for example, in aeronautical, maritime, and amateur radio bands. The fee reflects the costs incurred in administering such a spectrum. ACR is applicable to spectrum that is not congested, used for mainly noncommercial reasons, and where the risk of interference is low.

The regulators usually auction a spectrum as technology specific, that is, certain frequency bands can only be used for GSM and not for LTE and so on. Secondly, the frequencies are assigned to specific networks (operators) and cannot be shared with other operators in the vast majority of the cases. The other item is traffic capacity which is directly correlated to spectrum, in other words, to gain higher spectral efficiency (bits per second per hertz), the technology needs to squeeze more bits into the same spectrum. LTE and LTE-Advanced (4G) downlink spectral efficiency is roughly in the range of 1.5–30 bps/Hz which is much higher than 2G (GSM) and 3G (HSPA) systems. Similarly, 5G has to provide better spectral efficiency than LTE and 4G systems.

An important element associated with spectrum licensing is its heavy cost. Governments across the globe see it as a fat cow to generate funds and pay their debts. On the other hand, there are many cases where operators have yet to get their returns even after 5–7 years of launch; one key reason is heavy spectrum cost.

Thus, elements like technology neutrality, sharing, trading, and definitely cost may need to be kept in mind for 5G spectrum allocation/assignment activities.

15.1.3　Spectrum Auction Case Study

This case study primarily looks into the Next Generation Mobile Services (NGMS) spectrum auction that took place in 2014. The auction took place after a period of ten years for a variety of reasons. The country's telecom regulator, Pakistan Telecommunication Authority (PTA), first attempted to auction a 3G license in 2008, but due to the security situation and lack of enthusiasm from the operator community, it did not fly. Another attempt was made in 2012, but it also failed due to lack of transparency. Also, Government of Pakistan while privatizing state owned PTCL (Pakistan Telecommunication Company Limited) in 2006 made an agreement with Etisalat that it will conduct auction any new spectrum for the following seven years.

15.1.3.1　Background

The mobile service in the form of AMPS (Advanced Mobile Phone System) was started by two operators (Instaphone and PakTel) in 1990. The third operator, Mobilink, launched its GSM network in 1994–95, while the fourth operator, Ufone, joined the bandwagon in 2001. After the Deregulation Policy of 2003, an auction took place in 2004 which brought Telenor Pakistan and Warid Telecom to the country. During 2008, the license of Instaphone was revoked due to nonpayment while China Mobile started its operations by acquiring PakTel. Thus, it became a five-operator market offering GSM service to millions of users.

From the perspective of spectrum management, there are three key players. Ministry of Information Technology and Telecommunication that issues policy directives, the country's regulator PTA that conducts auctions, and Frequency Allocation Board (FAB) which is the custodian of this scarce resource.

At the time of auction, there were five cellular operators in Pakistan having a total GSM user base of around 130 million (at the end of March 2014). WiMAX (3.5 GHz) and EV-DO (1900 MHz) were the other two wireless broadband technologies operated by four and one players, respectively with a total subscriber base of around 140 million.

Cellular Spectrum: Each of the five players had technology neutral licenses in 900 and 1800 MHz with a total of 13.6 MHz as shown in Table 15.2. Each player offers GSM services in these bands.

TABLE 15.2
Frequency Allocations

Operator	900 MHz	1800 MHz	Total (MHz)
Warid	4.8	8.8	13.6
Telenor	4.8	8.8	13.6
Ufone	7.6	6.0	13.6
Zong	7.6	6.0	13.6
Mobilink	7.6	6.0	13.6

Fixed Wireless Broadband Spectrum: The bands in 1900 MHz and 3.5 GHz were auctioned as a wireless local loop that only allowed limited mobility (mobility within a cell site, no handoffs). However, the mobility element has not been strictly regulated and handoffs were usually taking place in WiMAX and EV-DO networks. It is expected that these technologies will eventually move to LTE.

Fixed Service Spectrum: The frequencies in the range of 10–38 GHz are used by both fixed wireless broadband and cellular operators to support backhaul. Operators run into frequency shortages which could be resolved by further assignments in the 42 GHz and 70–100 GHz bands. The 2004 Mobile Cellular Policy piggy-backed the use of these frequencies for backhaul with the radio access spectrum and operators were allowed to use it at almost no cost. This decision was made at that time to instigate growth, however, in the 2015 telecom policy, a proper fee structure was introduced, that is, an AIP for such frequencies.

15.1.3.2   Cellular Spectrum Auction

The PTA auctioned 3G (2100 MHz) and 4G (1800 MHz) in April 2014 after a period of 10 years. The auction took place under the banner of NGMS while maintaining technology neutrality. Spectrum blocks of 2 × 30 MHz in 2100 MHz band, 2 × 20 MHz in 1800 MHz band, and 2 × 7.38 MHz in 850 MHz were offered in this auction as shown in Table 15.3. The 850 MHz was only offered to new entrants (if any).

15.1.3.2.1   Results

The activity was finally successfully concluded in April 2014 after a gap of 10 years, earning U.S. $1.1128 Billion (U.S.D 1.2241 including withholding tax) for the national exchequer.

The four incumbent operators, namely Zong, Mobilink, Ufone, and Telenor, participated and obtained spectrum while the smallest player, Warid, kept itself out of the race. The government made several attempts to lure foreign operators into the picture. However, due to the security situation, low ARPU, and market stagnation, no foreign operator joined the race.

TABLE 15.3
Frequency Lots

Spectrum Band	Spectrum Quantity	Spectrum Floor	Spectrum Cap
2100 MHz	2 × 30 MHz	2 × 10 MHz	2 × 15 MHz
1800 MHz	2 × 20 MHz	2 × 10 MHz	2 × 10 MHz
850 MHz	2 × 7.38 MHz	2 × 7.38 MHz	2 × 7.38 MHz

Source: Pakistan Telecommunication Authority 2014. Information Memorandum, The Award of 2100 MHz, 1800 MHz and 850 MHz Spectrum: The Next Generation Mobile Services Award (NGMSA) [9].

TABLE 15.4

Spectrum Auction Results

Operator	Winner in 2100 MHz Band	Winner in 1800 MHz Band
Zong	2 × 10 MHz (306,920,000 $)	2 × 10 MHz (210,000,000 $)
Mobilink	2 × 10 MHz (300,900,000 $)	-
Ufone	2 × 5 MHz (147,500,000 $)	-
Telenor	2 × 5 MHz (147,500,000 $)	-
Total	$ 902,820,000	$ 210,000,000

Licenses were issued to Zong and Mobilink on payment of 100% of the auction winning price that was made within 30 days of the auction. Ufone and Telenor were issued licenses on payment of a minimum of 50% of the auction winning price. These two operators will pay the remaining payment in five equal installments including a certain interest fee. The breakdown of spectrum quantity and respective payments are shown in Table 15.4.

Zong was the only bidder in the 1800 MHz band and thus was easily able to secure a 2 × 10 MHz block. Ufone also expressed interest in 2 × 10 MHz in 1800 MHz band, but they were not able to participate as they did not win the required 2 × 10 MHz spectrum in 2100 MHz band. Only those operators who won 2 × 10 MHz spectrum in the 2100 MHz band were allowed to bid in the 1800 MHz auction.

15.1.3.2.2 Pros and Cons

The outcome of the auction is reasonably good for the sitting government as well as for the consumers. However, the main fallout of the spectrum is that the probability of ROI is very low for the foreseeable future for the operators.

Key Successes: The three key successes were the availability of MBB, a huge increase in the number of MBB subscribers from zero to over 35 million in 2.5 years, and the presence on the international floor of 3G/LTE.

Key Enabler—Survival of the Fittest: All the auction participants had one thing in mind before jumping on this bandwagon, that is, if they want to continue operating in the market, they need to win the license (following the principle of *survivability of the fittest*). The Pakistani market is very perception-oriented. If an operator does not have 3G, the existing subscribers will slowly start to switch to the ones that have it, even if they never planned on using the broadband service. The majority of the population is poor and price cautious, so they would prefer a 2G-3G combo service at about the same price of a 2G only service.

To avoid the above-mentioned pitfall, Warid Telecom started to offer LTE in its existing 8.8 MHz of 1.8 GHz band. PTA allowed Warid to offer LTE due to the clause of technology neutrality in their existing 2004 license, but with revised quality of service metrics. However, this is a very limited channel bandwidth to offer both GSM and LTE. The results were mixed and in late 2015, Mobilink acquired Warid and this merger was approved by the state in 2016.

Controversial 2 × 5 MHz Lot: One of the most controversial elements was the downsizing of the spectrum floor to 2 × 5 MHz at the eleventh hour in the 2100 MHz band [10]. When and how this decision was made is still up in the air, which in some sense, took out the spirit of the auction. When the demand became equivalent to supply, that is, four blocks of spectrum (two of 2 × 10 MHz and two of 2 × 5 MHz) offered to four operators, there was not much auction activity left. The base price of U.S.D 295 million for 2 × 10sMHz barely reached U.S.D 306.92 million, giving only a 4% increase from the original demand.

Interference-Ridden Spectrum Block: State has the responsibility to provide a clean and interference-free spectrum. However, after the launch of 3G service, Zong found out that its 20–1930 MHz uplink band had heavy interference. This interference was due to the presence of DECT (Digital Enhanced Cordless Telecommunications) 6.0 cordless phones which were heavily present in the country. Primarily, these phones were smuggled into the country and no proper records were available with the authorities. Afterward, a campaign was jointly run by PTA and FAB to make the public aware of its illegal use and ask for their surrender. To compensate for this negligence, state temporarily provided 6.6 MHz in the 1800 MHz band to Zong, which was still with the operator until the time of writing of this case study (end of 2016).

Other Factors: Pakistan was one of the lowest ARPU markets of the world (about U.S.D 2.0) before 3G/4G. Furthermore, some of the GSM operators have not started to make a profit and some have not even reached a break-even point on their initial investment. For example, the initial investment that was by made by the parent company (Telenor Group) in its Pakistani subsidiary (Telenor Pakistan) in 2004–05 did not reach the break-even point until the said auction activity and it is not expected to produce a profit for another few years. After two and half years of the 3G/4G launch, the ARPU has not changed much and ROI cannot be seen at the end of the tunnel due to the following key factors [11]:

- *Poor Population*: Pakistan is a poor country of about 200 million where 40% of the population lives below the poverty line. Another 40% have a basic education and have some means to buy to broadband, while the remaining 20% is comprised of the upper middle class and rich who have most of the means.
- *Literacy*: Data requires that people can read and write. Pakistan has a very low literacy rate and it follows a descending order from regional languages, Urdu (national language) to English.
- *Local Content*: The national language is Urdu and there are more than 100 dialects in Pakistan. Special emphasis has to be placed on the development of local content. Broadband will get a boost if the right local content is developed, implemented, and offered at the right price. Even with the development of content in local languages and Urdu, ARPU is not expected to change much, at least in the short term.
- *Public Transportation*: The success of 3G/4G is also linked to a good and safe public transportation system. In Europe and other developed Asian nations, broadband is heavily used in public transportation. Unlike most developed countries, the Pakistan public transportation infrastructure is extremely despicable. There are no intracity train services, no underground subways, and the conditions of road transportation are extremely intolerable. There are mainly two sections of the population that use the public transportation system, namely the poor and illiterate (or they just have a basic education), and a second group of the middle educated class which already has access to Internet either at work/school or home or at both places. There are also serious concerns of theft in public transportation so people do not dare to have or use broadband savvy devices in public transportation.
- *Operators' Offerings*: Operators also share the blame due to their cheaper and below the cost offering. The emergence of 3G/4G has also resulted in data dumping offers which is leading to low margins for the industry.

15.1.3.2.3 Final Thoughts

The delay in the auction caused considerable harm to the national exchequer and to some extent to the public and telecom sector. If this spectrum had been auctioned between 2008 and 2010, the situation would have been much better. The lack of spectrum strategy from the government, reluctance from the operators to get a 3G license, availability of some fixed broadband, unexpected energy crisis, nonstop steep rupee devaluation, and poor security situation all will jeopardize the return on 3G/4G, at least in the coming years. It can be safely said that no operator had forecasted such drastic rupee devaluation (from U.S.D 1 to PKR 60 in 2005 to U.S.D 1 to PKR 95 in 2014) and unthinkable energy crisis in their respective business plans.

Furthermore, the success of broadband in terms of revenue will remain questionable as long as the illiteracy rate remains high, availability of extra cash remains short, and content in Urdu and regional languages remains low. For an existing broadband user, 3G/4G operators do not have too much to offer except for some cheaper pricing plans and perhaps bundles with savvy devices.

At the same time, the auction was successful in providing a huge cash influx to the national exchequer, mobile broadband to the people of Pakistan, and a presence on the international stage of 3G/4G. At the same time, Pakistan has a huge younger population of around 40 million in the age group of 15–24 which is very attractive for mobile broadband services. It can be safely said that this age group and younger ones which will fall in the bracket of the 15–24 age group in the next few years will help to increase the broadband usage and make the cellular industry more durable in the longer run.

15.2   ENERGY MANAGEMENT

Energy management in cellular networks is a major concern for operators as it affects their profitability. Energy management mainly involves reducing OPEX and CO_2 emissions in networks.

According to a widely cited report of Gartner [12], the ICT sector was responsible for 2% of global emissions or about 830 $MtCO_2e$ (megaton carbon dioxide equivalent). If the trend continues, ICT will contribute about 1.43 $GtCO_2e$ by 2020. Out of this 1.43 $MtCO_2e$, telecom devices' contribution is expected to be 51 $MtCO_2e$ while the contribution of telecom infrastructure would be 299 $MtCO_2e$. Thus, the telecom sector may contribute about 25% to the total ICT emissions in 2020 [13]. On the other hand, a decrease in power consumption of telecom networks is expected due to the use of renewable energy sources, energy efficient devices, and infrastructure.

This section will briefly look into several aspects of energy management including problem definition, energy saving techniques, metrics for measuring greenness, alternative sources of energy, key activities around the world, and a case study of Telenor Pakistan.

15.2.1   PROBLEM DEFINITION

The main concern of cellular operators is the high OPEX cost associated with the nonstop increase in the number of base stations and small cells. In the developing/emerging economies like India, Nepal, Pakistan, and in many African countries, the impediment gets more problematic due to the frequent outages and unreliability of the electric grid power. For example, in Africa, where many countries have less than half of their towers connected to an electric grid, the problem gets worse. Thus, the energy-related OPEX makes up roughly 20%–40% of the total network OPEX for operators in such countries.

According to various studies [14,15] and practical deployments, the main energy consuming element of cellular networks is the BS along with associated elements such as air conditioning units and cables/feeders. As shown in Figure 15.1, PAs consume the majority of the energy of the BS, followed by signal processing, cooling units, and power supply.

PAs in base stations consume about 50%–80% of the BS power as shown in Figure 15.1. Startlingly, the total efficiency of a PA is generally anywhere in the range from 5% to 20% (depending on the standard - GSM, UMTS, CDMA, and the equipment's condition) [16]. Beside BS, mobile switching, and core transmission elements, data centers and retail outlets are the other elements that consume a considerable amount of power as shown in Figure 15.2. The switching sites house elements such as RNCs, MSCs, and so on, while data centers may retain SSGN and GGSNs [17].

15.2.2   ENERGY SAVING TECHNIQUES

The energy consumption of a BS can be reduced by improving the hardware design, software features or a combination of both. It can also be improved by using small cells in large numbers. For

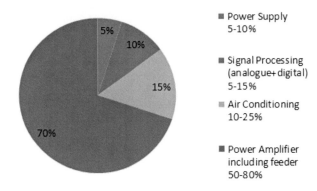

FIGURE 15.1   Distribution of energy consumption at a macro base station site.

example, a typical femtocell draws a total of 20–60 milliwatts of power as compared to a typical BS that consumes 60–240 W. Small cell deployments are continuously on the rise and are expected to gain more momentum in the coming 5G era. Some energy saving techniques are as follows [15–22]:

Power Amplifier Efficiency: Several techniques such as Doherty design, use of GaN amplifiers, DRX, discontinuous transmission (DTX) techniques, and others have been proposed/ implemented to improve PA efficiency. Doherty design can improve energy efficiency by 30% to 35% over a narrow bandwidth [18]. The enhancement can be further improved to over 50% using a digital predistorted Doherty architecture and GaN amplifiers as claimed in [16]. PA efficiency can also be improved by lessening its operating time during low traffic or in the idle mode. This is achieved by optimizing the number of transmit/receive control signals during such modes. For example, cellular technologies like WCDMA/HSPA and LTE realize power efficiency by using DRX and DTX modes for the mobile handsets. These techniques save power by momentarily powering down the devices while remaining connected to the network with reduced throughput [14].

Cell Zooming: This is a technique that allows BSs to adjust their coverage areas according to the network or traffic situation while reducing the energy consumption. When a cell gets congested with an increased number of users, it can zoom itself in, whereas the neighboring cells with less traffic can zoom out to cover those users that cannot be served by the congested cell or which are in coverage holes. Cells that are unable to zoom can sleep as needed to reduce energy consumption, while the neighboring cells can zoom out to help serve the mobile users in a cooperative fashion [15,19–21].

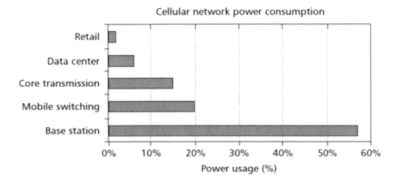

FIGURE 15.2   Energy consumption composition of a mobile network. (From Armour, S. et al. 2011. *IEEE Communications Magazine*, 49(6):46–54 [17].)

Cooperative Relay: Cooperative relaying technique provides gains in throughput and energy efficiency for wireless communications. In simpler terms, when a device transmits a data signal to a destination, a third device (which is authorized) overhears this transmission and relays the signal to the destination as well. The device at the destination finally combines the two received signals to improve decoding [22]. Delivering green communication via cooperative techniques can be achieved by two different approaches. The first approach is to install fixed relays within the network coverage area in order to provide service to more users using less power. The second approach is to exploit the users to act as relays.

15.2.3 RENEWABLE ENERGY RESOURCES

Cellular network operators in many developing nations rely on diesel powered generators to run base stations, which is not only expensive, but also generates CO_2 emissions. CO_2 emissions are naturally a key measure of greenness, but the overall share from the telecommunications networks is fairly low (about 2%) in the whole of scheme of things. However, the key motivation for green wireless communications includes economic benefits (lower energy costs) and better practical usage (for example, less energy consumption from network equipment and mobile devices). According to some estimates, there are more than five million cell phone towers worldwide, around 13% of which are not connected to an electrical grid and largely run on diesel power. For the last few years, the industry has explored and deployed a handful of renewable energy resources [23–25]. Primarily, solar and wind power technologies have provided an alternative to diesel.

- *Solar cells* with battery backup are becoming a primary alternative to off-grid sites and new solar cell technologies are moving the price steadily down. The solar arrays require minimal maintenance and their occasional cleaning prevents a gradual loss of panel efficiency from dust or bird droppings. Cloud and rainfall can reduce power output for a period, but rains can also help in cleaning the solar panel surface.
- A number of trade-offs need consideration in selecting a *wind power turbine* for a cell site. The peak wind speeds determine the size of the structure required to mount the turbine because of the direct relationship between the height above ground and wind speed. A smaller turbine could be mounted on a high tower whereas in other cases, a large turbine on a lower tower may be more cost effective. Thus, the choice of turbine and structure has to be considered in the design of cell sites.
- An initiative called Green Power for Mobile by GSMA is in progress in many African countries [25]. This program is meant to aid the mobile industry to deploy solar, wind or sustainable *biofuel technologies* to power 118,000 new and existing off-grid BSs in such developing countries. Powering that many BSs on renewable energy would save up to 2.5 billion L of diesel per annum (0.35% of global diesel consumption of 700 billion L per annum) and cut annual carbon emissions by up to 6.8 million tonnes.

15.2.4 ACTIVITIES IN SDOs/INDUSTRY/ACADEMIA

During the last few years, a number of green projects have been funded to facilitate the research, evaluation, and experimental deployment of green techniques in mobile networks. SDOs for example, 3GPP, IETF, and others, have introduced energy management parameters in their respective standard specifications. For example, 3GPP Rel-11 that was functionally frozen in March 2013 spelled out the adaptation of inter-RAT, inter-eNobeB, and intra-eNobeB energy savings management (including OAM aspects) in networks. In 3GPP Rel-12 (frozen in March 2015), studies on energy efficiency related performance measurements and system enhancements to support energy efficient deployments and potential solutions enabling energy saving within the GSM BTS and E-UTRAN were initiated [26]. The 3GPP in its Rel-13 (frozen in March 2016) revised the previously defined energy efficiency related performance measurements. Along the same lines, IETF is working on energy efficient networking techniques while specifying energy consumption monitoring requirements [27] and managed objects [28] for creating interoperable standards for green networks. Table 15.5 provides a summary of key projects and some details can be found in [29].

At the same time, various measures have been developed to calculate the efficiency of energy utilization by other organizations as well. For example, the ETSI has defined two network-level energy efficiency metrics. The first metric is defined as the ratio of the total coverage area to the power consumed at the site and is measured in units of $km^2/Watt$. The second metric is described as the ratio of the number of subscribers to the power consumed at the site and is measured in units of users/Watt [18]. A nonexhaustive list of greenness metrics is shown in Table 15.6.

It can be seen that a number of metrics are available and are in use as required. However, there is a need to come up with an agreed set of measurements that can be used across networks around the globe.

15.2.5 CASE STUDY–TELENOR PAKISTAN

Beside measurement metrics and standard specifications, the key and ongoing challenge in many developing countries is the absence of a reliable electrical grid. There are hundreds of networks and millions of BSs where operators have to rely on alternatives for nonstop and continuous operations. This section will primarily look into one case which was successfully managed by the incumbent operator.

Telenor Pakistan started its operations in March 2005 when the country's electrical grid was quite stable with only a few hours of load shedding. With the passage of time, this has worsened and in 2010 it reached 10 hours on average. Thus, the operator had to provide an alternative electric supply to its network of 8000+ S in the absence of state provided electric power. This unreliable supply has considerably increased the cost of doing business in the country and in particularly for operators like Telenor.

15.2.5.1 Background

The country has been facing an energy shortage for the last several years, but it has turned into a crisis since the beginning of the current decade. The key reasons for this aggravated situation are a continuous increase in the cost of electricity and diesel (Figure 15.3) and an increase in the grid outages due to an unreliable distribution system (Figure 15.4). The reduction in diesel and power costs is attributed to the global phenomena.

Diesel and electricity costs are the major contributors to the OPEX. Diesel price is regulated by the government and is affected by the taxes imposed and a reduction of subsidy. Electricity prices are also regulated by the government; however, electricity prices are also dependent on fuel prices. Fuel is a major import for the production of electricity and its price is subsidized as well. For the last few years, the grid outages are mainly attributed to the circular debt* phenomenon prevalent in the country. Due to seasonal impact, the electricity outages increase beyond 10 hours in the summers and decrease in the winters. In rural areas, 15–18 hours of power outages are not uncommon during summer. However, on average, yearly outages have gone up from 8 hours in 2010 to 10 in 2014 and are expected to remain at 10 hours in the coming 2–5 years due to a higher inflow of remittances from overseas Pakistanis which not only help in controlling the circular, but also allow their loved ones to buy electric kitchen appliances, air conditioners, and other consumer electronics. And, it is also due to the continued ineffective power distribution system.

On top of this, diesel and electricity pilferage is a ubiquitous problem in many developing markets like Pakistan. It is a mafia-style phenomenon and has become a lucrative business due to poverty,

---

* Circular debt refers to the debt prevailing among different departments of the government over the power usage charges.

TABLE 15.5

Summarization of Key Green Mobile Networks Projects

Project	Organizer	Region	Participants	Targets	Working Emphasis
5GrEEn	EIT ICT Labs	Europe	Ericsson AB, KTH Royal Institute of Technology, Aalto University and Telecom Italia	5G Mobile networks	• Design of environmentally friendly mobile networks of the future for a connected society
Earth	European Commission FP7 1P (3 years/15 million €)	Europe	European main mobile operators and research organizations	Mobile networks	• Energy aware radio and network technology • Energy efficient deployment, architecture, adaptive management • Multicell cooperation
Green IT	METI & JEITA (Ministry of Economy, Trade and Industry/ Japan Electronics and Information Technology Industries Association) (Japan)	Japan	Over 100 companies, institutes, and organizations	IT	• Power efficiency at data centers, networks, displays • Policy & mechanisms to encourage green IT • Collaboration of industry, academia, and government
Green Touch	Green Touch Consortium	Global	Experts form industry and academia	Telecom networks and mobile networks	• Reinvention of telecom networks • Sustainable data networks • Optical, wireless, electronics, routing, architecture, etc.
Opera-Net	CELTIC/EUREKA (3 years/ 5 million €)	Europe	European main mobile operators	Mobile networks	• Heterogeneous broadband wireless network mobile radio access network • Link-level power efficiency, amplifier, test bed
Green-T	CELTIC (3 years/ 6 million €)	Europe	European main mobile operators	Mobile networks (particularly 4G)	• Multistandard wireless mobile devices • Cognitive radio and cooperative strategies • QoS guarantee
Green Radio	Mobile VCE (Virtual Centre of Excellence) (3 years)	UK	UK universities	Base station and handsets of mobile data service	• Power amplifier, power efficient processing • Backhaul redesign, multihop routing, relaying • Resource allocation, dynamic spectrum access
Cool Silicon	Silicon Saxony Management	Global	Over 60 ICT companies and institutes	ICT	• Micro-/nano-technology • Media communication • Sensor network
Green Grid	8 main contributor companies	Global	Global ICT companies	Data centers	• Data center energy efficiency (design, measurement, metrics)
GSMAMEE (GSMA Mobile Energy Efficiency)	GSM Association Congress	Global	Over 800 mobile operators and 200 companies	Mobile networks	• Benchmarking of mobile energy efficiency networks

Source: Min, C. et al. 2011. *A Survey of Green Mobile Networks: Opportunities and Challenges. Mobile Networks and Applications.* Springer Science+Business Media, LLC. DOI 10.1007/s11036-011-0316-4 [29].

TABLE 15.6

Taxonomy of Green Metrics (Nonexhaustive List)

Metrics	Full Names	Creator	Targets	Calculation	Units	Remarks
PUE	Power usage effectiveness	Green grid	Data center	PUE = Total facility power/IT equipment power	Ratio	Ranging from 1 to infinite
DCiE	Data center infrastructure efficiency	Green grid	Data center	DCiE = 1/PUE × 100%	Percentage	Ranging from 0 to 100% (a reciprocal of PUE)
DCP	Data center productivity	Green grid	Data center	DCP = Useful work/total facility power	Ratio	Ranging from 1 to infinite
ECR	Energy consumption rating	ECRInitiate, (IXIA, juniper)	ISP, ICT enterprises	ECR = Energy consumption/effective system capacity	Watt/Gbps	Energy normalized to capacity
ECRW	ECR-weighted	ECRInitiate, (IXIA, juniper)	ISP, ICT enterprises	ECRW = (0.35 × E_f + 0.4×E_h + 0.25 × Ei)/T_f	Watt/Gbps	Ef, Eh and Ei are the energy consumption in full-load, half-load, and idle modes respectively. Tf is the effective throughput.
TEER	Telecommunications energy efficiency ratio	ATIS	General, server, transport		Gbps/Watt	Di is the data rate of each interface i; $P0j$, $P50j$, and $P100j$ are the power of module j at data utilization of 0%, 50%, and 100%, respectively
TEEER	Telecommunications equipment energy efficiency ratio	Verizon NEBS	Transport, switch, router, access, amplifier	TEEER = log(0.35xP_{MAX} + 0.4xP_{50} + 0.25xP_{sleep})/throughput	log(Watt/Gbps)	Referring from ECRW and TEER, $PMAX$, $P50$ and $Psleep$ are the power consumption at 100%, 50%, and 0% load utilization. (Formula may change based on different types of devices)
CCR	Consumer consumption rating	Juniper	Consumer network devices		Rad (dimensionless)	E is power rating of a consumer network device; A is energy allowance per function; j is the set of all allowances claimed. Value 1 matches an average device.

(Continued)

TABLE 15.6 (Continued)
Taxonomy of Green Metrics (a Nonexhaustive List)

Metrics	Full Names	Creator	Targets	Calculation	Units	Remarks
EPI	Energy proportionality index	Hp Labs	Network devices	$EPI = (PM - PI/PM)*100\%$	Percentage	*PI is the power consumption at idle mode, PM is the power consumption at maximum workload*
WattsPerVLL	Watts per VLL (virtual leased line)	Ericsson	IP networks	WattsPerVLL = Power consumption/ number of VLLs	Watt/line	*Used for Virtual Leased Line (point-to-point Ethernet-line) services based on the Number of VLLs*
WattsPerMAC	Watts Per MAC port	Ericsson	IP networks	WattsPerMAC = power consumption/ Number of MAC ports	Watt/port	*Used for MAC address (multipoint Ethernet-LAN) services based on the Number of MAC ports*
P_{BBline}	Power consumption per line of broadband	ETSI	Broadband telecommunication networks equipment	$P_{BBline} = P_{BBeq}$/No. of Sub lines	Watt / subscriber line or Watt / port	*P_{BBeq} is the Power consumption of fully equipped broadband equipment. No. Of Sub Lines is the maximum number of subscriber lines supported*
NPC	Normalized power consumption	ETSI	Broadband telecommunication networks equipment	$NPC = 1000 * P_{BBline}$/Bit Rate $\times$ Line length	Watt/(Mbps x km)	*Normalized power consumption per line for broadband network equipment based on the line length*

Source: Hasan, Z., Boostanimehr, H. and Bhargava, V.K. 2011. Green Cellular Networks: A Survey, Some Research Issues and Challenges. Cornell University, arXiv:1108.5493v3 [cs.NI] 24 Sep 2011.; Min, C. et al. 2011. *A Survey of Green Mobile Networks: Opportunities and Challenges. Mobile Networks and Applications.* Springer Science+Business Media, LLC. DOI 10.1007/s11036-011-0316-4. [15,29].

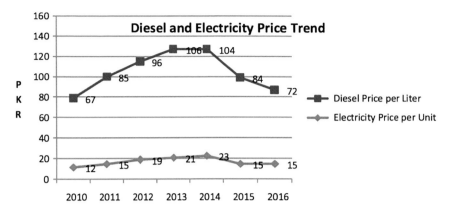

FIGURE 15.3 Diesel and electricity price trend.

low wages, and rundown ethics. According to some estimates, the diesel pilferage is around 5% in the telecom sector.

The most important element that needs to be kept in mind is that no operator had thought that energy shortages would turn into a crisis so quickly. It can be said with confidence that no operator included it as a high expense item in their initial business cases.

15.2.5.2 Initiatives

To reduce energy-related OPEX and have a greener network, Telenor Pakistan has been taking certain initiatives. These have started to make a positive impact on the financial health of the company and on the overall network. The list of some key initiatives which either have been introduced or are under development is as follows:

- *Generator Manager*: In March 2005 when Telenor launched its network, the diesel price was at PKR 24.37 per L with not more than 2–3 hours of average load shedding. It was viable to run generators in the absence of grid power at the required sites. As the price started to get significantly higher (PKR 85 per L) with 8+ hours of daily average grid outage, it no longer remained a sensible business preference. Thus, Generator Managers (GEMs) were introduced to minimize the generator run time, and start genset at 47 VDC (volts direct current), that is, to delay the start of genset. The generator manager is a device installed in the ATS that provides a delayed start-up of a generator during commercial power shutdowns. In simpler

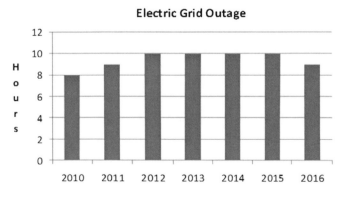

FIGURE 15.4 Grid outage trend. IndexMundi. 2016. (From Diesel Monthly Price—Pakistan Rupee per Gallon. http://www.indexmundi.com/commodities/?commodity=diesel&months=300&currency=pkr; INCPak 2016. Petrol Prices in Pakistan. http://www.incpak.com/info/petrol-prices-pakistan/ [30,31].)

terms, it keeps the system on batteries during grid power shutdown, while monitoring the battery voltage. Once the voltage reaches the set threshold, it starts the generator. This has resulted in savings on an average of 3 or so hours of diesel cost.

- *Generator-Less Operation*: Along the same lines, Telenor is increasing the battery backup time to remove gensets at some sites. The network modernization, which has resulted in the introduction of energy efficient BTS, has made it possible to increase the existing battery backup time.
- *Remote Monitoring System*: Telenor has deployed RMS to protect itself from abnormal billing by authorities. It will also help to a certain extent in minimizing and quantizing the percentage of diesel pilferage.
- *Power Integration Unit*: Power Interface Unit or Power Integration Unit (PIU) is an electrical interface and control unit used for GSM/Telecom installation. PIU is used to contain separate units like UPS (uninterrupted power supply), AC distribution panel, DC power distribution unit, battery backups, alarm panel, and rectifier (to convert AC to DC) in one cabinet. PIU controls the environment within different temperature zones in a shelter minimizing the use of air conditioners and diesel generators to reduce operating expenditure. Telenor has been deploying PIU at several low voltage sites. This has resulted in a savings of an average of two or so hours of diesel cost.
- *Power Saving Controller*: Telenor has deployed PSC at several sites to intelligently monitor alternative power inputs such as solar panels and coordinate the start-up of back up diesel generators, optimizing system power use (OPEX) and improving the reliability at grid/solar hybrid cell sites.
- *Solar Power System Solution*: Telenor has upgraded a number of sites on a solar enabled power system and this trend will continue in the coming years. In some areas where grid power is not available, Solar/Gen-set hybrid solutions have been used. However, the solar/Gen-set hybrid solutions require a larger space to accommodate the solar panels to fulfill the power requirement of a BTS.
- *Wind Energy*: Telenor is exploring the use of wind power at certain sites that are located in good wind corridors. The wind turbine will reduce dependency on the grid and fuel.
- *Network Modernization*: The company completed a network modernization (replacement with energy efficient equipment) exercise in the recent past. It has reduced electricity consumption by 15%–20% (compared to the level of 2011).

15.2.5.3 Results

In the initial years, it was viable to follow a Grid → Generator → Battery back-up process. However, due to coverage expansions, considerable increase in outages, and rising fuel costs, it later became an unviable option. Due to the above-mentioned initiatives, the company is following the cost effective and greener Grid → Battery → Generator path and at sites where solar panels are available, the sequence of Solar → Grid (if available) → Battery → Generator.

Before these initiatives, the energy-related OPEX ranged between 20%–25% of the total OPEX. As the result of these initiatives, the OPEX has been reduced to less than 20%*.

15.3 PATENT PORTFOLIO MANAGEMENT

Mobile communications heavily rely on standardized technologies which inevitably make use of numerous inventions or patents. Patents† have become a market differentiator in the high tech mobile world. Designers, manufacturers, and even the operators are constantly in a race to increase

---

* The precise figures and savings breakdown due to each initiative cannot be shared due to confidentiality.
† A patent is a set of exclusive rights granted by law to applicants for inventions that are new, nonobvious, and commercially applicable. It is valid for a limited period of time (generally 20 years).

their patent portfolio to gain a market advantage. Further on, the equipment manufactures and semiconductor designers do their utmost to make their IPR (intellectual property rights) become part of the standards. Once these are included in standards, the return on future investments becomes relatively predictable.

In the last few years, this unavoidable race has erupted in the form of patent wars costing the losing party and benefiting the winner by millions of dollars. The technologies such as LTE and products such as iPhones have been in the limelight of patent wars. As product innovation continues to grow in a competitive marketplace, it will reciprocate not only in intellectual capital protection space, but also be used more as a strategic tool in the quest for market leadership [32].

This section will primarily look into the distribution of 2G, 3G, LTE, and LTE-Advanced patents across the industry. It will also provide an overview of litigation examples and some strategies to avoid the same. Finally, U.S. based operator Sprint's commitment toward patent innovation will be shared as a case study.

15.3.1   2G, 3G, and LTE Patent Portfolio

In recent years, LTE and 4G (LTE-Advanced) technologies have become the dominant enablers of mobile broadband data and have become household names. On a similar scale, these are also encompassing the technology patent space worldwide. However, the share of 2G and 3G mobile connections, which is currently over 60%, along with a sizeable patent portfolio cannot just be ignored. The technology patent portfolio is presently dominated by only a handful of companies.

This section will briefly provide some statistics on such patents and companies, however, before diving into the patent distribution, it is worthwhile to understand certain key definitions.

Standard Essential Patent: A patent which is declared to an SDO like ETSI and so on, can be considered an essential one to a specific technology or to a standard. In other words, a standard essential patent is one in which the practice of one or more of the inventions set forth therein is required in order to comply with a technical standard published by an SDO. A patent's essentiality is tested in detailed technical discussions mapping patent claims to the technical specification of the standard on a clause by clause basis [33–35].

Seminal Patents are those that are determined to be strong based on a set of parameters and that can create or shift a particular technology space. These may be cited in new inventions.

Table 15.7 provides the patent distribution of the top 20 companies based on the patent data from U.S. PTO (U.S. Patent & Trade Office) and ETSI (European Telecommunications Standards Institute). The table* was populated from two separate reports provided by iRunway [36] and Cyber Creative Institute [37]. The table covers 2G, 3G†, LTE, and LTE-Advanced patents that were declared before 2012. It can be witnessed that Qualcomm leads the overall race followed by Nokia and Samsung.

If 2G and 3G are only considered, then Nokia Corporation (excluding Nokia Siemens Network) is number one followed by Qualcomm. Considering LTE and LTE-Advanced cases only, Qualcomm leads the race with 655 patents, followed by Samsung with 652 patents, and Huawei is in the third place with 603 patents.

15.3.2   Patent Wars

A patent war is a legal battle between corporations or individuals to secure patents for litigation, whether offensively or defensively. These legal battles between some of the world's largest technology

---

* The 100% accuracy of this table is not guaranteed. It only provides a snapshot of the overall patent portfolio of mobile communications.
† 2G-GSM, GPRS, EDGE, and CDMAOne; 3G-UMTS and CDMA2000.

TABLE 15.7

2G, 3G, LTE, and LTE-Advance Patent Counts as of 2012

#	Company	2G & 3G Standard-Essential Patents (ETSI+U.S. PTO)	2G & 3G Seminal Patents (U.S. PTO Only)	LTE/LTE-Advanced Standard-Essential Patents (Estimation on Those Declared to ETSI)	Other LTE/ LTE-Advanced Patents (ETSI Only)	Total Patent Portfolio
1	Qualcomm	146	113	318	337	914
2	Nokia Corporation	227	143	245	260	875
3	Samsung	57	21	233	419	730
4	Huawei	15	0	273	330	618
5	Ericsson	85	51	177	222	535
6	InterDigital	35	48	206	212	501
7	LG	38	118	237	80	473
8	ZTE	0	0	253	115	368
9	Motorola	16	0	111	199	326
10	Sharp	0	0	86	103	189
11	Nokia Siemens Networks	33	15	63	44	155
12	Texas Instruments	0	0	90	35	125
13	Panasonic Corporation	0	0	53	54	107
14	Alcatel-Lucent	0	45	24	38	107
15	NEC	0	0	64	37	101
16	Apple	0	0	25	53	78
17	Innovative Sonic Limited	19	0	35	9	63
18	Research in Motion	14	14	16	15	59
19	Nortel Networks	22	10	9	12	53
20	ETRI (Electronics and Telecommunications Research Institute, Korea)	0	0	35	16	51

Source: iRunway 2013. 2G & 3G Mobile Communication; 3GPP TR 36.887 (V12.0.0) Evolved Universal Terrestrial Radio Access Network (E-UTRAN); Study on Energy Saving Enhancement for E-UTRAN. Technical Report (Release 12), Technical Specification Group Radio Access Network, 3GPP, March; Cyber Creative Institute Co. Ltd 2013. Evaluation of LTE Essential Patents Declared to ETSI (Version 3.0), June [34,36,37].

and software giants have become a global phenomenon. The most common ones involve smart phone manufacturers such as Apple Inc., Nokia, Motorola, and Samsung [38–40].

It can be safely said that the current pattern of suing and countersuing began in the late 2000s as the smart phone market started to grow rapidly. Apple has been very aggressive in protecting its intellectual property and has been suing and getting countersued by others in this domain [41].

This section will briefly describe the two patent wars that were waged years ago. One such conflict started in 2010 and ended in 2013, while the one that started in 2011 is still nowhere near conclusion.

15.3.2.1 Case # 1: Apple Inc. versus Samsung Electronics Co., Ltd. (2011)

Apple Inc. versus Samsung Electronics Co., Ltd. was the first of a series of ongoing lawsuits between the two corporations regarding the design of smart phones and tablet computers.

The battle started on April 15, 2011, when Apple sued Samsung, alleging in a 38-page federal complaint in the U.S. District Court for the Northern District of California that a number of Samsung's Android phones and tablets had infringed on its intellectual property. Samsung countersued Apple on 3G technology patents and took the fight to Germany, Japan, and Korea.

Some key timelines associated with this case are as follows [41,42]:

May-December 2011: The battle increased to 30 cases across four continents-North America, Asia, Europe, and Australia [43,44].

July 2012: The two companies were fighting more than 50 lawsuits across the globe, with billions of dollars in damage control [45].

August-Dec 2012: In the U.S., Apple had a $1.049 billion victory while Samsung won rulings in South Korea, Japan, and the UK. An appeals court lifted an injunction on U.S. sales of the Samsung-made Galaxy Nexus. Judge Koh found that the U.S. jury calculated damages incorrectly. She invalidated $400 million out of the $1.049 billion awarded to Apple. She then ordered a retrial to determine proper damages.

Since then both Samsung and Apple have been trying hard to tilt the rulings and the damages in their favor.

June–August 2013: Samsung won a limited ban from the U.S. International Trade Commission (ITC) on sales of certain Apple products after the commission found Apple had violated a Samsung standards essential patent [46]. However, this was vetoed by U.S. Trade Representative Michael Froman before going into effect. Then, a few days later, the ITC blocked some older Samsung phones from sale in the U.S. for violating two Apple patents.

November 2013: The retrial on damages invalidated by Judge Koh got started. Apple sought $379.8 million while Samsung argued the amount should be $52 million. A Samsung representative admitted that some of its devices had used some elements of the Apple design. Judge Koh awarded Apple $290 million in damages which brought Samsung's total penalty in the first U.S. case to $939 million from $1.049 billion. The following day, Samsung filed a formal appeal against the judgment.

March 2014. The second U.S. trial got underway on March 31st where Apple was seeking roughly $2 billion in damages. The second trial involved different patents and different products than the first trial.

December 2014: Samsung urged a U.S. appeals court to invalidate the $939 million in the first case verdict won by Apple stating it did not copy the iPhone design and the damage award was too high.

The United States District Court for the Northern District of California awarded $120 million to Apple in October 2016. Samsung appealed to the Supreme Court, but the Court announced in November 2017 that it would not hear the appeal, leaving the District Court's ruling in Apple's favor in place.

15.3.2.2 Case # 2: Microsoft Corp. versus Motorola Inc. (2010)

Motorola is the holder of IEEE 802.11 (WiFi) and ITU-T H.264 (video encoding) standard essential patents. This means that it would be necessary to use the concepts incarnated in these patents in order to build an 802.11/H.264 compliant device. Motorola agreed in writing to both IEEE and ITU that it would license these patents on reasonable and nondiscriminatory (RAND) terms on a worldwide, nondiscriminatory basis. Microsoft had a number of products that used these standards, most importantly the Xbox gaming console, Windows phone software, and Microsoft Windows.

Here are some key timelines associated with this case [41,47,48]:

October 2010: Motorola sent letters to Microsoft arguing that it should pay royalty payments on all its end products that utilized these standards at the rate of 2.25%.

November 2010: Microsoft filed a lawsuit against Motorola in the Western District Court of Washington claiming that Motorola violated RAND licensing obligations.

July 2011: Motorola countersued Microsoft in Germany, accusing Microsoft of infringing on two German H.264 essential patents.

February 2012: The Western District Court of Washington rejected Motorola's claims that a licensee like Microsoft would have to negotiate to be eligible for RAND terms.

April 2012: Motorola was awarded an injunction by the German court against Microsoft, prohibiting Microsoft from selling allegedly infringing products in Germany.

May 2012: The U.S. district court granted Microsoft an antisuit injunction. This antisuit injunction prevented Motorola from enforcing a foreign (Germany) patent infringement injunction.

September 2012: The case was brought to the Appellate court in an interlocutory appeal by Motorola. The Ninth Circuit Court of Appeals affirmed the District Court's decision, thus stopping Motorola from preventing the sale of several Microsoft products in Germany.

September 2013: The jury finds that Motorola breached RAND obligations, unanimously awarding $14.5 million in damages to Microsoft [49].

15.3.3 Solution to Avoid Litigation

This section provides two possible options for avoiding patent litigations, namely patent pooling and RAND accuracy.

15.3.3.1 Patent Pooling

In patent law, a patent pool is a consortium of at least two companies agreeing to cross-license patents relating to a particular technology. The creation of a patent pool can save patentees and licensees time, earn and reduce royalty amounts, and reduce litigation battles [50].

Patent pools are conducive to spreading standards by making licensing easy and royalties reasonable. An LTE patent pool, administered by Via Licensing, was established in October 2012 to serve the same objectives. The LTE Patent License Agreement from Via Licensing provides access to all of the patents from the participating licensors which are essential to the implementation of the 3GPP LTE standard. Royalties are not calculated as a percentage of the product price but are levied at a fixed amount (dollars/unit) regardless of the product price [51,52].

The program has less than 40 licensors and has not been very successful as many important LTE patent holders still are not on the list. The key LTE patent holders such as Apple, Ericsson, Huawei, and Qualcomm, still have not joined the program [53].

The success of this approach is still to be determined [54].

15.3.3.2 RAND Effectiveness

Many SDOs, in an effort to reduce the possibility of patent holdup* and increase competition, asked that the holders of any standards essential patents agree to license these patents on a fair, nonrestrictive basis. The organizations, under their own willingness, can license standard essential patents on RAND terms on a worldwide, nondiscriminatory basis to SDOs.

However, as shown in the Microsoft versus Motorola case, litigation still finds its place. Thus, Judge James L. Robart 207 page order setting RAND royalty terms in this case can be effective for similar cases. His following recommendations can be adopted to mitigate the litigations involving RAND [55].

- A RAND royalty should be set at a level consistent with the SDOs' goals of promoting widespread adoption of their standards.
- A proper methodology for determining a RAND royalty would mitigate the risk of patent hold-up.

---

* Hold-up refers to the ability of a holder of a standard essential patent to demand more than the value of its patented technology and to attempt to capture the value of the standard itself.

- A proper methodology for determining a RAND royalty would address the risk of royalty stacking by considering the aggregate royalties that would apply if other standard essential patents' holders made royalty demands of the implementer.

15.3.4 CASE STUDY–SPRINT CORPORATION

This section will present a brief case study on Sprint's commitment toward patent innovation. This study will only compare Sprint with other mobile operators around the globe to have an apple to apple comparison.

15.3.4.1 Background

Sprint is the fourth largest wireless network operator in the U.S., serving more than 60 million customers. Japan's telecommunications Softbank Corporation owns 80% of the company. Sprint has always been at the forefront of technology development and innovation since its inception in1899. This attribute is well endorsed throughout its history, for example [56]:

- Founders of Sprint (or Brown Telephone Company) joined with 14 other Kansas independents to incorporate the Union Telephone and Telegraph Company that provided long distance service to Kansas City (1903).
- Post Great Depression, it started to show profits again and formed United Utilities, Incorporated (1942).
- Completion of the first nationwide, 100% digital, fiber optic network (1987).
- Completion of the first 3G CDMA2000 1X call in North America (March 2000).
- Completion of first and fastest data call using EV-DO technology in U.S. (April 2001).
- Named to the CW (connected world) 100 list of the most important and influential providers of M2M services for the tenth consecutive year in September 2013.
- Sprint was granted 1213 U.S. patents in 2012 and 2013 by the U.S. Patent and Trade Office.

15.3.4.2 Methodology

The data for this case study was acquired from IEEE Spectrum magazine's 'Patent Power' sections, Chetan Sharma Consulting 2013 report [57], and from the author's personal contribution in this area to Sprint. The data from IEEE is based only on operators' U.S. patents (i.e., that were granted by U.S. PTO), while the Chetan Sharma report includes data from U.S. PTO and EPO (European Patent Office). In the U.S. and Europe, a patent is valid for 20 years from the date of filing and it roughly takes 3–5 years to get a decision from the U.S. PTO and EPO.

15.3.4.3 Success Story

In the 1990s, Sprint fully acquired Sprint PCS (Personal Communications Service) from the rest of the shareholders. In the coming years, it became the wireless division of Sprint beside its long distance and local divisions. As most of the activities were taking place in the wireless arena, Sprint PCS from the very start encouraged the culture of innovation. The employees were encouraged to file patent applications and during those days, it paid $1000 to each inventor as an incentive.

As with any other initiative, the number of patent filings was low, but it started to pick up in the early 2000s. Now in the 2010s, it can be considered a major player in this area (in the service provider community). Sprint maintains one of the largest patent portfolios in the category of communications/Internet service providers. The number of patents it has been granted is constantly on the rise along with a key API score since 2010 as shown in Figure 15.5. The metric API shows the impact of an organization's patent portfolio on subsequent technological developments[*].

---

[*] According to IEEE the Adjusted Pipeline Impact is an estimate of a company's overall patent power [58].

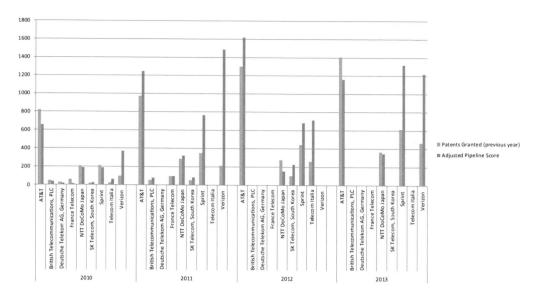

FIGURE 15.5   Mobile service providers' patent portfolio.

In 2012, USPTO granted 613 patents and IEEE gave Sprint an API score of 1316. This score made Sprint the 6th ranked player in this overall category and number 2 as a mobile operator around the globe as listed in the IEEE Patent Power 2013 Scorecard, surpassing its main rivals AT&T and Verizon [59]. If only patents (not API) that were granted by EPO and the U.S. PTO from 1996–2013* are included, Sprint still ranks relatively high. It ranks number 3 ahead of T-Mobile and Verizon and only behind AT&T and DoCoMo as shown in Figure 15.6.

Thus, this shows Sprint's effective management toward research and patent innovation.

15.4   FINAL REMARKS

Keeping 5G in mind, spectrum management, energy efficiency, and patent portfolio management aspects will become more valuable due to the maturity and complexity of networks.

- Spectrum Management: This chapter suggested some techniques for managing the spectrum along with a case study of auction.
- Energy Management: The energy saving techniques discussed are considerably relevant, particularly for gaining ROI on 3G/4G and future 5G networks in developing countries.
- Patent Portfolio Management: The complexity will increase with the inclusion of more players and due to potential convergence of media, telecom, and IT. The solutions to avoid litigation as discussed will help in reducing future patent wars.

Finally, the case studies presented provide lessons which could be used in the standardization and implementation phases of 5G.

PROBLEMS

1. Define spectrum management?
2. What is the difference between spectrum allocation and assignment?
3. List the key reasons that inhibited the reasonable allocation for future IMT services?
4. What are the key challenges in the proliferation of broadband services in Pakistan?

---

* The search included patents granted until 2/15/2013.

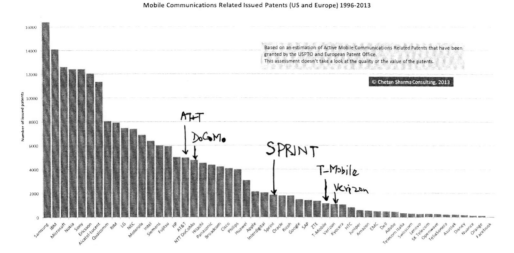

FIGURE 15.6 Mobile related issued patents in U.S. and Europe (1995–2012). (From Sharma, C. 2013. *Mobile Patents Landscape—An In-depth Quantitative Analysis*, 2nd Edn. Chetan Sharma Consulting, Washington, DC, USA [57].)

5. Which network element consumes most of the energy and how is it consumed?
6. Name some of the key techniques to improve PA efficiency?
7. Explain the concept of cell zooming?
8. What are some of the alternative energy resources?
9. Why was running cell sites on diesel in the absence of grid power no longer suitable for Telenor Pakistan and which measurement was used to address the challenge?
10. What are standard essential patents?
11. What are seminal patents?
12. What is a patent war?
13. Describe patent pooling?
14. What is RAND?
15. How can Judge James L. Robart order assist in mitigating the litigations involving RAND?
16. How many patents were granted to Sprint Corporation by the end of 2012 and what was its API score? Define API as well?

REFERENCES

1. Asif, S. 2011. *Next Generation Mobile Communications Ecosystem: Technology Management for Mobile Communications.* Wiley Inc., UK.
2. ITU-R 2013. Spectrum Management Fundamentals, Part1—International, The Need for Spectrum Management, ITU Regional Radiocommunication Seminar for Americas.
3. Manikkoth, S. 2014. Spectrum Scarcity—An Alternate View (in Special Interest Group Cognitive Radio in 5G White Paper Novel Spectrum Usage Paradigms for 5G). Cognitive Networks Technical Committee, IEEE Communications Society.
4. ITU-R 1997. Final Acts WRC-97. *World Radiocommunication Conference*, Geneva, Switzerland, 27th October—21st November 1997.
5. ITU-R 2015. Final Acts WRC-15. *World Radiocommunication Conference*, Geneva, Switzerland, 02-27 November 2015.
6. ITU-R 2012. Radio Regulations—Articles—Edition of 2012.
7. Asif, S.Z. 2017. World Radiocommunication Conference 2015 Results and Unresolved UHF Case. *(Under Consideration)*.
8. Ministry of Information Technology, Government of Pakistan 2015. Telecommunications Policy 2015.

9. Pakistan Telecommunication Authority 2014. Information Memorandum, The Award of 2100 MHz, 1800 MHz and 850 MHz Spectrum: The Next Generation Mobile Services Award (NGMSA).
10. DAWN Newspaper 2016. AGP (Auditor General of Pakistan) to Look into Auction of 3G, 4G Spectrums.
11. Asif, S.Z. 2008. Broadband Technology Management & Evaluation. *2008 International Conference on Computer and Electrical Engineering*, Phuket, Thailand, Dec. 20–22, 2008, pp. 271–275.
12. Maio, A.D. 2007. Green IT: A New Industry Shockwave. *Presentation at Symposium/ITXPO Conference*, April.
13. SMART 2020 - The Climate Group 2008. Enabling the Low Carbon Economy in the Information Age. http://www.smart2020.org/
14. Goran, P., Garma, T. and Lorincz, J. 2012. Measurements and Modelling of Base Station Power Consumption under Real Traffic Loads. *Sensors*, 12:4281–4310. DOI:10.3390/s120404281.
15. Hasan, Z., Boostanimehr, H. and Bhargava, V.K. 2011. Green Cellular Networks: A Survey, Some Research Issues and Challenges. Cornell University, arXiv:1108.5493v3 [cs.NI] 24 Sep 2011.
16. Claussen, H., Ho, L.T.W. and Pivit, F. 2008. Effects of Joint Macrocell and Residential Picocell Deployment on the Network Energy Efficiency. *IEEE 19th International Symposium on Personal, Indoor and Mobile Radio Communications (PIMRC)*, Cannes, France, Sep. 15–18, 2008.
17. Armour, S. et al. 2011. Green Radio: Radio Techniques to Enable Energy-Efficient Wireless Networks. *IEEE Communications Magazine*, 49(6):46–54.
18. Chen, T., Kim, H. and Yang, Y. 2010. Energy Efficiency Metrics for Green Wireless Communications. *Proceedings of the International Conference on Wireless Communications and Signal Processing (WCSP)*, Suzhou, China, October 21–23.
19. Alsharif, M.H., Nordin, R. and Ismail M. 2013. Survey of Green Radio Communications Networks: Techniques and Recent Advances. Hindawi Publishing Corporation. *Journal of Computer Networks and Communications*, Volume 2013, Article ID 453893, 13 pages. DOI:10.1155/2013/453893.
20. Murthy, C.R. and Kavitha, C. 2012. A Survey of Green Base Stations in Cellular Networks. *IRASCT— International Journal of Computer Networks and Wireless Communications*, 2(2):232–236.
21. Gong, J. et al. 2010. Cell Zooming for Cost-Efficient Green Cellular Networks. *IEEE Communications Magazine*, 48(11):74–79.
22. Bettstetter, C. 2015–16. Cooperative Relaying in Wireless Networks. http://bettstetter.com/research/relay
23. Tweed, C. 2013. Why Cellular Towers in Developing Nations Are Making the Move to Solar Power. Scientific American http://www.scientificamerican.com/article/cellular-towers-moving-to-solar-power/
24. Motorola Solutions 2007. Alternative Power for Mobile Telephony Base Stations, Solutions Paper.
25. GSMA 2010. Green Power for Mobile Community Power. http://www.gsmworld.com/greenpower
26. 3GPP TR 36.887 (V12.0.0) Evolved Universal Terrestrial Radio Access Network (E-UTRAN); Study on Energy Saving Enhancement for E-UTRAN. Technical Report (Release 12), Technical Specification Group Radio Access Network, 3GPP, March.
27. Quittek, J. (ed). 2010. Requirements for Power Monitoring. Internet Draft, Network Working Group, IETF. http://tools.ietf.org/html/draft-quittek-power-monitoring-requirements-02
28. Quittek, J. (ed). 2010. Definition of Managed Objects for Energy Management. *Internet Draft, Network Working Group, IETF.* http://tools.ietf.org/html/draft-quittek-power-mib-02
29. Min, C. et al. 2011. *A Survey of Green Mobile Networks: Opportunities and Challenges. Mobile Networks and Applications.* Springer Science+Business Media, LLC. DOI:10.1007/s11036-011-0316-4
30. IndexMundi. 2016. Diesel Monthly Price - Pakistan Rupee per Gallon. http://www.indexmundi.com/commodities/?commodity=diesel&months=300&currency=pkr
31. INCPak 2016. Petrol Prices in Pakistan. http://www.incpak.com/info/petrol-prices-pakistan/
32. Thomson Reuters 2012. Inside the iPhone Patent Portfolio, September.
33. GSMA Intelligence 2016. Global Mobile Trends.
34. iRunway 2013. 2G & 3G Mobile Communication.
35. Ericsson 2012. Essential Patents and LTE, September.
36. iRunway 2012. Patent & Landscape Analysis of 4G-LTE Technology.
37. Cyber Creative Institute Co. Ltd 2013. Evaluation of LTE Essential Patents Declared to ETSI (Version 3.0), June.
38. Rowinski, D. 2012. Patent Wars Turn Tech into a Battlefield. ReadWriteWeb.
39. Francis, T. 2012. Can You Get A Patent On Being A Patent Troll? NPR.
40. Sascha, S. 2012. Infographic: Smartphone Patent Wars Explained. PC Magazine.
41. Wikipedia http://en.wikipedia.org/wiki/Main_Page
42. Digital Trends 2014. Why Are Apple and Samsung Throwing Down? A timelines of the Biggest Fight in Tech. http://www.digitaltrends.com/mobile/apple-vs-samsung-patent-war-timeline/

43. Barrett, P.M. 2012. Apple's War on Android. Bloomberg Businessweek. Bloomberg.
44. Albanesius, C. 2011. Every Place Samsung and Apple Are Suing Each Other. PC Magazine (Ziff Davis).
45. Pyett, A. (Reporting), Lincoln, F. (Writing) and Davies, E. (Editing). 2011. Australian Court to Fast-Track Samsung Appeal on Tablet Ban. Reuters (Sydney: Thomson Reuters).
46. Mueller, F. 2012. Apple Seeks $2.5 Billion in Damages from Samsung, Offers Half a Cent per Standard-Essential Patent. FOSS Patents.
47. Essential Patent Blog 2014. Microsoft v. Motorola, Inc. http://www.essentialpatentblog.com/?s=Microsoft+v.+Motorola
48. Microsoft v. Motorola, 696 F.3d 872 (United States Court of Appeals for the Ninth Circuit 2012).
49. Shu, C. 2013. Federal Jury Orders Motorola Mobility To Pay Microsoft $14.5 Million In Patent Case. TechCrunch.
50. International Mfg. Co. v. Landon, 336 F.2d 723, 729 (9th Cir. 1964) The Pooling of the Patents, Licensing All Patents in the Pool collectively, and Sharing royalties Is Not Necessarily An Antitrust Violation. In a Case Involving Blocking Patents Such An Arrangement Is the Only Reasonable Method for Making the Invention Available to the Public.
51. Kosaka, T. and Kerr, C. 2013. Establishment of LTE Patent Pool. *NTT DoCoMo Technical Journal*, 14(4):54–59.
52. Via Licensing 2012. Long Term Evolution—Standards-Essential Patent Licensing.
53. Via Licensing 2016. Licensor Partners. http://www.via-corp.com/licensing/partners.html
54. Center on Law and Information Policy (At Fordham Law School) 2012. The Impact of the Acquisition and Use of Patents on the Smart Phone Industry.
55. Leech Tishman Fuscaldo & Lampl, LLC 2013. Latest High-Profile Software Decision Tackles Standard-Essential Patents. http://usip.com/Publications/EyeonIP/130514.html
56. Sprint Corporation http://www.sprint.com/
57. Sharma, C. 2013. *Mobile Patents Landscape—An In-depth Quantitative Analysis*, 2nd Edn. Chetan Sharma Consulting, Washington, DC, USA.
58. IEEE 2012. Patent Power 2012. http://spectrum.ieee.org/at-work/innovation/patent-power-2012/constructing-the-patent-power-scorecard
59. IEEE Spectrum 2013. Interactive: Patent Power 2013. http://spectrum.ieee.org/static/interactive-patent-power-2013#anchor_cie

16 The Way Forward
Fast Forward to Year 2040

What has been described in this book for the most part pertains to the traditional way of doing mobile communications business. What needs to be done perhaps is to look at a bigger picture and the long term future of the overall ICT* sector. The self-declared target is the year 2040.

16.1 CURRENT PICTURE

Forty years ago, who would have predicted the world of mobile communications that we live in today; a world of full cellular towers and handheld gadgets, which is satisfying the need of billions.

How long this way of communications will last is anybody's guess. However, it is for sure that we will communicate much differently in the next 10–20 years than the way we do today.

The fast pace of commercialization is taking a toll on every of facet of life. In the case of mobile communications, we like to access the required information and results in the blink of an eye. However, the urge for instant gratification is not a very healthy trend.

2G brought mobile voice telephony and digitization in the 1990s, 3G brought mobile data in the 2000s, while 4G is providing faster data connections and better user experience in the 2010s. There was roughly a 10-year transition period from one generation to the next. However, the mobile communications sector brought LTE (between 3G and 4G) in the late 2000s to compete with IEEE WiMAX technology and it successfully eliminated it.

LTE to 4G (LTE-Advanced) and to LTE-Advanced Pro has not brought anything out of the blue and that may need to be considered for the future of the ICT sector. Quick transitions due to competing technologies and commercial needs do not always have a happy ending.

5G, which we expect to see in 2020, needs to look at a bigger picture and may bring out of the box solutions. Searching beyond 5G and not to the distant future, that is, 2040, it is important that the sector collectively look to transform the ICT arena.

16.2 WHY 2040?

The first question that comes to mind is why 2040, why not some different target year. Why not follow the world's fair of 2064 (if there will be one!), following the tradition that happened 100 years before in 1964 [1].

The year 2040 is not a magical year, but the sector needs a target and the following two things are in its favor:

- First, Generation Z, which is the first generation that was born having mobile phones and Internet. There is no agreement as such on the start and end dates for the generation, however, it reflects those who were born from 1995 until 2012 [2–4]. The following human generation (Generation Z+1) will be the first generation whose biological parents were the true recipients and enablers of digitized communications. This generation could be

---

* ICT is an umbrella term encompassing telecommunications including mobile cellular telephony, computer, Internet, and information systems technologies.

those who were/will born from 2013 until perhaps 2030. Thus, the focus of the 6th/7th generation of mobile communications may be on Generation Z and its succeeding generation (Generation Z+1), which roughly takes us to the 2040s/2050s.

• Second, after the launch of 5G networks in the 2020s, there would be a sufficient time of another 10–15 or so years for an ROI on LTE/4G and 5G and to capture learnings.

Overall, it will provide ample time to look back and understand the good, bad, and ugly endeavors of the last 30 or so odd years (2000s–2030s) and bring technologies and solutions that can transform the ICT sector which we have not seen since the 2000s. A hypothetical future ICT network is shown in Figure 16.1.

16.3   KEY BIG TRENDS

The aim of this section is to list some of the large-scale trends that are happening and that may happen during our journey toward 2040. Some development can be predicted with high certainty and some with less.

IT and Telecom Total Convergence: The convergence between IT and Telecom toward ICT is already at a mature stage. However, today we have separate companies providing IT and telecom services. Even in many telecom operator firms, there are separate telecom and IT organizations lead by the CTO (Chief Technical Officer) and CIO (Chief Information Officer), respectively. In the next 5–10 years, we can expect to have more consolidation and convergence. The buyout of the Nokia mobile phone business by Microsoft is a step in that direction.

ISP Acceleration in Telecom: The role of ISPs in telecom is increasing. Google has made many inroads to join the bandwagon and the trend will continue.

Telecom and Media: The boundaries between telecommunications and media/broadcasting are fading. Today, content is produced by media firms and is delivered over the Internet through

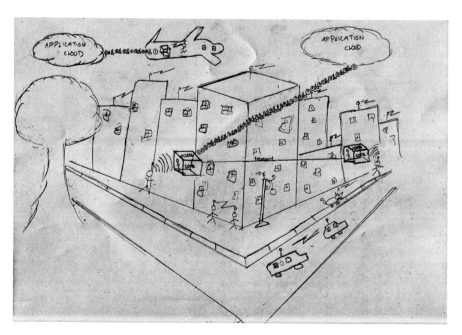

FIGURE 16.1   ICT future (courtesy of my young daughters).

networks of telecommunication service providers. Mobile operators are making headway to acquire media firms. AT&T acquired DIRECTV, which was the largest pay TV provider in the U.S.

Network Minimization: LTE eliminated one key element, that is, the radio controller, from the network. 5G may bring more minimization in the network. By 2040, the focus may be to reduce it to three domains, access, transport, and applications (content), where perhaps a single box can provide radio access network, core networks and operation support functions.

Alternative to Silicon: Research is already underway to find and build a more suitable material than silicon for computing, electronics, and networking chips. Emphasis can continue to be on finding an alternate that is widely naturally available and has excellent chemical properties.

Renewable Energy: Energy costs are increasing, energy resources are decreasing, and energy dissipation is not environmentally friendly. The focus may need to continue to be on cheaper alternate sources and on energy friendly devices and equipment.

Too Much Data: The annual global IP traffic is expected to reach 3.3 Zettabytes by 2021 [5]. This trend may increase as more and more users and Generation Z+1 get on board. The sector needs to prepare for a world where everything and everyone is connected and consuming and emitting information. How to capture, store, analyze, and produce meaningful results is still a big and unanswered question.

Cyber Security: It is a buzz word and will be continue to be for the coming years. National policies in some countries are still being developed. Overall, security objectives may not impede the investment, innovation and provision of telecom services.

16.4   FINAL SUGGESTIONS

The goal of this section is to have the global forums like ITU, ICT administrations, and regulators start thinking along these lines in addition to their normal routines, annual events, and 3–5-year long plans.

So, let us begin.

Basic Research: One of the most important elements for progression in the field of science and technology including ICT is the devotion toward basic research. Basic research has been hit hard by the fast pace of commercialization and industry short-sightedness.

Laser is just one example, as explained in the second chapter, that came out of basic research which has a revolutionary impact on ICT. To find the next killer application, it may be worthwhile to invest in basic research.

To bring the next revolution in ICT, it is of the utmost importance that funds are fed into basic research with a negligible weight on ROI. This long term investment is not easily feasible for the industrial players. Thus, ITU as the common platform may perhaps take the leading role. The ITU may set some funds aside for it and recommend that COMSTECH, European Commission, the U.S. federal government, and other rich economies do the same.

Applied research is ongoing, but basic research is the need of the day.

6th Generation: In the next 20 or so years, instead of following the 10 year trend, it may be worthwhile to only advance by one generation, that is, to 6G and not to 7G. It would also help in slowing down the fast pace of commercialization and provide an economic sigh of relief for the overall sector.

World with NO Cell Towers: People need to communicate with others and access and disperse information. However, this does not mean that cell towers are required. These can be replaced with environmentally friendly small size access points. As the network grows across the globe, so does the number of cellular towers and so does the pollution. Thus, it may be worthwhile to replace these towers with an alternative.

Spectrum: The world will be more wirelessly connected and less cabled. The data usage and number of subscribers are continuously increasing which demands more and more spectrum. ITU's WRC, as the responsible party for worldwide frequency allocation, may need to actively look into terahertz for wireless connectivity.

Smart City: The smart city is one that adopts scalable solutions that take advantage of ICT to increase efficiencies, reduce costs, and enhance quality of life [6]. This trend may not be restricted to only developed nations and needs harvesting in developing cities as well.

Protection for Us: Finally, privacy and security of data is and will be a fundamental need of everyone. Effective means are needed to protect every bite of it at reasonable costs.

And that's all, everyone, for now, and have a peaceful and prosperous present and future.

REFERENCES

1. Fortune 2014. The World's Fair of 2064. *Fortune Asia Pacific Edition*, 169(1):60.
2. Horovitz, B. 2012. After Gen X, Millennials, What should Next Generation be? USA Today.
3. WJSchroer Company (Generations X, Y, Z and the Others. http://socialmarketing.org/archives/generations-xy-z-and-the-others/
4. Urban Dictionary. Generation Z. http://www.urbandictionary.com/define.php?term=Generation%20Z
5. Cisco 2017. The Zettabyte Era: Trends and Analysis.
6. Falconer, G. and Mitchell, S. 2012. Smart City Framework—A Systematic Process for Enabling Smart + Connected Communities. Cisco.

Index

Note: Page numbers followed by "*fn*" indicate foot notes.

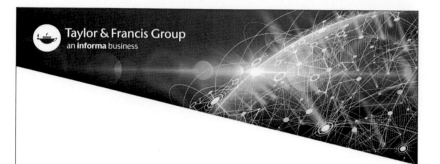